Horse Health Management

Horse Health Management

Edited by Eliza Melton

SYRAWOOD
PUBLISHING HOUSE
New York

Published by Syrawood Publishing House,
750 Third Avenue, 9th Floor,
New York, NY 10017, USA
www.syrawoodpublishinghouse.com

Horse Health Management
Edited by Eliza Melton

International Standard Book Number: 978-1-64740-076-7 (Hardback)

Cataloging-in-Publication Data

Horse health management / edited by Eliza Melton.
p. cm.
Includes bibliographical references and index.
ISBN 978-1-64740-076-7
1. Horses--Health. 2. Horses--Diseases--Treatment. 3. Veterinary medicine. I. Melton, Eliza.
SF951 .H67 2021
636.108 9--dc23

TABLE OF CONTENTS

PREFACE

Horse health management is one of the vital processes required for optimal health and long life of horses. Horse health management involves processes like veterinary care, leg care and bandaging, feeding, grooming, hoof care and shoeing, etc. Veterinary care is an integral part of horse health management as horses are affected by many disorders such as colic, laminitis, and internal parasites. There are many infectious diseases in horses that can be cured by regular vaccination. Rabies, tetanus and equine encephalitis are the common vaccines given to horses depending on the area they live in. Leg care and bandaging is another vital process in horse health management which involves routine observation for lacerations or swelling. Many riders wrap the horse's legs with protective boots or bandages to prevent injury in horses while working or exercising. Leg wraps are useful in preventing injury or for treating horses with leg injuries. This book unravels the recent studies in the field of horse health management. It discusses the fundamentals as well as modern approaches of horse health management. Those in search of information to further their knowledge will be greatly assisted by this book.

All of the data presented henceforth, was collaborated in the wake of recent advancements in the field. The aim of this book is to present the diversified developments from across the globe in a comprehensible manner. The opinions expressed in each chapter belong solely to the contributing authors. Their interpretations of the topics are the integral part of this book, which I have carefully compiled for a better understanding of the readers.

At the end, I would like to thank all those who dedicated their time and efforts for the successful completion of this book. I also wish to convey my gratitude towards my friends and family who supported me at every step.

Editor

1

The prevalence of ocular diseases in polish Arabian horses

Katarzyna Paschalis-Trela[1,2*], Anna Cywińska[3], Jan Trela[2], Michał Czopowicz[1], Jerzy Kita[1] and Lucjan Witkowski[1]

Abstract

Background: Equine ocular diseases pose a medical challenge due to long-lasting and cost-consuming therapies as well as economic issues associated with potential decrease in value of affected horses. The scale of the problem is significant but difficult to precisely define because epidemiological data is limited and lacks consistency in presentation. To date, no retrospective studies specifically investigating Arabian horses have been published.

Results: The aim of this cross-sectional study was to investigate the prevalence of ocular lesions and define the ocular diseases present in Arabian horses from breeding farms in Poland. Clinical and ophthalmic examination of 615 Arabian horses at Polish breeding farms (15% of Arabian population in Poland) were performed and medical history from the previous 5 years was analyzed. Data was obtained from review of veterinary archives and epidemiological interview of the resident veterinarian at each farm. The prevalence of ocular diseases was 9.75%. The following pathologies were diagnosed (with their respective prevalence): equine recurrent uveitis (ERU; 5.5%); cataract not related to ERU (3.3%); non-visual eyes (1.13%); posttraumatic lesions (0.8%); glaucoma (0.16%).

Conclusions: In this study, ERU was the most common ocular disease identified in Arabian horses in Poland. Its prevalence was lower than usually reported in Europe and the United States. There was no sex or farm predisposition but ocular disease prevalence increased with age. Other severe ocular pathologies were also observed, confirming that ocular diseases remain an important clinical problem.

Keywords: Epidemiology, Equine, Eye, ERU, Horse, Uveitis

Background

The Polish Arabian horse breeding program is considered one of the largest and the most successful worldwide, as evidenced by the considerable material value and global interest in the horses being sold. For example, in 2015 at the high-profile Pride of Poland Arabian Horse Auction the mare Pepita sold for the high dollar amount of €1.4 million ($US 1.6 m; $NZ 2.5 m). Three hundred years of tradition and many years of international success have contributed to the importance of Polish horses for the Arabian and Warmblood breeds [1, 2].

It is widely accepted that ocular diseases and related complications are a serious health problem worldwide with great impact on quality of life and value of affected horses, as well as athletic and show potential or use for other purposes. In addition to the pain associated with these conditions that can impair welfare, eye disorders may lead to partial or full blindness, which may exclude the horse from performance occupations and cause potential risk for the rider. Ocular diseases often require long-lasting and expensive treatments, and particularly if vision-threatening, may lead to disruption of training and potential disqualification from competition due to ongoing therapies and impairment. Thus, these diseases lead to large financial losses in the equine industry, estimated at millions of dollars per year in the US [3].

It is difficult to define the scale of the problem and associated occurrence of ocular diseases in Poland where equine ophthalmology is a relatively new veterinary discipline and epidemiological data is lacking. To date, no reports have been published on the prevalence and type of ocular lesions in Arabian horse breeding farms. A limited number of epidemiological studies have been

* Correspondence: katarzyna.paschalis@gmail.com
[1]Laboratory of Veterinary Epidemiology and Economics, Faculty of Veterinary Medicine, Warsaw University of Life Sciences-SGGW, Warsaw, Nowoursynowska 159c, 02-786 Warsaw, Poland
[2]TRELA VETs Referrals, Warsaw, Poland

published but none have focused on a large group of horses of one specific breed [3–12].

Thus, the aim of this study was to determine the prevalence of ocular pathologies in the Arabian horse population of Poland.

Methods

This study was approved by the 3rd Local Commission for Ethics in Animal Experiments (Decision No. 24/2011) and carried out in 2011–2013. All horses included in this cross-sectional study were recorded in the Polish Arabian Stud Book and their pedigree confirmed with an official passport. At the time the study was designed, the Arabian horse population in Poland numbered approx. 4000 individuals, however most of them were kept in several hundred small stables. There were only three large breeding farms constituting 615 Arabian horses total (roughly 15% of the general population) – 292, 270 and 53 animals from each farm, respectively. A minimum required sample size to estimate prevalence of ocular diseases at expected true prevalence of 50%, precision of 5% and 95% level of confidence was determined to be 352 individuals using EpiTools [13]. As 615 horses were available, they all were enrolled in the study. Sex and age were recorded for each animal and they were assigned to three age/use classes per E.C.A.H.O (European Conference of Arabian Horse Organization) recommendations: junior (up to 5 years), main reproduction age (5–12 years) and older horses (over 12 years).

Each horse underwent general clinical and ophthalmological examination of both eyes according to commonly accepted standards [3, 5, 14–16]. Medical histories from the preceding five years were reviewed. Clinical reports provided by each staff veterinarian were utilized to determine relevant medical details such as the recurrence of uveitic episodes. Ophthalmological examination was performed in horses without sedation or nerve blocks, and all examinations were conducted by a single investigator in a dark stable or in the predawn morning to ensure proper dark adaptation. Both eyes were examined with a penlight and slit-lamp (Kowa SL15). Intraocular pressure was measured (Tono-Pen XL, Mentor, USA) after use of topical anesthetic (Alcaine® proparacaine hydrochloride ophthalmic solution, USP 0.5%). Both pupils were dilated with Tropicamide 1% (Polfa, Poland) followed by a direct ophthalmoscopy (Heine EN 20-1d). When indicated (e.g. partially opaque cornea), ocular ultrasonography was carried out to evaluate the state of the globe and internal structures (Portable Ultrasound Machine GE LOGIQ I 10 MHz). Horses with suspected differences in globe size underwent ultrasonography as well. The measurement results allowed evaluation for conditions such as buphthalmos. Glaucoma diagnosis was made based on both case history and clinical signs, including corneal edema with endothelial striae, elevated IOP >30 mmHg, and globe enlargement.

Normality of age was assessed using a Shapiro-Wilk test and as it proved non-normally right-hand asymmetrically distributed ($p < 0.001$; coefficient of skewness of 1.7), it was reported with a median and interquartile range (IQR) and compared between groups using a Mann-Whitney U test. Ninety-five percent confidence intervals (CI 95%) for proportions were calculated using the Wilson score method [17]. A Pearson chi-square test and the associated version for a linear trend [18] were used for comparing prevalence of ocular diseases in the three age classes. All statistical hypotheses were two-tailed and significance level (α) was set at 0.05. Statistical analyses were performed in Statistica 12 (StatSoft Inc.).

Results

The study population consisted of 425 females (69.1%) and 190 males (30.9%) ranging in age from 6 months to 33 years. The median age of males was significantly younger than that of females (median 2 years, IQR 2 to 4 years vs. median 5 years, IQR 2 to 8 years; $p < 0.001$). There were 348 young horses (58.9% female), 209 adult horses (87.0% female) and 58 elderly horses (67.2% female).

A total number of 1228 eyes in 615 horses were examined. Two horses had one eye enucleated previously, one because of trauma and the other due to glaucoma and chronic pain. Abnormalities were found in 60 horses (9.8%; CI 95%: 7.7 to 12.4%) (Fig. 1.) with prevalence increasing with age: 6.0% (CI 95%: 4.0% to 9.0%) in young horses, 11.5% (CI 95%: 7.9% to 16.6) in adults and 25.9% (CI 95%: 16.3% to 38.4%) in elderly horses ($p < 0.001$) (Fig. 2.). The prevalence was uniform in all three breeding farms ($p = 0.303$), as well as in mares versus stallions ($p = 0.103$). In most examined horses with ocular disease, more than one pathology was noted. Therefore, the data is presented in figures and tables with reference to anatomical structures of the eye rather than to horses. The prevalence of observed symptoms is shown in Fig. 1, with the number of specific pathological changes in Table 1. Four horses presented with phthisis bulbi, one microphthalmia and one buphthalmia. Periocular structure deformations were detected in two horses: posttraumatic orbital fracture (one horse) and osteophyte formation at the base of the lacrimal bone (one horse). Nasolacrimal system pathologies were also diagnosed, including nasolacrimal duct obstruction and, rarely, lack of distal nasal lacrimal punctum due to imperforate nasal punctum. Other detected anomalies included: sarcoids affecting the upper eyelid in two horses, unilateral third eyelid paralysis, and in one horse, a scar resulting from a lower eyelid laceration. Conjunctivitis was diagnosed in six cases, all secondary to other coexistent

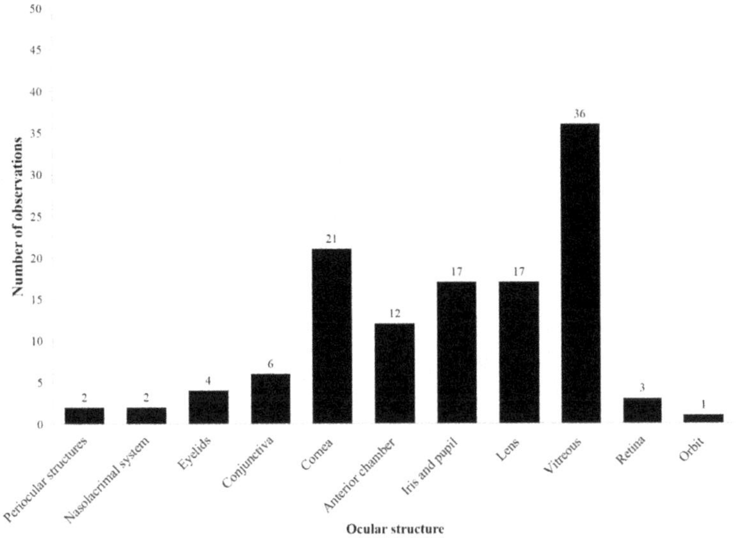

Fig. 1 Prevalence of the lesions in eye structures in Arabian horses in breeding farms in Poland

pathologies present on the day of examination. The most common pathologies affecting the anterior globe were corneal scarring and ulcers (potentially indicative of previous keratitis) in 21 horses.

Four older horses had senile mature cataracts and 15 horses had immature cataracts. Opacities of the lens capsule (3 anterior lens capsule, 1 posterior) or lens cortex (10 horses with perinuclear and 3 with nuclear) were present in 17 cases.

One horse was diagnosed with hyphema and anterior lens luxation. Furthermore, 12 horses had hyphema and fibrin clots in the anterior chamber, 24 horses had anterior and posterior synechiae, 2 horses had iris cysts, and iris atrophy was observed in 1 horse. The most common abnormalities in the posterior segment of eyeball were vitreal opacities (36 cases) and subretinal/retinal haemorrhage with detached retina (3 cases). During fundus examination in one horse, papilloedema was observed.

Seven horses had non-visual eyes; one was completely blind because of bilateral mature cataracts, retinal detachment and vitreal opacities, and the remaining 6 had unilateral ocular abnormalities including mature cataracts, retinal detachment, phthisis bulbi, and in one case microphthalmia. In 36 horses pathologically vitreous changes were detected. In some older horses this was due to senile degeneration, but in general, changes were primarily caused by inflammation.

Thirty four horses demonstrated clinical signs characteristic for ERU: blepharospasm, corneal edema, aqueous flare, hypopyon, miosis, vitreous haze, iris fibrosis and hyperpigmentation, posterior synechia, corpora nigra

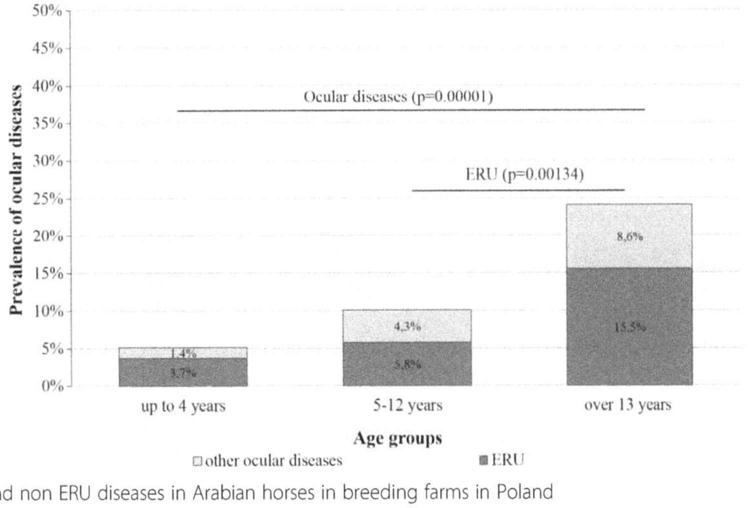

Fig. 2 Prevalence of ERU and non ERU diseases in Arabian horses in breeding farms in Poland

Table 1 Prevalance of pathological changes observed during ophthalmological examination

Pathological changes (some of them coexisting in one horse)	Number of findings
Phtisis bulbi	1
Microophthalmia	1
Buphthalmia	
glaucoma	1
Structure deformations	2
Conjunctivitis	6
Eyelid	
laceration	1
sarcoid	2
third eyelid paralisis	1
Nasolacrimal duct	
obstruction	1
lack of distal punctum	1
Corneal changes	
opacity	9
scar	17
ulcer	4
Anterior chamber	
hyphema	12
fibrin clots	17
Posterior synechiae	24
Iris	
atrophy	1
cysts	2
Lens	
luxation	1
mature cataract	4
immature cataract	15
Vitreal opacities	36
Optic nerve	
papilloedema	1
Retinal detachment	3

degeneration, miosis, cataract formation, vitreous degeneration and discoloration, and/or retinal degeneration coupled with the history of recurrent or persistent uveitis in one or both eyes [13, 17, 18]. Thus, the clinical diagnosis of ERU was confirmed in 34 out of 60 (56.7%; CI 95%: 44.1% to 68.4%) horses with ophthalmic lesions and was the most common ocular disease (Table 2.). This corresponds to 5.5% (CI 95%: 4.0 to 7.6%) of Polish Arabian horse population. The prevalence did not differ between the three breeding farms (p = 0.779), nor did it differ between mares and stallions (p = 0.606). Prevalence of ERU was significantly higher in older horses (15.5%; CI 95%: 8.4% to 26.9%) compared to those younger than 13 years (4.5%; CI 95%: 3.1% to 6.6%) (p = 0.002). The difference between young and adult horses (3.7% and 5.8%, respectively) was not statistically significant (p = 0.286).

Table 2 Prevalence of ocular diseases in Arabian horses in breeding farms in Poland

Ocular disease	Number (%) of horses affected	Confident interval 95%
ERU	34 (5.5%)	4.0% - 7.6%
Post-traumatic lesions	5 (0.8%)	0.4% - 1.9%
Not ERU related cataract	20 (3.3%)	1.9% - 4.7%
Mature cataract	4	
Inmature cataract	15	
Congenital cataract	1	
Glaucoma	1 (0.16%)	
Total	60 (9.8%)	7.7% - 12.4%

Discussion

The main goal of the study was to investigate and raise awareness of the prevalence of ocular diseases in horses in Poland. Since it is a new area of study in this country, the authors chose a valuable and well-known breed, the Polish Arabian, as the target population.

Data collection for this study was a significant challenge, requiring collaboration with study farms and veterinarians to gather clinical and historical information, including archived data, for 615 horses. The lack of published studies for reference also posed some difficulty. Previous studies describing ocular findings in horse populations were conducted as disease surveys limited to specific group of horses such as working horses [5], army horses [6] and horses in riding clubs [19]. Breed-specific studies have been limited to Thoroughbred racehorses [20] and Rocky Mountain Horses [21].

Several limitations of this study have to be taken into consideration. The primary issue is the structure of the Arabian horse population in Poland. There are approx. 4000 individuals (including 598 breeding mares and 227 stallions) kept at several hundred farms. Farm size varies from a just a few to over 200 individuals. The three farms enrolled in this study were the only locations at which more than 20 horses were kept. This represents a non-probability (convenience) sampling method, however the authors believe this is a sound approach as it allowed inclusion of the largest number of horses (15% of the Polish Arabian population) with detailed medical histories. It is also considered likely that this sample set was genetically representative of the entire population as 44% (264 of 598) of breeding mares were kept on the three farms and over 90% of the remaining 334 breeding mares on other farms were born on one of these 3 farms. Moreover, the ocular diseases under investigation were non-contagious, meaning that the concentrated population distribution of the sample set was unlikely to bias the results if compared to the larger majority of horses kept in tiny clusters at smaller farms. Neither general ocular disease prevalence nor ERU prevalence differ between the three study farms, therefore it was also unlikely to differ between other stables based on management differences.

The prevalence of ocular lesions reported in this study (9.8% of the investigated population), as well as the increasing prevalence observed in older horses, is consistent with most similar studies [4–7, 19–23]. However, it is difficult to compare the results obtained in our study with literature data due to substantial differences in study design (detection of the disease, description of eye lesions, variations in horse population), methods used (detailed ophthalmological examination, inspection only, or ophthalmological examination with the analysis of medical history), distinct forms of classification of ophthalmic disorders (usually categorized based on the type or place of lesions) as well as data presentation and description (lesions vs. horse with disease). Furthermore, other factors including climate, environment, husbandry practices, social and economic background, and the occurrence of infectious diseases and vectors (e.g. flies) also may influence the results.

Some epidemiological reports have been designed as retrospective studies of a specific disease, such as glaucoma [24] or ulcerative keratitis [25]. These studies were performed in patients at equine clinics selected for one disease, which could be diagnosed and described in detail. However, this type of study design does not give insight into the prevalence of disease in the general horse population. Most horses in the current study were not exhibiting acute signs of disease at the time of examination and were observed in their everyday living situation versus a clinic environment, adding validity to the results.

Epidemiological studies on the non-hospitalized population reveal multiple disease stages, although very rarely the acute phase. Instead, ongoing chronic pathological processes or evidence of previous ophthalmological disorders are commonly observed [4–7, 19–21, 23]. Untreated disease may extend to other eye structures, leading to visual impairment, which is often not identified by the owner due to patient adaptation [5, 19]. Use of the horse and the owner's knowledge is critical in this situation. In an Australian study, vision-threatening eye diseases surprisingly were detected in 7.5% of horses without any reported history of vision impairment [19]. This was greater than previously perceived, and highlighted the importance of ocular examination as a part of any routine physical examination of horses. This has important implications for human safety as well, as riding horses with unknown vision impairments pose a considerable risk for humans as well as themselves.

Other studies have reported general eye lesions in specific groups of horses, such as horses undergoing pre-purchase examination or riding school inspection [6], patients of equine veterinary clinics [7, 21], geriatric horses [5, 20], healthy neonatal foals [26] or race horses [19]. These studies were focused on lesion description, which can result from multiple diseases. Progressive or cumulative effects of several processes were potentially present and specific diseases were not precisely diagnosed. Moreover, they were based on general inspection versus a detailed ophthalmological examination. Such study design allows evaluation of the risk for vision loss and possible consequences [6, 19], but only provides limited information on disease prevalence. This study was designed to focus on proper examination and diagnosis of ocular diseases using a group of horses with common genetic background (Arabian breeding horses) across a wide age distribution. Moreover, the detailed medical records and consistent veterinary oversight of this study population allowed for selection of horses without other health issues. To the authors' knowledge there was no similar database source utilized for other published studies.

Trauma, glaucoma, and cataracts are the primary reported eye abnormalities [4, 6, 7, 19, 22, 23]. However, the most common cause of blindness is equine recurrent uveitis (ERU) which seems to be the most important ophthalmological problem in horses worldwide [3, 4, 8–11, 27–29]. Consistent with this, the current study demonstrates the importance of ERU in Arabian horses in Poland. Prevalence of ERU has been reported as 0% to 25% and even 40% in endemic areas [3, 4, 6, 8–10, 12], although other studies indicate that 8–10% occurrence is typical for most countries [5, 7, 8, 10]. Generally, the prevalence of eye disease in horses with potential threat to vision has been reported as below 1% up to 15% [4, 6, 7, 19, 21], rarely higher than 20% [22, 23]. However, subclinical ophthalmic lesions detected during ocular examination, but not noticed by owners or during routine physical examination, has been reported in up to 60–90% of examined horses [5, 6, 19, 20].

A cross-sectional study conducted in Ethiopia observed eye abnormalities in 23% of horses. Over 50% of affected eyes had visual impairment and ocular discharge. Moreover, ocular pain was evident in approximately 30% of horses but only a small number were receiving treatment or had been presented to a veterinary clinic with an eye problem [22].

Age predisposition for ocular diseases is widely accepted. In geriatric horses ocular lesions are common (approx. 60–90%). Vitreous degeneration, cataracts, and senile retinopathy or retinal degeneration are described most often [5, 20, 30]. The current study confirmed the higher prevalence of eye abnormalities in older horses, however, it did not exceed 25%. Interestingly, one published study claims that vision problems are less important for geriatric animals as they are no longer used for work or breeding [5]. One mare from our study population was blind for ten years due to mature bilateral cataracts, retinal detachment, and vitreous opacities. However, she was still a successful broodmare, likely due to management accommodations provided by her caretakers.

In foals, congenital ocular abnormalities pose the most serious problem. A high prevalence of such lesions has been reported in Rocky Mountain Horses [31] but appear rather uncommon in Standardbred foals [14, 26]. Acquired ophthalmic abnormalities as an ocular hemorrhage are relatively frequent [14, 26]. All foals examined in our study were older than 6 months, therefore details in the neonatal foal population are not described.

Our data indicate that ERU is the most prevalent disease in the Polish Arabian population, identified in 5.5% of 615 individuals examined, which corresponds to over 50% of horses with ophthalmic lesions. Breed predisposition for this disease has been demonstrated primarily in Appaloosas, but also to a lesser extent in European Warmbloods, draught breeds, Standardbred trotters and color-dilute horses [9, 12, 28, 29, 32–35]. As a breed, Arabians represent a population of average risk for ERU development, which must be taken into consideration for comparison with data from mixed breed populations in Europe.

Variation in ERU has been reported previously, ranging from 0% to 25% and even 40% in endemic areas. Although most studies indicate that prevalence of 8–10% is typical for Europe, we could not identify the primary source of this information and only sparse current epidemiological data are available [3–12]. Most recent studies describe lower ERU prevalence, such as 0.2% (2 from 805) of horses in Brazil [7], 0.8% (6 from 901) of horses in Iran [4], and 0.9% of horses (6 from 693) in United Kingdom [6]. Only the studies from United Kingdom [5] and Germany [36] reported similar prevalence to that indicated by our study; 6% (5 from 83 geriatric horses) and 7.6% (78 from 1014 animals) respectively. Some epidemiological studies describe uveitis in horses but without any information about the type of disorder, e.g. in 24 of investigated 500 horses in India (4.8%), uveitis cases constituted 24% of horses with ophthalmologic disorders [23]. ERU may be considered in these cases but cannot be diagnosed based only on the detected lesions without proper documentation of medical history as lesions caused by episodes of ERU and other reasons cannot be differentiated.

Taking into account the specific diagnosis, ERU is the most or second most common (after ocular trauma) reported ophthalmologic disease. In one Iranian study, ERU was diagnosed in 15% of horses with eye abnormalities and only eye injuries with various sequelae were more commonly reported [4, 7, 22]. However, the high occurrence of ERU in Polish Arabian horses with eye abnormalities of 56.7% is surprising, particularly given the research was conducted in breeding farms rather than veterinary clinics.

Lack of age and gender predilections for ERU have been reported previously [37]. Our study confirmed this finding. We found no significant differences related to sex and environment (farms), however, the prevalence of ERU increased with age and disease was diagnosed most frequently in horses older than 12 years. This could be explained by the type of the study, as a cross-sectional study decreases the probability of finding a horse experiencing an acute phase of the disease. Diagnosis was based on detected lesions and historical data. Because of the recurrent and progressive nature of ERU, the lesions develop with age, reflecting multiple episodes. Occurrence of ERU is spontaneous and recurrence is unpredictable. The disease tends to increase in severity with repeated episodes resulting in cumulative damage to the eye which can then be detected during ophthalmoscopic examination [10–12, 27, 28, 37, 38].

Similar lesions of the cornea can be caused by trauma, inflammation, glaucoma or episodes of uveitis [39]. Infectious disorders of eye may also lead to identical lesions if not managed correctly [40]. Trauma was reported as the most important cause of ocular lesions in working horses (36% of cases) [4, 7, 22]. In this study we observed post-traumatic lesions (according to medical history gathered from farm vets) in 0.8% of horses. This is comparable with other studies [5, 6, 19], however lesions indicating a history of chronic keratitis caused by other factors and coexisting with other lesions were detected in 20 horses. In racehorses in Japan ulcerative keratitis caused by different factors was the most frequent corneal and ocular disease but the incidence rate was very low 0.015% [25]. In contrast, corneal edema, opacity and scarring were diagnosed in 97.5% of horses older than 15 years [5]. This indicates that data from questionnaires and medical history is crucial for proper recognition and diagnosis of any abnormalities.

In this study cataracts unrelated to ERU were diagnosed in 3.3% of investigated horses and congenital cataracts were diagnosed in only one case. In most cases cataracts were diagnosed in horses with concurrent ocular abnormalities and in older horses, consistent with previous studies. Cataracts have been reported to affect 5–7% of horses [6, 15, 41] but some studies reveal significantly different prevalence of less than 1% [4,

7, 23], approx. 20% [5, 19] and even 97% in geriatric horses [20].

The current study shows low prevalence of glaucoma in horses (a single case, 0.16%). Primary glaucoma in horses has been reported rarely, and is most often secondary to ERU. Age greater than 15 years is the main risk factor [15, 23, 24].

Conclusions

Ocular diseases in Arabian horses in Poland represent a significant welfare and economic concern. Equine recurrent uveitis appears to be the most common, however its prevalence in this population is lower than has been previously reported in Europe and the United States. Various ocular pathologies were observed, including severe conditions, confirming the presence and describing the prevalence of this important clinical problem among the study population. There was no observed sex or farm predisposition but the prevalence increases with age.

Abbreviations

CI: Confidence intervals; ERU: Equine recurrent uveitis; IQR: Interquartile range

Acknowledgements

The authors thank Aleksandra Tomkowicz from Department of Small Animal Diseases with Clinic SGGW, for editing help.

Funding

The study was partially supported by a grant from National Science Centre, grant number DEC-2011/03/B/NZ6/04682.
Publication was funded by KNOW (Leading National Research Centre) Scientific Consortium "Healthy Animal - Safe Food", decision of Ministry of Science and Higher Education No. 05-1/KNOW2/2015.
The funders had no role in study design, data collection and analysis, decision to publish, conclusions and preparation of the manuscript.

Authors' contributions

KPT conceiving and designing the study, clinical examination of the horses and drafting the manuscript. AC participation in data analysis and assistance in drafting the manuscript. JT participation in clinical examination of the horses and assistance in drafting the manuscript. MCz statistical analysis. JK assistance in conceiving and designing the study and critical revision of the manuscript. LW acquisition of funding, assistance in conceiving, designing and coordination of the study and assistance in drafting the manuscript. All authors read and approved the final manuscript.

Consent for publication

Not applicable.

Competing interests

The authors declare that they have no competing interests

Author details

[1]Laboratory of Veterinary Epidemiology and Economics, Faculty of Veterinary Medicine, Warsaw University of Life Sciences-SGGW, Warsaw, Nowoursynowska 159c, 02-786 Warsaw, Poland. [2]TRELA VETs Referrals, Warsaw, Poland. [3]Division of Pathophysiology, Department of Pathology and Veterinary Diagnostics, Faculty of Veterinary Medicine, Warsaw University of Life Sciences-SGGW, Nowoursynowska 159c, 02-786 Warsaw, Poland.

References

1. Strumiłło A.: Al Jawad, Konie Janowa: Dom Wydawniczy Benkowski; 2007. http://www.benkowski.home.pl.
2. Bielański W. Historia i współczesne zagadnienia hodowli i użytkowania koni w Polsce. Kraków: Wydawnictwo PAU; 2008.
3. Gilger BC, Deeg C. Equine ophthalmology. St. Louis, Missouri: Elsevier Saunders; 2011.
4. Bazargani TT, Moaddab SH, Raoofi A, Masoudifard M, Bahonar AR. Study of the prevalence and type of ophthalmic diseases among different breeds of horses in Tehran riding clubs. International Journal of Veterinary Research. 2011;5(1):13–6.
5. Chandler KJ, Billson FM, Mellor DJ. Ophthalmic lesions in 83 geriatric horses and ponies. Vet Rec. 2003;153(11):319–22.
6. Grundon RA. A retrospective comparison of eye lesions in two populations of horses: horses at pre-purchase examination and at riding school inspections. BrAVO Proceedings. 2008;2(1):1–3.
7. Reichman P, de Olivera Dearo AC, Rodrigues TC. Occurence of ophthalmologic diseases in horses used for urban cart hauling in Landrina. Ciencia Rural. 2008;38(9):2525–8.
8. Deeg CA, Hauck SM, Amann B, Pompetzki D, Altmann F, Raith A, Schmalzl T, Stangassinger M, Ueffing M. Equine recurrent uveitis - A spontaneous horse model of uveitis. Ophthalmic Res. 2008;40(3–4):151–3.
9. Lowe RC. Equine uveitis: a UK perspective. Equine Vet J Suppl. 2010;37:46–9.
10. Spiess BM. Equine recurrent uveitis: the European viewpoint. Equine Vet J Suppl. 2010;37:50–6.
11. Witkowski L, Cywinska A, Paschalis-Trela K, Crisman M, Kita J. Multiple etiologies of equine recurrent uveitis – a natural model for human autoimmune uveitis: a brief review. Comp Immunol Microbiol Infect Dis. 2016;44:14–20.
12. Gilger BC, Deeg CA. Equine Recurrent Uveitis. In: Gilger BC, editor. Equine ophthalmology. St. Louis: Elsevier Science; 2011. p. 317–49.
13. Thrusfield MV. Veterinary Epidemiology 3rd edition: Wiley-Blackwell; 2007.
14. Labelle AL, Hamor RE, Townsend WM, Mitchell MA, Zarfoss MK, Breaux CB, Thomasy SM, Hall T. Ophthalmic lesions in neonatal foals evaluated for nonophthalmic disease at referral hospitals. Javma-Journal of the American Veterinary Medical Association. 2011;239(4):486–92.
15. Matthews AG. Lens opacities in the horse: a clinical classification. Vet Ophthalmol. 2000;3(2–3):65–71.
16. Wilkie DA. Equine glaucoma: state of the art. Equine Vet J Suppl. 2010;37: 62–8.
17. Altman D, Machin D, Bryant T, Gardner M: Statistics with Confidence: Confidence Intervals and Statistical Guidelines, 2nd edn: BMJ Books; 2000.
18. Zar JH. Biostatistical analysis. 5th ed. Prentice Hall: Upper Sadle River; 2010.
19. Hurn SD, Turner AG. Ophthalmic examination findings of thoroughbred racehorses in Australia. Vet Ophthalmol. 2006;9(2):95–100.
20. Ireland JL, Clegg PD, McGowan CM, McKane SA, Chandler KJ, Pinchbeck GL. Disease prevalence in geriatric horses in the United Kingdom: veterinary clinical assessment of 200 cases. Equine Vet J. 2012;44(1):101–6.
21. Mellor DJ, Love S, Walker R, Gettinby G, Reid SWJ. Sentinel practice-based survey of the management and health of horses in northern Britain. Vet Rec. 2001;149(14):417–23.

22. Scantlebury CE, Aklilu N, Reed K, Knottenbelt DC, Gebreab F, Pinchbeck GL. Ocular disease in working horses in Ethiopia: a cross-sectional study. Vet Rec. 2013;172(4):99.
23. Thangadurai R, Sharma S, Bali D, Rana BP, Mahajan V, Samanta I, Hazra S. Prevalence of ocular disorders in an Indian population of horses. J Equine Vet Sci. 2010;30(6):326–9.
24. Curto EM, Gemensky-Metzler AJ, Chandler HL, Wilkie DA. Equine glaucoma: a histopathologic retrospective study (1999-2012). Vet Ophthalmol. 2014; 17(5):334–42.
25. Wada S, Hobo S, Niwa H. Ulcerative keratitis in thoroughbred racehorses in Japan from 1997 to 2008. Vet Ophthalmol. 2010;13(2):99–105.
26. Barsotti G, Sgorbini M, Marmorini P, Corazza M. Ocular abnormalities in healthy Standardbred foals. Vet Ophthalmol. 2013;16(4):245–50.
27. Curling A. Equine recurrent uveitis: classification, etiology, and pathogenesis. Compend Contin Educ Vet. 2011;33(6):E2.
28. Gerding JC, Gilger BC. Prognosis and impact of Equine Recurrent Uveitis. Equine Vet J. 2015. doi:https://doi.org/10.1111/evj.12451.
29. Gilger BC. Equine recurrent uveitis: the viewpoint from the USA. Equine Vet J Suppl. 2010;37:57–61.
30. Nell B, Walde I. Posterior segment diseases. Equine Vet J Suppl. 2010;37: 69–79.
31. Ramsey DT, Ewart SL, Render JA, Cook CS, Latimer CA. Congenital ocular abnormalities of Rocky Mountain horses. Vet Ophthalmol. 1999;2(1):47–59.
32. Deeg CA, Marti E, Gaillard C, Kaspers B. Equine recurrent uveitis is strongly associated with the MHC class 1 haplotype ELA-A9. Equine Vet J. 2004;36(1): 73–5.
33. Fritz KL, Kaese HJ, Valberg SJ, Hendrickson JA, Rendahl AK, Bellone RR, Dynes KM, Wagner ML, Lucio MA, Cuomo FM, et al. Genetic risk factors for insidious equine recurrent uveitis in appaloosa horses. Anim Genet. 2014; 45(3):392–9.
34. Kulbrock M, Lehner S, Metzger J, Ohnesorge B, Distl O. A Genome-Wide Association Study Identifies Risk Loci to Equine Recurrent Uveitis in German Warmblood Horses. PLoS One. 2013;8(8):e71619.
35. Dwyer AE, Crockett RS, Kalsow CM. Association of Leptospiral Seroreactivity and Breed with Uveitis and blindness in horses - 372 cases (1986-1993). J Am Vet Med Assoc. 1995;207(10):1327–31.
36. Szemes PA, Gerhards H. Study on the prevalence of equine recurrent uveitis in the cologne-Bonn area. Praktische Tierarzt. 2000;81(5):408-+.
37. Deeg CA, Kaspers B, Gerhards H, Thurau SR, Wollanke B, Wildner G. Immune responses to retinal autoantigens and peptides in equine recurrent uveitis. Invest Ophthalmol Vis Sci. 2001;42(2):393–8.
38. Faber NA, Crawford M, LeFebvre RB, Buyukmihci NC, Madigan JE, Willits NH. Detection of Leptospira spp. in the aqueous humor of horses with naturally acquired recurrent uveitis. J Clin Microbiol. 2000;38(7):2731–3.
39. Cutler TJ. Corneal epithelial disease. Vet Clin North Am Equine Pract. 2004; 20(2):319–43.
40. Brooks DE. Inflammatory stromal keratopathies: medical management of stromal keratomalacia, stromal abscesses, eosinophilic keratitis, and band keratopathy in the horse. Veterinary Clinics of North America-Equine Practice. 2004;20(2):345–60.
41. Matthews AG. The lens and cataracts. Veterinary Clinics of North America-Equine Practice. 2004;20(2):393–415.

2

An epidemiologic study of antimicrobial resistance of *Staphylococcus* species isolated from equine samples submitted to a diagnostic laboratory

Ronita Adams[1], Jackie Smith[2], Stephen Locke[2], Erica Phillips[2], Erdal Erol[2], Craig Carter[2] and Agricola Odoi[1*]

Abstract

Background: Antimicrobial resistance limits traditional treatment options and increases costs. It is therefore important to estimate the magnitude of the problem so as to provide empirical data to guide control efforts. The aim of this study was to investigate the burden and patterns of antimicrobial resistance (AMR) among equine *Staphylococcus* samples submitted to the University of Kentucky Veterinary Diagnostic Laboratory (UKVDL) from 1993 to 2009. Retrospective data of 1711 equine *Staphylococcus* samples submitted to the UKVDL during the time period 1993 to 2009 were included in the study. Antimicrobial susceptibility testing, that included 16 drugs, were performed using cultures followed by the Kirby-Bauer disk diffusion susceptibility test. The proportion of resistant isolates by animal breed, species of organism, sample source, and time period were computed. Chi-square and Cochran-Armitage trend tests were used to identify significant associations and temporal trends, respectively. Logistic regression models were used to investigate predictors of AMR and multidrug resistance (MDR).

Results: A total of 66.3% of the isolates were resistant to at least one antimicrobial, most of which were *Staphylococcus aureus* (77.1%), while 25.0% were MDR. The highest level of resistance was to penicillins (52.9%). Among drug classes, isolates had the highest rate of AMR to at least one type of β-lactams (49.2%), followed by aminoglycosides (30.2%). Significant ($p < 0.05$) associations were observed between odds of AMR and horse breed, species of organism and year. Similarly, significant ($p < 0.05$) associations were identified between odds of MDR and breed and age. While some isolates had resistance to up to 12 antimicrobials, AMR profiles featuring single antimicrobials such as penicillin were more common than those with multiple antimicrobials.

Conclusion: Demographic factors were significant predictors of AMR and MDR. The fact that some isolates had resistance to up to 12 of the 16 antimicrobials assessed is quite concerning. To address the high levels of AMR and MDR observed in this study, future studies will need to focus on antimicrobial prescription practices and education of both practitioners and animal owners on judicious use of antimicrobials to slow down the development of resistance.

Keywords: Staphylococcus, Antimicrobial resistance, Multidrug resistance, Horses, Equine, Kentucky, Burden, Predictors, Logistic regression model

* Correspondence: aodoi@utk.edu
[1]Department of Biomedical and Diagnostic Sciences, University of Tennessee, College of Veterinary Medicine, 2407 River Dr., Knoxville, TN 37996, USA
Full list of author information is available at the end of the article

Background

The development of antimicrobial agents has been one of the most critical advances in both human and veterinary medicine within the last century. However, due to a combination of factors, but most notably to the rise in the use of antimicrobials for treating both human and domestic species, antimicrobial resistance has become a global scientific and public health concern in both human and veterinary medicine [1, 2]. The quantity of antimicrobials used in both human and veterinary medicine as well as in aquaculture have contributed to the selection for antimicrobial resistance [3]. High rates of antimicrobial resistant bacterial infections increase morbidity, be it to a single agent, or multiple drug classes, hindering the ability to effectively treat infections. As a result, both morbidity and mortality of antimicrobial resistant infections have increased in affected populations [1]. Identification of the resistance profiles of microorganisms is a critical step in understanding antimicrobial resistance and is useful in providing information to guide treatment options and to combat the problem.

According to the World Health Organization, the frequency of resistance to first-line drugs that have traditionally been used to treat infections caused by *Staphylococcus* has increased globally [4]. Unfortunately, this resistance is not limited to human medicine, but is being seen more frequently in domestic species, and in equine medicine in particular [1]. Although the widespread use of antimicrobials among equine species in the U.S. has been addressed in multiple forums, the epidemiology of antimicrobial resistance in bacteria found in horses has not been assessed [2]. Identifying and describing the burden of antimicrobial resistance among domestic species has become even more important due to evidence of potential cross transmission of certain bacteria between humans and domestic species [5]. Both the Centers for Disease Control and Prevention (CDC) and the United States Department of Agriculture (USDA) have reported such findings in past years [6, 7]. Outcomes from these investigations found evidence of a potential zoonotic transfer of *Staphylococcus* bacteria and/or their genetic material between healthy humans and horses [6, 8]. Other reports suggest that resistant *Staphylococcus* infections in domestic animals may contribute to transmission seen in human contacts [9].

Understanding the burden of antimicrobial resistant *Staphylococcus* infections in horses is critical in not only being able to understand the risk to those in immediate contact with these animals, but also in effectively providing information to guide efforts for the development of antimicrobial stewardship programs. Although a number of studies have investigated mainly methicillin-resistant *S. aureus* in horses [10–14], many other *Staphylococcus* species not only exhibit resistance to antimicrobials, but are clinically relevant to understanding the epidemiology of antimicrobial resistance in horses and its zoonotic spread to humans [2]. Thus, the objective of this study was to estimate the proportion of antimicrobial resistant staphylococcal isolates among equine samples submitted to the University of Kentucky Veterinary Diagnostic Laboratory between 1993 and 2009 and to identify potential predictors of antimicrobial resistance and multidrug resistance.

Methods

Data sources, preparation & study area

Laboratory records of all samples from horses submitted to the University of Kentucky Veterinary Diagnostic Laboratory were included in this study. The records included a combination of antimicrobial sensitivity test results and animal demographic information. For the isolation of bacteria, specimens were cultured on blood agar and eosin methylene blue agar plates at 37 °C in 5–10% CO_2, for a minimum of 24 h. If the specimen was from a likely contaminated site such as nasal swab, a Columbia colistin and nalidixic acid (CNA) plate with blood was also inoculated. The plates were examined for pathogenic bacteria and were incubated for an additional 24 h at 37 °C in aerobic incubators and examined again for pathogenic bacteria. The criteria used for reporting a microorganism was the isolation of the microorganism in pure culture or significant numbers from specimens (as the predominate microorganism). *Staphylococcus* isolates were identified by using colony morphology, dark-field examination, β-hemolysis on blood agar and CNA plates, and conventional biochemical tests, including coagulase, catalase, maltose, mannitol, and trehalose. Additionally, selective and differential plates with antimicrobials and indicator were used to differentiate between *S. aureus* and *S. hyicus*.

Antimicrobial susceptibility testing, that included 16 drugs, were performed using Kirby-Bauer disk diffusion susceptibility test. The laboratory followed procedures of the Clinical Laboratory Standards Institute (CLSI) testing and classification to determine the susceptibility of isolates [15–19]. Sizes of the zones of inhibition were measured and interpreted as susceptible, intermediate, or resistant. Sizes of zones of susceptible and resistant in millimeters were as follows: bacitracin (≥ 13, ≤ 8), cephalothin (≥ 18, ≤ 14), erythromycin (≥21, ≤15), neomycin (≥ 17, ≤ 12), kanamycin (≥ 18, ≤ 13), streptomycin (≥ 15, ≤ 11), oxacillin (≥ 13, ≤ 10), lincomycin (≥ 19, ≤ 15), enrofloxacin (≥ 21, ≤ 17), amoxicillin/clavulanic acid (≥ 20, ≤ 19), nitrofurantoin (≥17, ≤14), gentamicin (≥ 15, ≤ 12), novobiocin (≥ 17, ≤ 14) penicillin (≥ 28, ≤ 19), tetracycline (≥ 23, ≤ 18), and trimethoprim and sulfamethoxazole (≥ 16, ≤ 10). Isolates were classified as either susceptible, intermediate or resistant based on the above classification procedure [15–19]. For the purpose of this study, only susceptible and resistant isolates were included for subsequent analyses. Only records from the state of Kentucky were included in the study.

Data analysis

All statistical analyses were performed in SAS 9.4 [20]. For the purpose of this study, the resistance status variable was reclassified into a binary outcome, resistant or susceptible. Thus, all isolates indicated as "intermediate" were not included in the analysis. Antimicrobial resistance (AMR) was defined as resistance to at least one antimicrobial. Additionally, multi-drug resistance (MDR) was defined as resistance to three or more antimicrobial classes [21]. The proportion of resistant isolates and 95% confidence intervals were computed by breed, sex, age, sample source, the species of *Staphylococcus*, antimicrobial agent, year (which was scaled by subtracting 1993 from each year), season and month. Season was classified as follows: summer (June–August), fall (September–November), winter (December–February), and spring (March–May). All specimen types that had frequencies of less than 1% were combined into a category called "Other". These were too many to list. Similarly, breeds with frequencies less than 1% were classified as "other breeds" and included Appaloosa, Belgian, Burro, Clydesdale, Donkey, Draft, French Warmblood, Hanover, Miniature Horse, Missouri Fox Trotter, Morgan, Other, Paint, Palomino, Percheron, and Pony.

Temporal graphs were generated in excel to visualize the temporal patterns of resistance. In addition, the Cochran-Armitage Trend test was used to identify significant temporal trends. Simple and multivariable logistic regression models were used to investigate if AMR had significant associations with breed, sex, age, sample source, species of *Staphylococcus* organism, year, season, and month. The model building process was done in two steps. In the first step, simple logistic regression models were fitted with "AMR, (1 = Resistant, 0 = Susceptible)" as the outcome and each of the variables in Table 1 as the explanatory variables. Variables with *p*-values less than 0.15 were considered for inclusion in the multivariable logistic regression model that was used in the second step. During this 2nd step, the multivariable logistic regression model was fitted using a manual backwards selection procedure. Confounding was assessed by comparing the change in parameter estimate of the variables in the model with and without the suspected confounding variable. A 20% change in the estimate of any of the variables already in the model was considered to be indicative of a confounder that was then retained in the final model. Odds ratios and their corresponding 95% confidence intervals were computed for all variables included in the final model. Goodness-of-fit of the final model was assessed using the Hosmer-Lemeshow goodness-of-fit test. No evidence of lack of fit was found. Steps 1 and 2 for the process above were repeated to investigate predictors of multidrug resistance (MDR). In this model, the outcome variable used was "MDR, (1 = Multidrug Resistant/0 = Not Multidrug Resistant)". Again, Goodness-of-fit of the final model was assessed using the Hosmer-Lemeshow goodness-of-fit test. No evidence of lack of fit was found.

Results

Summary statistics

A total of 1711 samples, from 26 horse breeds, were included in the study. The most common breeds were Thoroughbreds (74.3%) followed by Tennessee Walking Horses (5.6%) (Table 1). Overall, more samples were submitted from female horses (83.7%) than male horses (16.3%) (Table 1). Similarly, horses > 4 years old contributed the highest proportion of samples (46.0%), followed by aborted fetuses (22.6%) and those < 1 year old (19.7%) (Table 1). Additionally, samples testing positive for *coagulase negative Staphylococcus* were most frequent (47.8%), followed by coagulase positive *Staphylococcus aureus* (40.3%). *S. hyicus* was the least frequent (4.4%).

Overall, 66.3% of the isolates were resistant to at least one antimicrobial. Of the samples with known breed information, the highest proportion of resistant isolates was from Thoroughbreds (70.5%) followed by the Standardbreds (68.6%) and Arabians (68.4%), while the lowest proportion of resistance was seen in mixed breeds (40.0%) (Table 1). Standardbreds had the highest proportion of MDR isolates (37.1%), followed by Thoroughbreds (31.1%), and Quarter Horse (18.3%). The lowest proportion of MDR was in the Tennessee Walking Horse (3.4%) (Table 1). Although females seemed to have a slightly higher level of AMR (68.1%) than males (64.0%), these differences were not statistically significant. However, the same does not apply to the levels of MDR between the sexes. In fact, males had a markedly higher proportion of MDR (32.9%) than females (25.4%) (Table 1).

Foals (< 1 years old) showed the highest levels of AMR (75.9%), followed by horses 2–4 years old (67.3%), and yearlings (1–2 years old) (65.6%). Adult horses (> 4 years old) had the lowest levels of antimicrobial resistance (60.0%) (Table 1). Foals again showed the highest levels of MDR (37.6%) when compared with other age groups (Table 1). MDR for horses 2–4 years old (28.9%) and those 1–2 years old (18.8%) were again the next highest. The highest proportion of AMR was observed among *Staphylococcus aureus* isolates (77.1%) followed by coagulase negative *Staphylococcus* strains (60.1%) (Table 1). Similarly, *Staphylococcus aureus* (38.3%) again

Table 1 Distribution and antimicrobial resistance of equine *Staphylococcus* samples submitted to the University of Kentucky veterinary diagnostic laboratory, 1993–2009

Variable	No. of samples tested	Percentage of samples tested[e] (%)	95% CI[a]	AMR[b] samples	AMR[b, f] (%)	95% CI[a]	MDR[c] samples	MDR[c, g] (%)	95% CI[a]
Breed	*n* = 1577			*n* = 1046					
Arabian	19	1.2	0.7, 1.7	13	68.4	43.5, 87.4	4	21.1	6.1, 45.6
American Saddlebred	63	4.0	3.0, 5.0	34	54.0	40.9, 66.6	10	15.9	7.9, 27.3
Mixed breed	30	1.9	1.2, 2.6	12	40.0	22.7, 59.4	3	10.0	2.1, 26.5
Quarter horse	60	3.8	2.9, 4.8	28	46.7	33.8, 60.0	11	18.3	9.5, 30.4
Rocky Mountain Saddlebred	16	1.0	0.5, 1.5	7	43.8	19.8, 70.1	1	6.3	0.2, 30.2
Standardbred	35	2.2	1.6, 3.1	24	68.6	50.7, 83.2	13	37.1	21.5, 55.1
Thoroughbred	1172	74.3	72.2, 76.5	826	70.5	67.8, 73.1	365	31.1	28.5, 33.9
Tennessee Walking Horse	88	5.6	4.5, 6.7	46	52.3	41.4, 63.0	3	3.4	8.5, 75.5
Other breeds	94	6.0	4.8, 7.3	56	59.6	49.0, 69.6	9	9.6	4.5, 17.4
Sex	*n* = 1377			*n* = 928					
Female	1152	83.7	81.6, 85.6	784	68.1	65.3, 70.7	293	25.4	22.9, 28.1
Male	225	16.3	14.4, 18.4	144	64.0	57.4, 70.3	74	32.9	26.6, 39.5
Age Groups	*n* = 717			*n* = 459					
> 4 years	330	46.0	42.4, 49.7	198	60.0	54.5, 65.3	68	34.3	16.4, 25.4
2–4 years	52	7.3	5.5, 9.4	35	67.3	52.9, 79.7	15	28.9	17.1, 43.1
1–2 years	32	4.5	3.0, 6.0	21	65.6	46.8, 81.4	6	18.8	7.2, 36.4
< 1 year	141	19.7	16.8, 22.6	107	75.9	68.0, 82.7	53	37.6	29.6, 46.1
Aborted fetus (0 years)	162	22.6	19.5, 25.7	98	60.5	52.5, 68.1	35	21.6	15.5, 28.8
Species of organism	*n* = 1711			*n* = 1131					
CoNS[d]	817	47.8	45.4, 50.1	491	60.1	56.7, 63.5	163	20.0	17.3, 22.9
Staphylococcus aureus	689	40.3	37.9, 42.6	531	77.1	73.7, 80.2	264	38.3	34.7, 42.1
Staphylococcus hyicus	75	4.4	3.4, 5.4	31	41.3	30.1, 53.3	1	1.3	0.03, 7.2
Staphylococcus intermedius	130	7.6	6.3, 8.9	78	60.0	51.1, 68.5	16	12.3	7.2, 19.2

[a]95% Confidence Interval
[b]AMR: Antimicrobial Resistance
[c]MDR: Multidrug Resistance
[d]Coagulase negative *Staphylococcus*
[e]The denominators are the number of samples tested for each variable and vary (e.g. breed = 1577; sex = 1377; etc) due to missing data
[f]The denominators for the percentage of AMR are the number of samples tested per row
[g]The denominators for the percentage of MDR are the number of samples tested per row

had the highest levels of MDR, followed by coagulase negative *Staphylococcus* strains (20.0%) (Table 1).

Distribution of resistance across antimicrobials

Overall, 16 antimicrobials from 10 antimicrobial classes were examined in this study (Table 2). Highest proportions of AMR isolates were seen among β-lactams (49.2%), with more isolates exhibiting resistance to Penicillin (52.9%) than oxacillin (15.6%) (Table 2 and Fig. 1). The drug class with the second highest proportion of AMR isolates was aminoglycosides (30.2%) (Fig. 1), with 28.9% and 22.8% of the isolates exhibiting resistance to Kanamycin and Gentamicin, respectively (Table 2). As for MDR, β-Lactams again had the highest levels (23.5%) of isolates that were MDR followed by Aminoglycosides (22.1%) (Table 2 and Fig. 1). Although the majority of resistant isolates (51.3%) were only resistant to 1 or 2 antimicrobial classes, 13.4% of the resistant isolates were resistant to 5 antimicrobial classes.

Of the isolates that were found to be MDR, 8.0% were resistant to 9 antimicrobials (Amoxycillin/clavulanic acid, Erythromicin, Gentamicin, Kanamycin, Oxacillin

Table 2 Distribution of antimicrobial resistance categorized by antimicrobial class among equine *Staphylococcus* samples submitted to the University of Kentucky veterinary diagnostic laboratory, 1993–2009

Antimicrobial class	Drug	AMR[a] samples	AMR[a] %	95% CI[b]	MDR[c] samples	MDR[c] %	95%CI[b]
Aminoglycosides		516/1710	30.2	28.0, 32.4	377/1710	22.1	20.1, 24.1
	Neomycin	53/1582	3.4	2.5, 4.4	46/1582	2.9	2.1, 3.9
	Kanamycin	486/1682	28.9	26.7, 31.1	369/1682	21.9	20.0, 24.0
	Streptomycin	59/287	20.6	16.0, 25.7	28/287	9.8	6.6, 13.8
	Gentamicin	369/1622	22.8	20.7, 24.9	270/1622	16.7	14.9, 18.6
β-lactams		841/1710	49.2	46.8, 51.6	402/1710	23.5	21.5, 25.6
	Penicillin	814/1539	52.9	50.4, 55.4	396/1539	25.7	23.6, 28.0
	Oxacillin	254/1634	15.6	13.8, 17.4	235/1634	14.4	12.7, 16.2
	Amoxicillin/Clavulanic Acid	115/1644	7.0	5.8, 8.3	107/1644	6.5	5.4, 7.8
Macrolides		292/1668	17.5	15.7, 19.4	249/1668	14.9	13.3, 16.7
	Erythromycin	292/1668	17.5	15.7, 19.4	249/1668	14.9	13.3, 16.7
Sulfonamides		463/1645	28.2	26.0, 30.4	372/1645	22.6	20.6, 24.7
	Sulfonamide	488/1702	28.7	26.5, 30.9	372/1702	21.9	19.9, 24.0
	Trimethoprim-sulfadiazine	330/1355	24.4	22.1, 26.7	297/1355	21.9	19.7, 24.2
Lincosamides		28/970	2.9	1.9, 4.2	25/970	2.6	1.7, 3.8
	Lincomycin	28/970	2.9	1.9, 4.2	25/970	2.6	1.7, 3.8
Aminocoumarins		141/1578	8.9	7.6, 10.6	31/1578	2.0	1.3, 2.8
	Novobiocin	141/1578	8.9	7.6, 10.6	31/1578	2.0	1.3, 2.8
Cephalosporins		63/1711	3.7	2.8, 4.7	63/1711	3.7	2.8, 17.1
	Cephalothin	48/1692	2.8	2.1, 3.7	48/1692	2.8	2.1, 3.7
Fluoroquinolones		1/25	4.0	0.1, 20.4	1/25	4.0	0.1, 20.4
	Enrofloxacin	1/24	4.2	0.1, 21.1	1/24	4.2	0.1, 21.1
Tetracyclines		451/1682	26.8	24.7, 29.0	326/1682	19.4	17.5, 21.4
	Tetracycline	451/1682	26.8	24.7, 29.0	326/1682	19.4	17.5, 21.4
Polypeptides		45/1649	2.7	2.0, 3.6	36/1649	2.2	1.5, 3.0
	Bacitracin	45/1649	2.7	2.0, 3.6	36/1649	2.2	1.5, 3.0

[a]AMR: Antimicrobial Resistance
[b]95% Confidence Interval
[c]MDR: Multidrug Resistance

(Methicillin), Penicillin, Sulfonamides, Tetracycline, and Trimethoprim-sulfadiazine) belonging to 5 antimicrobial classes (Aminoglycosides, β-Lactams, Macrolides, Sulfonamides and Tetracyclines) (Table 3). Another 7.0% were resistant to the same profile of antimicrobials except Erythromycin. In fact, 46.0% of the isolates that were MDR, and had a sample size greater than 10, had resistance profiles that contained penicillin, kanamycin, sulfonamides, and trimethoprim-sulfadiazine (Table 3). Additionally, 34.0% of the MDR samples with sample sizes greater than 10 showed resistance to oxacillin (Table 3).

Temporal trends

There was a significant ($p = 0.023$) decreasing temporal trend in AMR over the study period (Fig. 2). The proportions of AMR isolates were highest in 2000 (76.0%) and reached their lowest levels by 2007 (52.4%) (Fig. 2). On the contrary, there was an increasing temporal trend in MDR ($p = 0.007$) over the study period (Fig. 2). The proportion of MDR isolates began at its lowest point in 1993 (14.4%) before reaching the highest level in 2000 (42.5%) (Fig. 2).

Predictors of antimicrobial resistance (AMR) and multidrug resistance (MDR)

Species of organism, breed, age, sex, season, and year all had significant simple associations with the odds of AMR at an $\alpha = 0.15$. (Table 4). Similarly, species of organism, breed, age, sex, and year had significant simple associations with the odds of MDR (Table 4). All variables found to be significant ($p \leq 0.15$) in the AMR or

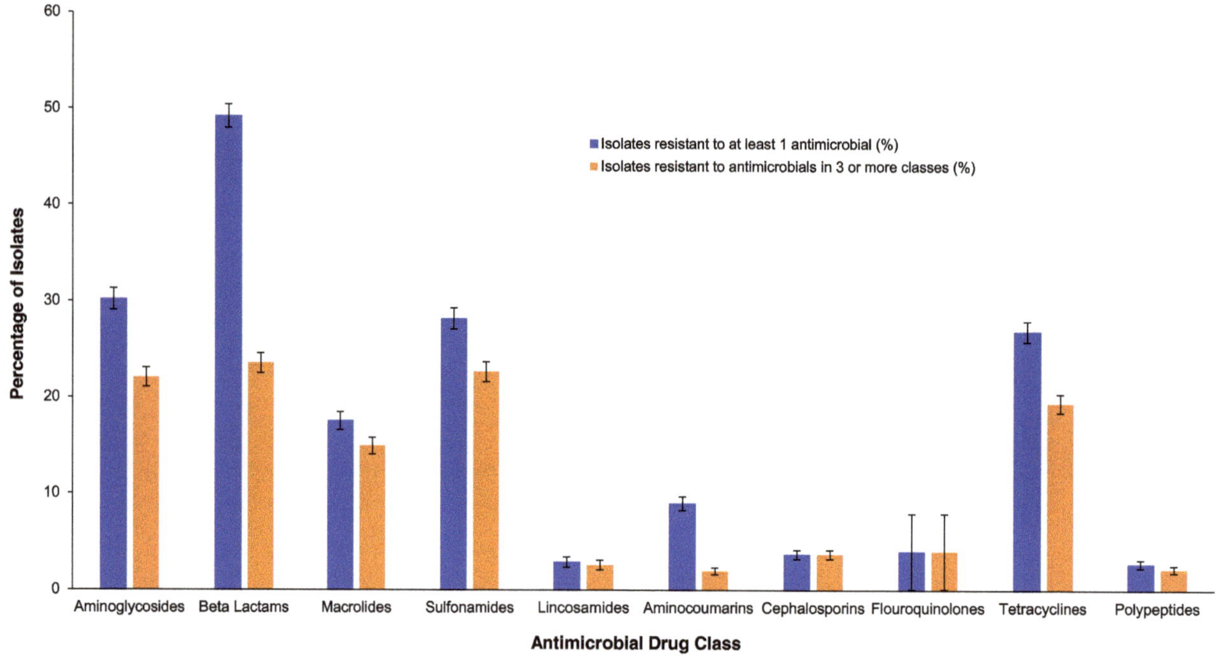

Fig. 1 Antimicrobial resistance and multidrug resistance by drug class from equine *Staphylococcus* samples submitted to the University of Kentucky veterinary diagnostic laboratory, 1993–2009

MDR simple models were considered for inclusion in their respective multivariable models.

Breed ($p = < 0.001$), species of organism ($p = < 0.001$) and year ($p = 0.023$) were significantly associated with the odds of antimicrobial resistant *Staphylococcus* infections in horses (Table 5). There was a significant ($p = < 0.001$) association between breed and AMR with Thoroughbreds having higher odds (Odds Ratio [OR] = 1.61; 95% Confidence Interval [CI] = 1.07, 2.42) of AMR than other breeds (Table 5). Interestingly, species of organism was a significant predictor for AMR but not MDR. The odds of AMR among *Staphylococcus aureus* isolates was significantly ($p < 0.0001$) higher (OR = 2.30; 95% CI = 1.81, 2.93) than that of coagulase negative *Staphylococcus* isolates (Table 5), while the odds of AMR among *Staphylococcus hyicus* isolates was significantly ($p < 0.0001$) lower (OR = 0.46; 95% CI = 0.27, 0.77) than that of coagulase negative *Staphylococcus* isolates. Year had a negative association with AMR (OR = 0.97, 95% CI = 0.95, 1.00).

Table 3 Antimicrobial resistance profiles of equine resistant *Staphylococcus* samples submitted to the University of Kentucky veterinary diagnostic laboratory, 1993–2009

Profile	No. of samples	Percent	95% CI[a]
Amo-Cep-Ery-Gen-Kan-Oxa-Pen-Sul-Tet-Tri	14	4.3	2.3, 6.9
Amo-Ery-Gen-Kan-Oxa-Pen-Sul-Tet-Tri	28	8.3	5.6, 11.8
Amo-Ery-Kan-Oxa-Pen-Sul-Tet-Tri	10	3.0	1.4, 5.4
Amo-Gen-Kan-Oxa-Pen-Sul-Tet-Tri	10	3.0	1.4, 5.4
Ery-Gen-Kan-Oxa-Pen-Sul-Tet-Tri	25	7.4	4.9, 10.8
Ery-Gen-Kan-Pen-Sul-Tet-Tri	14	4.2	2.3, 6.9
Ery-Kan-Oxa-Pen-Sul-Tet-Tri	12	3.6	1.9, 6.1
Gen-Kan-Oxa-Pen-Sul-Tet-Tri	17	5.0	3.0, 8.0
Gen-Kan-Pen-Sul-Tet-Tri	16	4.8	2.7, 7.6
Gen-Kan-Pen-Sul-Tri	10	3.0	1.4, 5.4

The denominator used for each percentage was (337) after missing values were removed

Amo Amoxiillin/clavulanic acid, *Cep* Cephalothin, *Ery* Erythromycin, *Gen* Gentamicin, *Kan* Kanamycin, *Oxa* Oxacillin, *Pen* Penicillin, *Sul* Sulfonamide, *Tet* Tetracycline, *Tri* Trimethoprim-sulfadiazine

[a]95% Confidence Interval

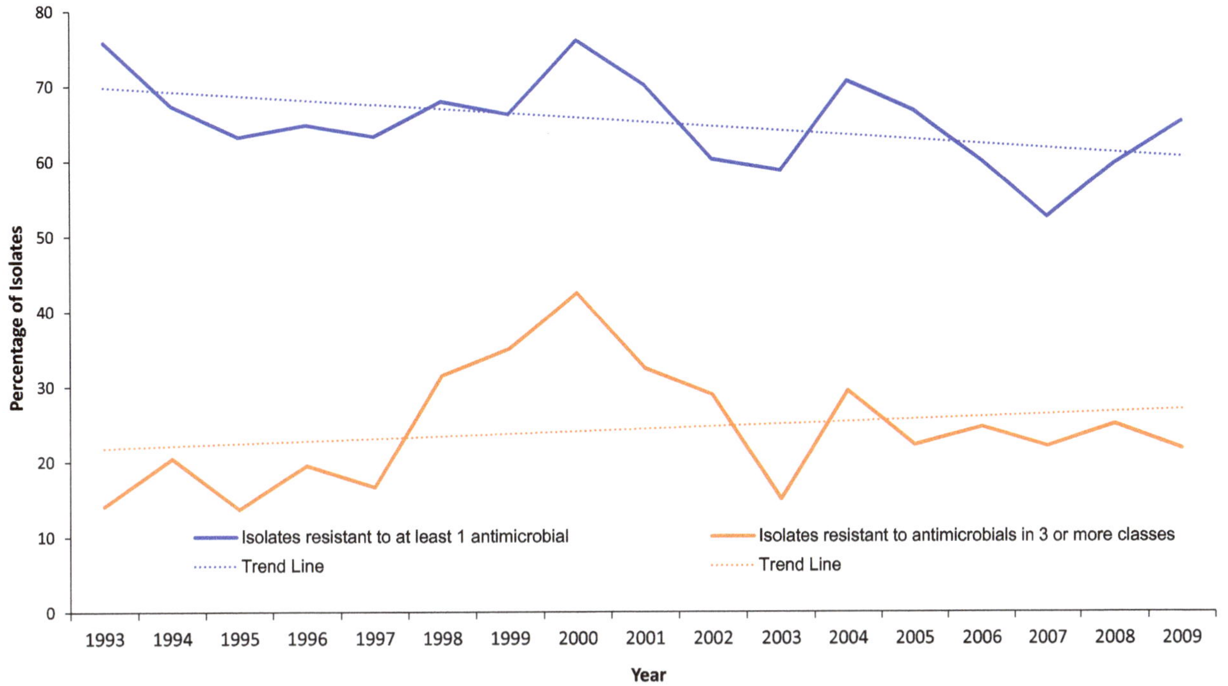

Fig. 2 Annual temporal distribution of antimicrobial resistance & multidrug resistance from equine *Staphylococcus* samples submitted to the University of Kentucky veterinary diagnostic laboratory, 1993–2009

Breed ($p = < 0.001$) and age ($p = 0.020$) were significantly associated with the odds of MDR of *Staphylococcus* (Table 6). The odds of isolates from Standardbreds being MDR were over 15 times (OR = 15.0; 95% CI = 3.7, 60.4) higher than those of isolates from other breeds, while the odds of MDR in isolates from Thoroughbreds were almost 7 times (OR = 7.0; 95% CI = 2.4, 19.8) higher than that of isolates from other breeds (Table 6). The odds of MDR among isolates taken from foals (< 1 year) were 63% (OR = 1.6; 95% CI = 1.0, 2.6) higher than that of horses > 4 years old (Table 6).

Table 4 Results of simple logistic models assessing predictors of antimicrobial resistance and multidrug resistance in equine *Staphylococcus* samples submitted to the Kentucky state diagnostic laboratory, 1993–2009

Variable	AMR[a] *P*-Value	MDR[b] *P*-Value
Breed	< 0.001	< 0.001
Age	0.019	0.002
Organism	< 0.001	< 0.001
Sex	0.107	0.021
Season	0.083	0.781
Month	0.379	0.519
Year	0.046	0.001
City	0.390	0.146

[a]Antimicrobial Resistance
[b]Multidrug Resistance

Discussion

This study was designed to investigate the burden and patterns of both AMR and MDR among equine *Staphylococcus* samples submitted to the University of Kentucky Veterinary Diagnostic Laboratory and to

Table 5 Significant predictors of antimicrobial resistant *Staphylococcus* in equines from samples submitted to the Kentucky state diagnostic laboratory, 1993–2009

Variable	Odds ratio	95% CI[a]	*P*-Value
Breed			< 0.001
Arabian	1.5	0.5, 4.3	0.331
American Saddlebred	0.9	0.5, 1.7	0.763
Mixed Equine	0.5	0.2, 1.2	0.057
Quarter Horse	0.6	0.3, 1.2	0.091
Standard Bred	1.5	0.6, 3.3	0.245
Thoroughbred	1.6	1.1, 2.4	<.001
Tennessee Walking Horse	0.8	0.5, 1.5	0.397
Other Breeds	.	.	.
Species of Organism			< 0.001
Staphylococcus aureus	2.3	1.8, 2.9	<.001
Staphylococcus hyicus	0.5	0.3, 0.8	<.001
Staphylococcus intermedius	1.1	0.7, 1.8	0.692
Coagulase negative Staphylococcus	.	.	.
Year	0.97	0.95, 1.00	0.023

[a]95% Confidence Interval

Table 6 Significant predictors of multidrug resistant *Staphylococcus* in equines from samples submitted to the Kentucky state diagnostic laboratory, 1993–2009

Variable	Odds ratio	95% CI[a]	*P*-value
Breed			< 0.001
Arabian	3.9	0.6, 25.1	0.159
American Saddlebred	2.5	0.5, 12.5	0.257
Mixed Equine	2.3	0.5, 11.5	0.308
Quarter Horse	2.2	0.5, 8.7	0.277
Standardbred	15.0	3.7, 60.4	0.001
Thoroughbred	7.0	2.4, 19.8	0.000
Tennessee Walking Horse	0.8	0.2, 3.6	0.730
Other	.	.	.
Age			0.020
Aborted Fetus 0 years	0.7	0.4, 1.2	0.171
< 1 year	1.6	1.0, 2.6	0.042
1–2 years	1.8	0.6, 4.9	0.266
2–4 years	1.5	0.7, 3.1	0.275
> 4 years	.	.	.

[a]95% Confidence Interval

investigate the predictors of AMR and MDR. The findings should provide information to guide future studies and ongoing surveillance of antimicrobial resistance. The proportion of antimicrobial resistant isolates seen in this study for both coagulase negative *Staphylococcus* infections (60.1%) and coagulase positive strains including *S. aureus* (77.1%), *S. intermedius* (60.0%), and *S. hyicus* (41.3%) suggest that the levels of AMR are high for both pathogenic and non-pathogenic *Staphylococcus* species.

Temporal trends

The temporal patterns observed in this study are interesting as a significant decreasing temporal trend was found for AMR, while an increasing temporal trend was observed for MDR. The reasons for this are unclear. However, a University of California (U.C.), Davis study that examined temporal trends in antimicrobial susceptibility patterns in equine case records from the William R. Pritchard Veterinary Medical Teaching Hospital (VMTH) from 1979 to 2010, found statistically significant increases over time in the percentage of *Staphylococcus* isolates susceptible to certain antimicrobials (chloramphenicol, ceftiofur, and penicillin) [22]. It is worth noting that, the U.C. Davis study investigated multiple organisms (*Pseudomonas* species, *Enterococcus* species, *E. coli*, *Salmonella* species., *Streptococcus* species, *Staphylococcus* species and *Actinobacillus* species) while our study was limited to *Staphylococcus* species. Findings from this study suggest that despite the significant decreasing AMR temporal trends, significant increasing MDR temporal trends in this population could have a negative impact on morbidity and mortality rates attributable to MDR infections [23, 24].

Antimicrobials

There is a paucity of published literature on antimicrobial resistance in equine *Staphylococcus* infections. Most of the work that has been published has focused only on *S. aureus* and especially MRSA. Thus the lack of literature addressing resistant *Staphylococcus* species in horses makes comparisons between the findings of this study and others difficult. Suffice it to say that although the overall proportions of AMR isolates in this study were high, MRSA levels were much lower (15.6%) than the percentage of MRSA (48%) found in a similar study done in Turkey [25]. A Belgian study, by Van den Eede et al., that assessed occurrence of MRSA in equine nasal samples found similar MRSA levels (10.9%) to those found in our study [26]. However, studies done in Australia, Canada, and Ireland that investigated *Staphylococcus aureus* colonization in healthy horses as well as isolation rates in horses with clinical presentation of MRSA found the percentage of AMR isolates to be much lower and ranging from 4% to 8% [27–29]. These differences could be attributed to the fact that we examined a higher number of antimicrobials and species of *Staphylococcus* in this study in comparison with the above studies that only investigated methicillin resistance in *S. aureus*.

The highest levels of resistance in this study was towards β-Lactams and Aminoglycosides. This may be due to the tendency of staphylococci to adapt to the selection pressure of antimicrobial use and become resistant to antimicrobials in general and the multiple mechanisms of resistance to aminoglycosides and β-Lactams in particular [30, 31]. These findings are comparable to those of a Swiss study which reported high levels of AMR not only to β-lactams and aminoglycosides, but to tetracyclines, lincosamides and macrolides as well when compared to other drug classes [32]. We also found the highest levels of AMR to be against penicillin (52.9%). Much higher levels of resistance were reported from equine hospital data in Zurich, where researchers identified AMR to penicillin in both coagulase negative staphylococci and *Staphylococcus aureus* to be around 82% and AMR to tetracycline to be 64% [14]. High levels of resistance to both penicillin (62.7%) and tetracycline (23.7%) were found in a retrospective study in France that investigated Staphylococci implicated in death or euthanasia in horses [33]. The higher levels of AMR *Staphylococcus* infections reported in hospitals could explain the higher AMR levels from the Zurich study.

A German study looking at resistance profiles of MRSA in horses from veterinary hospitals and large animal clinics found that gentamicin resistance was high

(85%) and mainly associated with isolates coming from equine clinics, while the majority of the isolates from all horses in the study were resistant to tetracycline (97.5%) and fluoroquinolones (79%) with only 15.6% being resistant to erythromycin [34]. Our study found much lower levels of AMR to gentamicin (22.8%), tetracycline (26.8%) and fluoroquinolones (4.0%), than the German study. Despite our MDR profiles containing gentamicin (16.7%) and tetracycline (19.4%) resistance, these levels were still not consistent with the findings of the German study. The differences in the levels of AMR and MDR seen in our study can be explained by the fact that the isolates from our study included multiple *Staphylococcus* species.

Of the resistant isolates in this study, 25% were MDR. This is double the percentage of MDR (13%) found in a Lithuanian study by Klimienė et al. [35] and a Zurich study [14] that both reported 13% MDR. However, it is more than double that reported by Toombs-Ruane et al. (10.1%) in New Zealand [36]. The Swiss study mentioned previously, also found that isolates were most likely to be MDR involving β-lactams, aminoglycosides, and tetracyclines [32]. That finding is similar to that of our study where the highest proportion of MDR infections involved aminoglycosides, β-lactams, sulfonamides, cephalosporins and tetracyclines. Interestingly, a recent companion animal study done in India found that not only were the incidences of *Staphylococcus aureus* wound infections higher in equines (57.14%), but that there was 100% MDR against kanamycin, colistin, clindamycin, penicillin-G, cotrimoxazole and cefotaxime [37]. However, it is worth noting that the current study only focused on *Staphylococcus* infections in horses and not multiple companion animals as was the case in the Indian study.

Antimicrobial resistance profile

Almost half of the MDR isolates in this study had antimicrobial resistance profiles that included penicillin, kanamycin, sulfonamides, and trimethoprim-sulfadiazine. These findings are consistent with those of a similar study that found that, in isolates identified to be MDR, *Staphylococcus* isolates that were oxacillin resistant, were also resistant to kanamycin, gentamicin and penicillin [38]. In our study less than 1% of the isolates were resistant to 12 antimicrobials and antimicrobial resistance profiles showed MDR to occur most frequently among isolates resistant to aminoglycosides, β-lactams, tetracyclines, sulfonamides, and cephalosporins. These findings were different from those of a study done in Switzerland by Schnelleman et al., [32], where 24% of the *Staphylococcus* isolates were resistant to all 12 of the antimicrobials tested, while the remainder of the isolates were resistant to a number of drug classes including β-lactams, combination β-lactam-β-lactamase-inhibitors, aminoglycosides, tetracyclines, chloramphenicol, macrolides, lincosamides and/or streptogramins [32]. It is important to note that isolates from the Swiss study were obtained only from horses undergoing colic surgery. A Lithuanian study by Klimienė et al. [35], found that the *Staphylococcus* isolates that were MDR showed high levels of resistance to penicillin G, erythromycin or tetracycline. Similar to the findings of our study, they reported that 66.7% of the isolates showed resistances to penicillin, erythromycin, tetracycline, ciprofloxacin, and gentamicin.

Distribution of resistance by host factors, species of organism and time

Thoroughbreds had the highest proportion of antimicrobial resistance (70.5%) in this study. This number is strikingly higher than the 5% AMR levels found in a similar study in Japan that examined MRSA colonization and infection in thoroughbreds [12]. However, because the Japanese study only looked at MRSA in thoroughbreds, while our study was able to examine both AMR and MDR in thoroughbreds, it is difficult to make direct comparisons between the AMR levels of the two. Nonetheless, a Canadian study looked at a mixture of draft, race, pleasure, breeding, school, and show horses and found no evidence of MRSA in thoroughbreds [29]. In this study, we found that the odds of AMR in thoroughbreds was higher than that of other breeds. The higher odds of AMR in thoroughbreds could be due to the extensive movement of this particular horse breed, increasing the risk for exposure to resistant *Staphylococcus* strains and contributing to higher resistance levels. Another Canadian study hypothesized that frequent contact with other horses, recurring and frequent travel to different sites, and the frequent use of antimicrobials in this set of horses could be associated with increased prevalence of MRSA in show and race horses [39]. Race horses, especially thoroughbreds, are moved frequently between Canada and the United States due to the large racing industry in both countries, which makes the risk of MRSA colonization and infection more widespread than seen in other breeds [40]. Horses, and thoroughbreds, in particular, are often moved between the United States, Australia, Canada, Japan, the UK, and Ireland, increasing the risk of importing infected carrier horses [12]. This could explain the high levels of resistance seen in thoroughbreds in this study.

A significant simple association was found between age and the odds of both AMR and MDR in this study. However, a significant association was only found between MDR and age in the multivariable model with age group less than 1 year showing significantly higher odds of MDR. Many past studies have focused on foals as an important population for studies of antimicrobial susceptibility [22, 41–44]. This is likely due to the higher susceptibility of younger animals to infection resulting in higher likelihood of antimicrobial treatment and hence selection for resistance. In this study *Staphylococcus aureus* was found to have significantly higher odds of AMR when compared with other *Staphylococcus* species, which is likely due to adaptability seen in *S. aureus*, [45], as well as the high prevalence of methicillin resistance in *Staphylococcus aureus* isolates, which indicates intrinsic resistance to all other β-Lactams, aminoglycosides and macrolides [46, 47].

Year was a significant predictor of AMR but not MDR in this study, where the odds of AMR isolates decreased over time. Decreases in AMR are likely due to changes in surveillance and reporting practices for resistant *Staphylococcus* infections, as well as adherence to sound antimicrobial prescription practices and policies. A study by Weese & Rousseau [48] found that after implementation of both active surveillance cultures and infection control procedures to address endemic MRSA, there was a rapid decrease in the proportion of horses colonized with MRSA. The study done by Weese & Rousseau focused on MRSA infections so direct comparisons cannot be made. However, it does indicate that appropriate control measures can affect the proportion of resistance infections observed and reported.

Study limitations

This retrospective laboratory-based study is not without limitations. Since data were not obtained using a statistical sampling technique, the study population should not be considered to be representative of the equine population in Kentucky. Only data available in the laboratory records could be investigated limiting the scope of investigation. For instance, information on past antimicrobial use was not available and therefore we could not assess its associations with levels of AMR or MDR. Furthermore, past medical history of the animals whose samples were used in this study was not reported.

Conclusion

The above limitations notwithstanding, the findings of this study provide useful information on the epidemiology of AMR and MDR in *Staphylococcus* infections in horses whose samples were submitted to the UKVDL. This information will be useful for guiding future primary base studies as well as efforts to address the problem. It is clear that equine *Staphylococcus* infections are exhibiting both AMR and MDR in horses. Factors such as breed and year are significant predictors of the odds of both AMR and MDR in this study, while species of staphylococci is also an important predictor of AMR and age of the horse was significantly associated with MDR. High levels of AMR and MDR could be indicative of problems in clinical prescription practices and procedures leading to selection for antimicrobial resistance. This highlights the need for a more comprehensive approach to investigating the epidemiology of AMR and MDR in horses. Future studies will need to focus on improving our understanding of antimicrobial use in horses as this will allow for more informed antimicrobial stewardship programs. Moreover, AMR surveillance in horses needs to include better record keeping and lab submission information (such as pre-treatment history). More information on risk factors may be gained through primary base observational studies that can more robustly identify risk factors that might otherwise not be investigated by retrospective lab-based studies.

Abbreviations

AMR: Antimicrobial resistance; CDC: Centers for Disease Control and Prevention; CI: Confidence interval; CLSI: Clinical Laboratory Standards Institute; CNA: Colistin and nalidixic acid; MDR: Multidrug resistance; MRSA: Methicillin-resistant *Staphylococcus aureus*; SAS: Statistical analysis system; UC: University of California; UKVDL: University of Kentucky Veterinary Diagnostic Laboratory; USDA: United States Department of Agriculture; VMTH: Veterinary Medical Teaching Hospital

Acknowledgements

The authors would like to thank the University of Kentucky Veterinary Diagnostic Laboratory for providing the study data.

Funding

The University of Tennessee, Center of Excellence in Livestock Diseases and Human Health (COE) Research Award.

Authors' contributions

RA was involved in data management and performed all statistical analyses and interpretation as well as preparation of the manuscript draft. AO was involved in conceptualization of research idea, study design, data analysis and interpretation as well as extensive editing of the manuscript. JS, SL, EP, CC, and EE were involved in all laboratory analyses, provision of laboratory resources, and data curation. All authors read and approved the final manuscript.

Consent for publication

Not Applicable.

Competing interests

The authors declare that they have no competing interests.

Author details

[1]Department of Biomedical and Diagnostic Sciences, University of Tennessee, College of Veterinary Medicine, 2407 River Dr., Knoxville, TN 37996, USA. [2]University of Kentucky, Veterinary Diagnostic Laboratory, 1490 Bull Lea Rd., Lexington, KY 40511, USA.

References

1. Bryan J, Leonard N, Fanning S, Katz L, Duggan V. Antimicrobial resistance in commensal faecal Escherichia coli of hospitalised horses. Ir Vet J. 2010;63(6):373–9.
2. Maddox TW, Clegg PD, Williams NJ, Pinchbeck GL. Antimicrobial resistance in bacteria from horses: epidemiology of antimicrobial resistance. Equine Vet J. 2015;47(6):756–65.
3. Nikaido H. Multidrug Resistance in Bacteria. Annu Rev Biochem. 2009;78: 119–46.
4. WHO: Antimicrobial Resistrance Fact Sheet. Geneva: Who Health Organizarion.
5. Epstein JH, Price JT. The significant but understudied impact of pathogen transmission from humans to animals. Mt Sinai J Med. 2009;76(5):448–55.
6. Johnson YJ, Myinit MS, Herrmann JA, Mathewson A, Maddox CW, Troutt HF, Opacich K. Preliminary findings on the prevalence of methicillin-resistant Staphylococcus species in healthy horses and their human contacts. Silver Spring: United States Food and Drug Administration; 2007. 128.
7. Ruthland BE, Weese JS, Bolin C, Au J, Malani AN. Human-to-dog transmission of methiciliilin-resistant Staphylococcus aureus. Emerg Infect Dis. 2009;15:1328.
8. Van Duijkeren E, Ten Horn L, Wagenaar JA, De Bruijn M, Laarhoven L, Verstappen K, et al. Suspected horse-to-human transmission of MRSA ST398. Emerg Infect Dis J. 2011;17:1137.
9. Bramble M, Morris D, Tolomeo P, Lautenbach E. Potential role of pet animals in household transmission of Methicillin-resistant Staphylococcus aureus: a narrative review. Vector Borne Zoonotic Dis. 2011;11(6):617–20.
10. Bergstrom K, Bengtsson B, Nyman A, Andersson UG. Longitudinal study of horses for carriage of methicillin-resistant Staphylococcus aureus following wound infections. Vet Microbiol. 2013;163(3–4):388–91.
11. Carfora V, Caprioli A, Grossi I, Pepe M, Alba P, Lorenzetti S, Amoruso R, Sorbara L, Franco A, Battisti A. A methicillin-resistant Staphylococcus aureus (MRSA) sequence type 8, spa type t11469 causing infection and colonizing horses in Italy. Pathog Dis. 2016;74(4):ftw025.
12. Kuroda T, Kinoshita Y, Niwa H, Shinzaki Y, Tamura N, Hobo S, Kuwano A. Meticillin-resistant Staphylococcus aureus colonisation and infection in thoroughbred racehorses and veterinarians in Japan. Vet Rec. 2016;178(19): 473–U444.
13. Schwaber MJ, Navon-Venezia S, Masarwa S, Tirosh-Levy S, Adler A, Chmelnitsky I, Carmeli Y, Klement E, Steinman A. Clonal transmission of a rare methicillin-resistant Staphylococcus aureus genotype between horses and staff at a veterinary teaching hospital. Vet Microbiol. 2013;162(2–4):907–11.
14. van Spijk JN, Schoster A, Wittenbrink MM, Schmitt S. A retrospective analysis of antimicrobial resistance in bacterial pathogens in an equine hospital (2012-2015). Schweiz Arch Tierheilkd. 2016;158(6):433–42.
15. National Committee for Clinical Laboratory Standards. Performance standards for antimicrobial disk susceptibility tests, approved standard M2-A5, vol. 14. 5th ed; 1993.
16. National Committee for Clinical Laboratory Standards: Performance standards for antimicrobial disk susceptibility tests – fifth informational supplement. Approved standard M100-S5, vol. 14; 1994.
17. National Committee for Clinical Laboratory Standards. Performance standards for antimicrobial susceptibility tests. Approved standard M2-A6. 6th ed; 1997.
18. National Committee for Clinical Laboratory Standards. Methods for dilution antimicrobial susceptibility tests for bacteria that grow aerobically. 3rd ed; 1997.
19. National Committee for Clinical Laboratory Standards: Performance standards for antimicrobial susceptibility testing. Approved standard, M100-S8. 1997.
20. SAS Institute. Base SAS 9.4 Procedures Guide. Cary: SAS Institute; 2014.
21. Magiorakos AP, Srinivasan A, Carey RB, Carmeli Y, Falagas ME, Giske CG, Harbarth S, Hindler JF, Kahlmeter G, Olsson-Liljequist B, et al. Multidrug-resistant, extensively drug-resistant and pandrug-resistant bacteria: an international expert proposal for interim standard definitions for acquired resistance. Clin Microbiol Infect. 2012;18(3):268–81.
22. Theelen MJP, Wilson WD, Edman JM, Magdesian KG, Kass PH. Temporal trends in in vitro antimicrobial susceptibility patterns of bacteria isolated from foals with sepsis: 1979-2010. Equine Vet J. 2014;46(2):161–8.
23. Leekha S, Diekema DJ, Perencevich EN. Seasonality of staphylococcal infections. Clin Microbiol Infect. 2012;18(10):927–33.
24. van Balen J, Mowery J, Piraino-Sandoval M, Nava-Hoet RC, Kohn C, Hoet AE. Molecular epidemiology of environmental MRSA at an equine teaching hospital: introduction, circulation and maintenance. Vet Res. 2014;45:17.
25. Aslantas O, Turkyilmaz S, Yilmaz MA, Erdem Z, Demir C. Isolation and molecular characterization of methicillin-resistant staphylococci from horses, personnel and environmental sites at an equine hospital in Turkey. J Vet Med Sci. 2012;74(12):1583–8.
26. Van den Eede A, Martens A, Lipinska U, Struelens M, Deplano A, Denis O, Haesebrouck F, Gasthuys F, Hermans K. High occurrence of methicillin-resistant Staphylococcus Aureus ST398 in equine nasal samples. Vet Microbiol. 2009;133(1):138–44.
27. Abbott Y, Leggett B, Rossney AS, Leonard FC, Markey BK. Isolation rates of meticillin-resistant Staphylococcus aureus in dogs, cats and horses in Ireland. Vet Rec. 2010;166(15):451–5.
28. Axon JE, Carrick JB, Barton MD, Collins NM, Russell CM, Kiehne J, Coombs G. Methicillin-resistant Staphylococcus aureus in a population of horses in Australia. Aust Vet J. 2011;89(6):221–5.
29. Burton S, Reid-Smith R, McClure JT, Weese JS. Staphylococcus aureus colonization in healthy horses in Atlantic Canada. Can Vet J. 2008;49(8):797–9.
30. Chambers HF, FR DL. Waves of resistance: Staphylococcus aureus in the antibiotic era. Nat Rev Microbiol. 2009;7(9):629–41.
31. Priyantha R, Gaunt MC, Rubin JE. Antimicrobial susceptibility of staphylococcus pseudintermedius colonizing healthy dogs in Saskatoon, Canada. Can Vet J. 2016;57(1):65–9.
32. Schnellmann C, Gerber V, Rossano A, Jaquier V, Panchaud Y, Doherr MG, Thomann A, Straub R, Perreten V. Presence of new mecA and mph (C) variants conferring antibiotic resistance in staphylococcus spp. isolated from the skin of horses before and after clinic admission. J Clin Microbiol. 2006; 44(12):4444–54.
33. Haenni M, Targant H, Forest K, Sevin C, Tapprest J, Laugier C, Madec JY. Retrospective study of necropsy-associated coagulase-positive staphylococci in horses. J Vet Diagn Investig. 2010;22(6):953–6.
34. Cuny C, Abdelbary MMH, Köck R, Layer F, Scheidemann W, Werner G, Witte W. Methicillin-resistant Staphylococcus aureus from infections in horses in Germany are frequent colonizers of veterinarians but rare among MRSA from infections in humans. One Health. 2016;2:11–7.
35. Klimienė I, Ružauskas M, Mockeliūnas R, Šiugždinienė R, Virgailis M, Špakauskas V. Antimicrobial resistance of *Staphylococcus* spp. isolated from riding-horses nasal mucosa. Veterinarija ir Zootechnika. 2015;71(93):32-38
36. Toombs-Ruane LJ, Riley CB, Kendall AT, Bolwell CF, Benschop J, Rosanowski SM. Antimicrobial susceptibilities of aerobic isolates from respiratory samples of young New Zealand horses. J Vet Intern Med. 2015;29(6):1700–6.
37. Tiwari R, Yadav SK, Singh S. Methicillin resistant Staphylococcus aureus isolated from wounds of livestock and companion animals of Uttar Pradesh India: a preliminary study. Int J Pharmacol. 2016;12(8):821–9.
38. Sieber S, Gerber V, Jandova V, Rossano A, Evison JM, Perreten V. Evolution of multidrug-resistant Staphylococcus aureus infections in horses and colonized personnel in an equine clinic between 2005 and 2010. Microb Drug Resist. 2011;17(3):471–8.
39. Tokateloff N, Manning ST, Weese JS, Campbell J, Rothenburger J, Stephen C, Bastura V, Gow SP, Reid-Smith R. Prevalence of methicillin-resistant Staphylococcus aureus colonization in horses in Saskatchewan, Alberta, and British Columbia. Can Vet J. 2009;50(11):1177–80.
40. Weese JS, Archambault M, Willey BM, Dick H, Hearn R, Kreiswirth BN, Said-Salim B, McGeer A, Likhoshvay Y, Prescott JF, et al. Methicillin-resistant Staphylococcus aureus in horses and horse personnel, 2000-2002. Emerg Infect Dis. 2005;11(3):430–5.
41. Giguere S, Lee E, Williams E, Cohen ND, Chaffin MK, Halbert N, Martens RJ, Franklin RP, Clark CC, Slovis NM. Determination of the prevalence of antimicrobial resistance to macrolide antimicrobials or rifampin in Rhodococcus equi isolates and treatment outcome in foals infected with antimicrobial-resistant isolates of R equi. J Am Vet Med Assoc. 2010;237(1):74–81.
42. Marsh PS, Palmer JE. Bacterial isolates from blood and their susceptibility patterns in critically ill foals: 543 cases (1991-1998). J Am Vet Med Assoc. 2001;218(10):1608–10.
43. Russell CM, Axon JE, Blishen A, Begg AP. Blood culture isolates and antimicrobial sensitivities from 427 critically ill neonatal foals. Aust Vet J. 2008;86(7):266–71.
44. Theelen MJ, Wilson WD, Edman JM, Magdesian KG, Kass PH. Temporal trends in prevalence of bacteria isolated from foals with sepsis: 1979-2010. Equine Vet J. 2014;46(2):169–73.
45. Lowy FD. Antimicrobial resistance: the example of Staphylococcus aureus. J Clin Investig. 2003;111(9):1265–73.

46. Guignard B, Entenza JM, Moreillon P. Beta-lactams against methicillin-resistant staphylococcus aureus. Curr Opin Pharmacol. 2005;5(5):479–89.
47. Sahebnasagh R, Saderi H, Owlia P. The prevalence of resistance to Methicillin in Staphylococcus aureus strains isolated from patients by PCR method for Detec-tion of mecA and nuc genes. Iran J Public Health. 2014; 43(1):84–92.
48. Weese JS, Rousseau J. Attempted eradication of methicillin-resistant Staphylococcus aureus colonisation in horses on two farms. Equine Vet J. 2005;37(6):510–4.

3

Comparison of three different methods for the quantification of equine insulin

T. Warnken[1,3*], K. Huber[2] and K. Feige[3]

Abstract

Background: Exact analysis of equine insulin in blood samples is the key element for assessing insulin resistance or insulin dysregulation in horses. However, previous studies indicated marked differences in insulin concentrations obtained from sample analyses with different immunoassays. Most assays used in veterinary medicine are originally designed for use in human diagnostics and are based on antibodies directed against human insulin, although amino acid sequences between equine and human insulin differ. Species-specific assays are being used more frequently and seem to provide advantages compared to human-specific assays. The aim of this study was to compare three immunoassays, one porcine-specific insulin enzyme-linked immunosorbent assay (ELISA), advertised to be specific for equine insulin, one porcine-specific insulin radioimmunoassay (RIA) and one human-specific insulin chemiluminescence immunoassay (CLIA), all three widely used in veterinary laboratories for the analysis of equine insulin. Furthermore, we tested their clinical applicability in assessing insulin resistance and dysregulation by analysis of basal blood and blood samples obtained during a dynamic diagnostic stimulation test (OGT) with elevated insulin concentrations.

Results: Insulin values obtained from the ELISA, RIA and CLIA, investigated for analyses of basal blood samples differed significantly between all three assays. Analyses of samples obtained during dynamic diagnostic stimulation testing with consecutively higher insulin concentrations revealed significantly ($p < 0.001$) lower insulin concentrations supplied by the CLIA compared to the ELISA. However, values measured by ELISA were intermediate and not different to those measured by RIA. Calculated recovery upon dilution, as a marker for assay accuracy in diluted samples, was 98 ± 4 % for ELISA, 160 ± 41 % for RIA and 101 ± 11 % for CLIA.

Conclusions: Our results indicate that insulin concentrations of one sample measured by different methods vary greatly and should be interpreted carefully. Consideration of the immunoassay method and reliable assay-specific reference ranges are of particular importance especially in clinical cases where small changes in insulin levels can cause false classification in terms of insulin sensitivity of horses and ponies.

Keywords: Horse, Equine, Insulin, Quantification, ELISA, RIA, CLIA, EMS

* Correspondence: tobias.warnken@tiho-hannover.de
[1]Department of Physiology, University of Veterinary Medicine Hannover, Foundation, Bischofsholer Damm 15, 30173 Hannover, Germany
[3]Clinic for Horses, University of Veterinary Medicine Hannover, Foundation, Bünteweg 9, 30559 Hannover, Germany
Full list of author information is available at the end of the article

Background

The equine metabolic syndrome (EMS) has attracted increasing attention in equine veterinary practice in recent years and is becoming more common due to today's horse husbandry and feeding practices. Different diagnostic tests to determine dysregulation of glucose and insulin homeostasis in horses are currently used to diagnose EMS. Tests are based on quantitative measures either of basal serum insulin concentration in fasted and non-fasted horses or of increased insulin concentration stimulated by oral [1, 2] or intravenous [3] dynamic diagnostic tests. Complex hyperinsulinaemic euglycaemic clamp (HEC) tests, which were considered to be the gold standard for assessment of insulin resistance (IR) [4], were usually reserved to research approaches due to their complex and expensive implementation. High insulin indicative of IR or insulin dysregulation (ID) in EMS horses is one of the features associated with this metabolic disorder [5]. Exact pathophysiology resulting in IR is not known to date. Several hypothesis leading to the clinical signs have been discussed in literature. Increased insulin degradation or neutralization, decreased binding of insulin to its receptor, as well as impaired downstream signaling were described. Whereas the term IR used in context with EMS is mainly characterized by reduced tissue response of insulin-dependent tissues the newly introduced term ID is used to describe in summary abnormalities of insulin metabolism [5]. Therefore, an exact quantification of insulin by laboratory analysis is the most important and challenging step to diagnose EMS-related changes in insulin regulation. Several immunoassay methods for the quantification of insulin in equine serum or plasma samples are commercially available. However, due to the use of different methods for quantifying equine insulin, discrepancies between results obtained from different studies have occurred [6, 7]. Since no assay with a specific antibody against equine insulin is available, most of the commercial immunoassays are based on antibodies directed against human or non-equine insulin. Only one enzyme-linked immunosorbent assay (ELISA) advertised by the manufacturer as being specific for measuring equine insulin is commercially available[1], and this is based on anti-porcine insulin antibodies. This assay has been validated for use in horses [8] and has already been used successfully in several studies [9, 10].

The aim of our study was to compare the results of this equine-optimized porcine-specific ELISA with results obtained from a radioimmunoassay[2] (RIA) and a chemiluminescence immunoassay[3] (CLIA) for measurements of equine insulin in serum samples. Furthermore, the second aim of the study was to evaluate the applicability of the three assays for samples obtained from horses and ponies under fasted conditions and from horses and ponies with stimulated insulin secretion during diagnostic procedure.

Methods

Animals and samples

Forty blood serum samples were collected from seventeen horses and ponies of different breeds, age (14 ± 6 years), weight (478 ± 179 kg) and body condition score (6.8 ± 1.1) to obtain samples with a broad range of insulin concentrations. Blood samples from healthy, university-owned horses were collected during a study, which has been approved by the ethics committee within the University of Veterinary Medicine, Hannover, and the State Office for Consumer Protection and Food Safety in accordance with the German Animal Welfare Law (LAVES–Reference number: 33.14 42502-04-13/1259). Blood samples from insulin dysregulated horses and ponies were collected during routine diagnostic procedures in the Clinic for Horses and informed consent was obtained from the clients for publication. Blood samples were collected after fasting the horses overnight (basal, $n = 20$), and additionally after stimulation by an oral glucose testing (OGT, $n = 20$) procedure [2]. OGT was carried out by administering 1 g/kg btw glucose powder dissolved in two liters water by naso-gastrical-tubation. Samples were collected via jugular vein catheter, transferred into plain tubes[4] for serum preparation and were incubated at room temperature for 1 h, centrifuged at 3000 g for 6 min and serum was stored at −80 °C. Prior to analyses samples were thawed once and split into different aliquots containing the required volume for each assay. Afterwards, they were re-frozen and sent on dry ice to the laboratories.

Assays

The porcine-specific insulin assay[1] is a solid phase two-side sandwich ELISA optimized for the quantification of equine insulin in serum or plasma. It is based on monoclonal mouse anti-porcine insulin antibodies coupled to plate and free monoclonal mouse anti-porcine insulin antibody conjugate labeled with horseradish peroxidase. Porcine insulin is used for calibrators to compute calibration curve via cubic spline regression. The limit of detection for this assay is 1.17 µIU/mL, but concentrations of samples with absorbance below the lowest calibrator (2.34 µIU/mL) should not be calculated. Cross reactivity of the assay is stated at 100 % for porcine insulin and 22 % with human insulin (according to the manufacturer's protocol) (Table 1).

For sample analysis by RIA, a porcine-specific one-site insulin RIA[2] was used. This assay is based on ^{125}I-labeled insulin and used guinea pig anti-porcine insulin antibodies and goat anti-guinea pig IgG antibodies. Purified human recombinant insulin preparations were used as standards

Table 1 Information about the three assays examined for quantification of equine insulin

Assay	Standards	Primary antibody	Second antibody
Mercodia Equine Insulin ELISA	Porcine insulin	Mouse monoclonal anti-porcine insulin	Mouse monoclonal anti-porcine insulin HRP-conjugate
Millipore Porcine Insulin RIA	Human recombinant insulin	Guinea pig anti-porcine insulin	Goat anti-guinea pig IgG
Siemens ADVIA Centaur Insulin Assay	-[a]	Monoclonal mouse anti-insulin AE-conjugate	Monoclonal mouse anti-insulin coupled to paramagnetic particles

[a]No information available

for calibration. The manufacturer's protocol referred to the assay's limit of linearity of 200 μIU/mL and recommended diluting the samples which have greater concentrations with the assay buffer. The limit of detection for this assay is 1.61 μIU/mL when 100 μl of sample is used. The cross-reactivity of the assay is stated at 100 % for porcine insulin, 100 % for human insulin and 90 % for bovine insulin. No information about cross-reactivity for equine insulin is given by the manufacturer. Measurements with the RIA were performed by a commercial veterinary endocrinology laboratory.

A human-specific insulin chemiluminescent immunoassay[4] with two-site sandwich technique and direct chemiluminescent technology was used for the sample analysis by CLIA. The assay is based on monoclonal mouse anti-insulin antibodies and chemiluminescent detection with monoclonal mouse anti-insulin antibodies labeled with acridinium ester. The limit of detection of this assay is 0.5 μIU/mL and the measuring range is from 1.0 to 300 μIU/mL. No information about cross-reactivity for equine insulin is given by the manufacturer. Measurements using the CLIA were performed by a commercial veterinary laboratory. Due to the fact that all tested assays use no equine standards, all the assays only provide an approximation of equine insulin concentrations.

Revalidation of the equine-optimized porcine-specific insulin ELISA

Intra-assay CV was calculated by division of the standard deviation by the corresponding mean of 15 replicates of equine serum samples with low (mean: 2.34 μIU/mL) and medium (mean: 108.81 μIU/mL) insulin concentrations. Inter-assay CV was calculated by division of the standard deviation by the corresponding mean of replicate samples, each measured on 25 different plates. Commercial controls[5] with low (5.03 μIU/mL), medium (18.84 μIU/mL) and high (60.84 μIU/mL) insulin concentrations (classifications according to manufacturer), and equine serum samples with low (mean: 10.53 μIU/mL) and medium (mean: 109.98 μIU/mL) insulin concentrations were used for the calculation of concentration-dependent inter-assay CVs. The recovery upon dilution (RUD) and the linearity of dilution were calculated to prove accuracy. A sample selected with a medium insulin concentration (99.45 μIU/mL) was measurable without any dilution and in five different dilution steps up to the 1:40 ratio dilution within the calibration range of the assay. Sample buffer[6] was used for dilution and at least 40 μl of the equine serum sample.

Comparison of methods

Basal serum samples and simulated samples from the OGT procedure were measured undiluted in all three assays for comparison of methods. Moreover, stimulated samples were additionally measured diluted in a ratio of 1:4 with sample buffer[6] to calculate the RUD for each assay to prove accuracy in the measurement of dilution procedure. The RUD was calculated as a percentage recovery of the insulin concentration in the diluted sample related to the corresponding undiluted samples. The RUD analyses were performed without the knowledge of the laboratories commissioned.

Stability of insulin after freezing, thawing and eight weeks of storage

In order to investigate the effect of repeated freezing and thawing on the stability of insulin in serum samples measured by equine-optimized, porcine-specific ELISA, 33 samples were thawed, measured and refrozen at −80 °C. Eight weeks later, samples were thawed at ambient temperature for 1 h and measured again. Samples covering a broad range of insulin concentrations were chosen to match the assay's analytical range from 2.34 to 175.5 μIU/mL. The samples were subdivided into three subgroups of low (3.51–15.21 μIU/mL), medium (30.42–90.09 μIU/mL) and high (92.43–125.19 μIU/mL) concentrations with eleven samples in each subgroup.

Statistics and calculations

Data analysis was performed using GraphPad Prism software[7]. The Shapiro-Wilk normality test was used to assess the normality of data distribution. Wilcoxon matched-pairs signed rank test, Spearman correlation and Deming regression analyses were used to compare results from different assays and to verify relationships between the three methods. In cases of assay-based non-detectable insulin concentrations, the corresponding samples were excluded from the statistics. Bland-Altman analysis was performed to calculate method-dependent bias and limits of agreements. Wilcoxon matched-pairs signed rank test was also used to examine the effect of repeated freezing and thawing on the stability of equine insulin. Statistical significance was set at $p < 0.05$.

Results

In re-validation experiments, intra-assay CV was 4.61 % at low insulin concentrations (2.34 μIU/mL) and 1.91 % at medium insulin concentrations (108.81 μIU/mL) using the equine-specific insulin ELISA. The inter-assay CV was 5.27 %, 3.24 % and 3.17 % for the low (reference value 5.03 μIU/mL insulin), medium (reference value 18.84 μIU/mL insulin) and high (reference value 60.84 μIU/mL insulin) commercial controls, respectively. Inter-assay CVs for equine serum samples were 7.34 % and 4.83 % for low (mean: 10.53 μIU/mL) and high concentrations (mean: 109.98 μIU/mL), respectively. The RUD in the first revalidation experiment was 95 % (range from 94 to 96 %), calculated in five dilution steps. The linearity upon dilution was excellent ($r^2 = 0.9994$; $p < 0.0001$), showing a strong relationship between the calculated and measured concentrations of the diluted samples (Fig. 1).

The ELISA provided results within the analytical range for 17 of 20 basal samples and for 11 of 20 stimulated samples (Table 2). The analytical range of the assay was from 2.34 to 175.5 μIU/mL (corresponding range in μg/L: 0.02 to 1.5; conversion factor: 117), as stated by the manufacturer. The mean RUD in samples analyzed by ELISA was 98 ± 4 % (Fig. 2a).

The RIA supplied results for 19 basal samples and for all 20 stimulated samples (Table 2). The analytical range for the RIA was stated as 1.61 to 200.00 μIU/mL. The RUD in the RIA analyses was 160 ± 41 % (Fig. 2b).

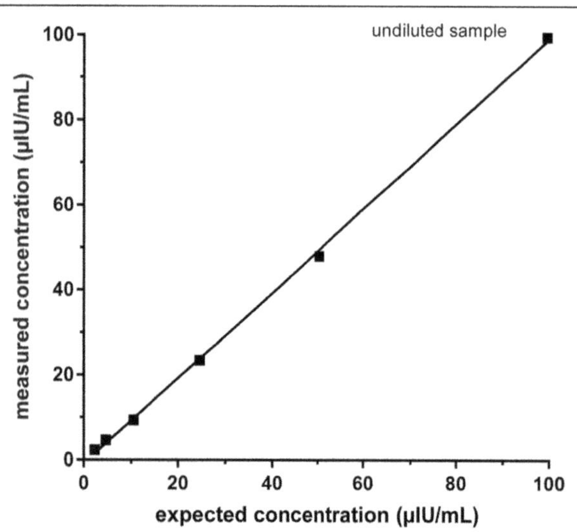

Fig. 1 Linearity upon dilution in porcine-specific insulin ELISA. Linear regression analysis indicates best-fit line of $Y = 0.9987\,x - 0.7385$ ($r^2 = 0.9994$, $P = < 0.0001$)

Table 2 Information about the number of samples which were measureable within the analytical ranges of the three assays investigated, without previously dilution

Assay	Number of samples	
	Basal, fasted samples	Stimulated OGT samples
ELISA	17/20	11/20
RIA	19/20	20/20
CLIA	20/20	19/20

One of the stimulated samples from the OGT procedure was not measureable within the analytical range (1.0 to 300 μIU/mL) in the CLIA (reported as > 300 μIU/mL) (Table 2). The RUD in the CLIA analyses was 101 ± 11 % (Fig. 2c).

A comparison of the results provided by the three methods of analysis of basal, fasted blood samples for equine insulin indicated statistically significant differences (Fig. 3). Results provided by the CLIA for basal blood samples were significantly ($p < 0.001$) lower compared to results supplied by ELISA or RIA (Fig. 3b, c). Furthermore, differences between ELISA and CLIA in stimulated samples were statistically significant (Fig. 4b), whereas results supplied by ELISA and RIA and RIA and CLIA did not differ significantly (Fig. 4a, c). Despite these differences between insulin concentrations measured by ELISA, RIA and CLIA, Spearman correlations and Deming regression analyses revealed significant correlations and linear relations between all three assays tested (Fig. 5a–c). Bland-Altman Plot analyses showed good agreement with a small bias of −0.44 μIU/mL (95 % limits of agreement: −18.12 to 18.99) comparing the results of ELISA and RIA (Fig. 6a), whereas variations in the results supplied by ELISA compared to CLIA (Fig. 6b; bias: 17.65 μIU/mL; 95 % limits of agreement: −13.92 to 49.22) and CLIA to RIA (Fig. 6c; bias: −8.39 μIU/mL; 95 % limits of agreement: −56.76 to 39.98) occurred. Additionally, the range of variation between results from ELISA and CLIA and between CLIA and RIA increased with elevated insulin concentrations (Fig. 6b, c).

The average recovery rate of equine insulin in 33 samples after one cycle of refreezing and thawing was 97 ± 17 %, but the corresponding range of recovery stretched from 71 to 158 %. In total, insulin concentrations were significantly ($p = 0.0232$) lower in the second measurement (54.99 ± 40.95 μIU/mL) compared to the first (58.5 ± 43.29 μIU/mL). When samples were subdivided into concentration range groups of low (3.51–15.21 μIU/mL), medium (30.42–90.09 μIU/mL) and high (92.43–125.19 μIU/mL) concentrations, the range as well as the mean recovery between the three concentration ranges differed, but there was no statistically significant difference between the first and second measurement when samples were clustered in subgroups depending on concentration. The mean

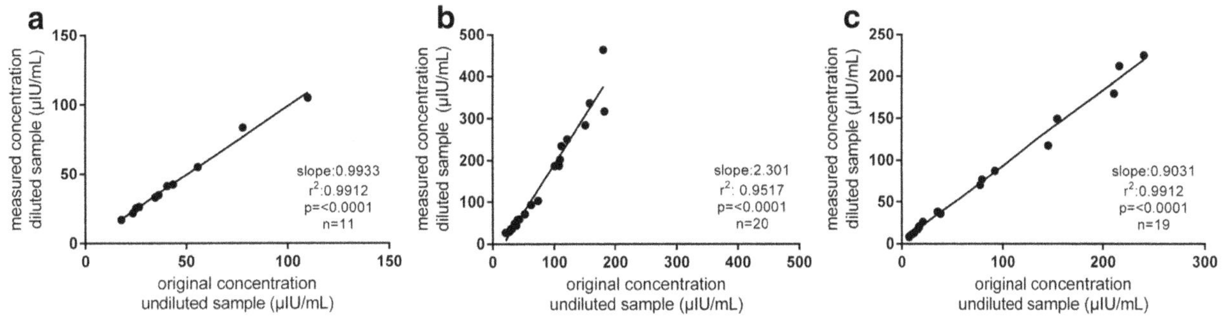

Fig. 2 Recovery upon dilution (RUD) of equine insulin in the three assays investigated: **a** ELISA (98 ± 4 %), **b** RIA (160 ± 41 %) and **c** CLIA (101 ± 11 %). The RUD was calculated as percentage recovery of the insulin concentration in the 1:4 diluted sample related to the corresponding undiluted sample. Samples were diluted with commercially available sample buffer[6]

recovery was 98 ± 16 % (range: 79 to 129 %) in the low, 99 ± 24 % (range: 74 to 158 %) in the medium and 92 ± 12 % (range: 71 to 107 %) in the high concentration group.

Discussion

Most immunoassays used in veterinary medicine for analyzing insulin concentrations in equine serum or plasma samples were originally designed for human diagnostics and research, and specific immunoassays for the quantification of equine insulin are not commercially available. A human-specific insulin radioimmunoassay[8] was used in a variety of equine studies [3, 11, 12] in which several IR and ID diagnostic tests and related reference ranges for plasma and serum insulin had been established. Unfortunately, this assay is no longer available, thus, there is an urgent need to reevaluate the reference ranges established on analyses from samples measured with other approaches for measuring insulin.

Reproducibility, recovery upon dilution and measuring range of assays

The CVs calculated for the ELISA showed that insulin concentration-dependent differences in CVs occurred and that the CVs calculated by the use of equine serum samples were higher than the CVs stated in the manufacturer's protocol. Very high or low insulin concentrations could not be detected by all assays due to the various analytical ranges of the assays. Therefore, as samples vary greatly in insulin concentration when obtained in fasted conditions or in dynamic diagnostic tests with induced insulin secretion, selection of an assay with an appropriate analytical range is important for the detection of very high or low insulin concentrations. Moreover, our results show that samples with high insulin concentrations, similar to those that can occur in patients with IR or ID, needed to be diluted in some cases to obtain results. Although the ELISA offered the smallest analytical range of the three assays tested, its excellent RUD results enabled the coverage of a broad range of insulin concentrations. Using this ELISA allowed valid measurements of earlier diluted samples of dynamic diagnostic tests. By contrast, using the RIA, all samples could be measured without dilution in the analytical range of the assay. However, RUD experiments did not provide satisfactory results. In a substantial amount of samples, too high insulin concentrations were obtained for the diluted samples compared to the corresponding undiluted samples. The CLIA offered the widest analytical

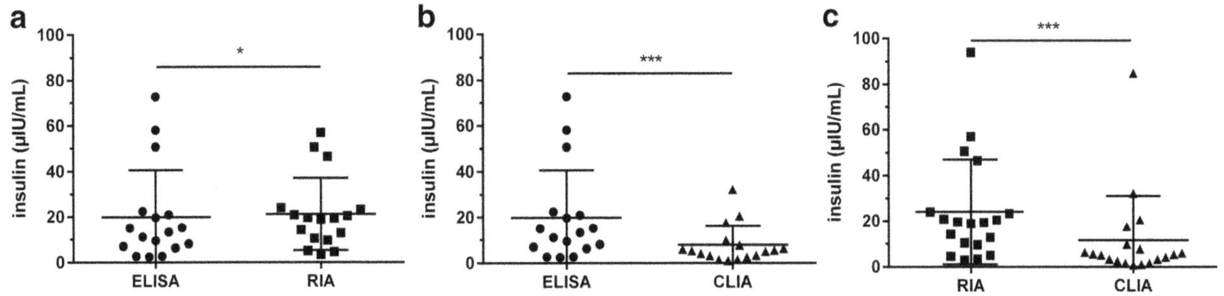

Fig. 3 Insulin concentrations measured in basal, fasted blood samples by ELISA, RIA and CLIA. Data were analyzed by Wilcoxon matched-pairs rank test. **a** ELISA and RIA ($n = 17$) **b** ELISA and CLIA (n =1 7) **c** RIA and CLIA ($n = 20$). *$p < 0.05$, **$p < 0.01$, ***$p < 0.001$

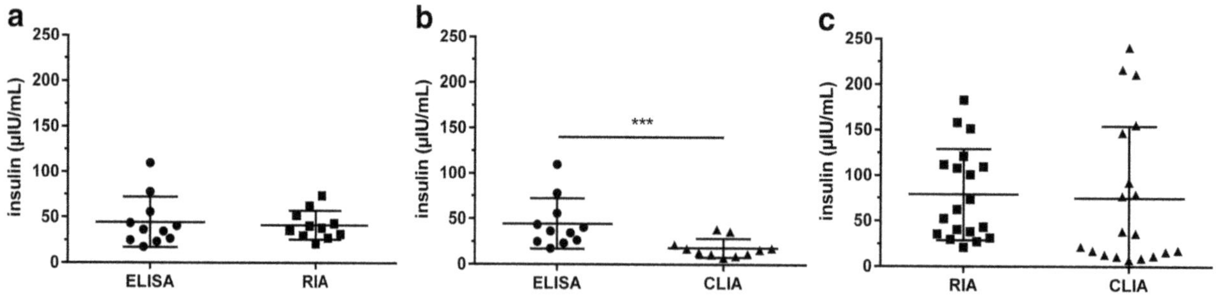

Fig. 4 Insulin concentrations measured in stimulated blood samples from OGT procedure by ELISA, RIA and CLIA. Data were analyzed by Wilcoxon matched-pairs rank test. **a** ELISA and RIA ($n = 11$) **b** ELISA and CLIA ($n = 11$) **c** RIA and CLIA ($n = 19$). $*p < 0.05$, $**p < 0.01$, $***p < 0.001$

range when compared to the other two assays, but also could not measure all high insulin concentration samples. This assay, similar to the ELISA, exhibited good RUD results, leading to the conclusion that appropriate dilution of samples in this assay do not falsify results.

Dilution buffers

We used commercially available sample buffer[6] to dilute the samples in this study. This buffer was designed for use in the ELISA investigated. Öberg and colleagues [8] evaluated the RUD and linearity of dilution in this ELISA with physiologic saline as a dilution medium in four dilution steps and found similar results regarding linearity, but poorer RUD of 92 to 122 %. Dilution of serum samples with zero calibrator provided in the ELISA yielded higher recovery rates of 102.4 ± 20.5 % in another study [13]. Therefore, dilution of samples with the manufacturer's commercially available sample buffer[6] seems to improve results obtained from diluted samples in this assay. Tinworth and colleagues [6] recommended that samples with high insulin concentrations which required dilution should be diluted with charcoal stripped equine plasma instead of the manufacturer's diluent in order to improve results obtained by analysis with a human-specific insulin RIA[8] mentioned previously. One limitation in our study is that all our samples have been diluted with commercially available sample buffer[6] designed for use in the equine-optimized, porcine-specific ELISA and were then used in the different assays. Therefore, results of this RUD analysis may also be impaired by interactions of the dilution medium and assay components. The RUD results may be improved by using an appropriate method-dependent dilution medium.

Insulin molecule stability

The stability of insulin in blood samples is a commonly discussed issue in both human and veterinary medicine. Optimal pre-analytical handling of samples is essential for valid analytical results, but is often not feasible in the daily routine of an ambulatory practice. The recovery range observed indicates quite variable results. Some of the refrozen and thawed samples measured had a lower insulin concentration in the second analysis, whereas others had higher insulin concentrations. Parts of the variations and effects between the first and second measurement may be due to inter-assay variations, but, as already presented, inter-assay CVs for low equine insulin concentration samples and high insulin concentration samples in the ELISA were within the generally accepted variation range for immunoassays [14]. A study

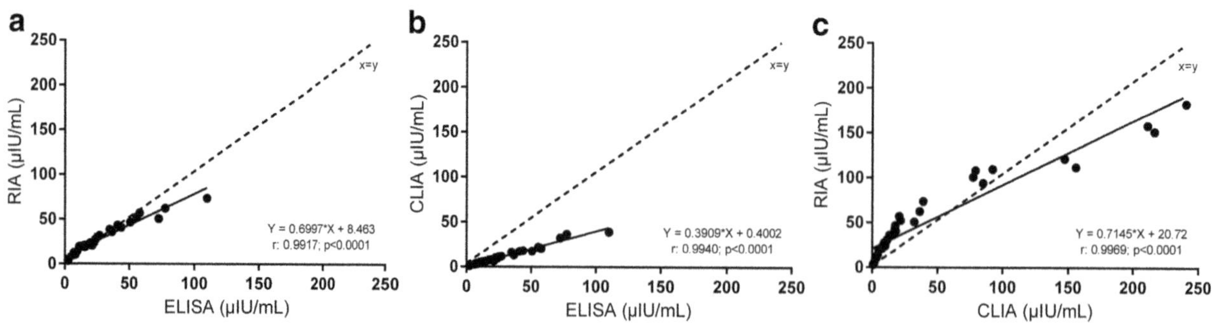

Fig. 5 Scatterplot of measurement results for basal and stimulated samples ELISA compared to **a** RIA and **b** CLIA. **c** Scatterplot of results supplied by CLIA compared to RIA

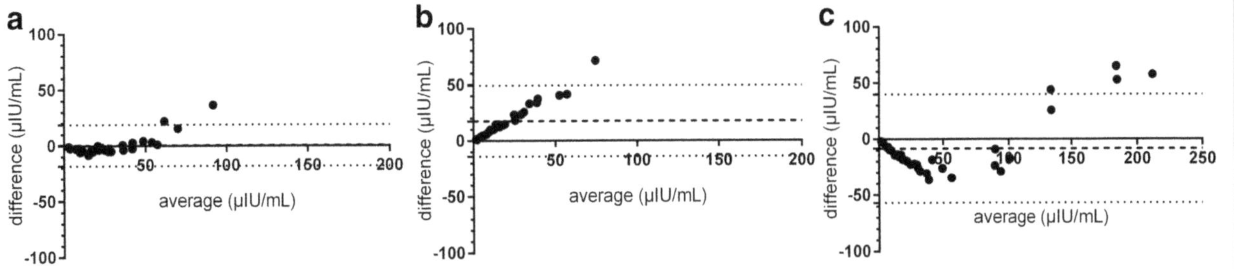

Fig. 6 Bland-Altman plot of average compared to difference **a** ELISA minus RIA ($n = 28$), **b** ELISA minus CLIA ($n = 28$) and **c** CLIA minus RIA ($n = 39$). *Dashed line* = bias, *pointed line* = 95 % limits of agreements

by Livesey and colleagues [15] indicated that repeated thawing and freezing affected the insulin concentration in human plasma samples collected during intravenous glucose tolerance testing procedures. Insulin concentrations had already declined after the first freezing and thawing cycle and exhibited a continuous decline during following freezing and thawing cycles. A more recent study by Mechanic and colleagues [16] examined the effect of short-term refrigerated storage at ≤ −70 °C on insulin in human blood samples and found no significant effect in a time period of 72 h and also no significant effect of delays in freezing following centrifugation 2 h post-collection, whereas they observed declined insulin concentrations following delays in centrifugation. Furthermore, repeated freezing and thawing of serum did not affect the concentrations of insulin in serum samples from beagle dogs [17]. Borer-Weier and colleagues [13] evaluated the stability of equine insulin after freezing and thawing in their study and showed similar results to our study, with large ranges of differences in the samples analyzed with human-specific RIA[8] and a high-range porcine insulin ELISA. One limitation in our study design, as well as in other previous studies, was that serum insulin concentrations were not measured directly after the sampling procedure, prior to first freezing, so the effect of the first, initial freeze remains unknown. Taking into account the variable effects of freezing and thawing on the insulin concentrations in the samples we analyzed, we recommend that samples are not frozen and thawed any more than necessary in order not to receive false results.

Conversion factors for insulin measures

Various conversion factors have been established in the past due to the differences in insulin binding efficacy. Data had to be converted from µg/L into the commonly used SI unit µIU/mL to compare the results from ELISA with results derived from RIA and CLIA. In previous studies in which insulin analysis had been performed with the equine-optimized porcine-specific ELISA, data was either presented in µg/L [9, 10, 18] or converted into mU/L [19]. In the latter case, conversion was performed without an exact statement regarding the conversion factor, which was suggested to be 10 in an earlier publication by the same author [9]. In addition, the factor 28.69 (1 µg/L = 28.69 for µIU/mL) [20] for the conversion of human insulin was discussed. However, for the ELISA used in our study, the cross-reactivity with the human insulin molecule is only 22 %, as declared by the manufacturer. Therefore, the use of both factors for conversion of the data generated by the equine-optimized, porcine-specific ELISA might be questionable. Thus, another equine-specific conversion factor (1 µg/L = 117 mU/L) was given by the manufacturer for this assay [21] and was used for the calculation of data in our study.

Comparison of methods and assay insulin antibodies

Equine insulin has a lower molecular weight (5.748 kDa) compared to human (5.808 kDa) and porcine insulin (5.778 kDa) and consists of 51 amino acids. The equine insulin differs from the human and porcine insulin molecule in the amino acid sequence [22, 23]. Even slight differences in the amino acid sequences between species and, thus, secondary structure can result in impaired antibody binding and can, therefore, cause variations in efficacy when measuring insulin by different immunoassay-based methods. Comparing the results of converted data, all three methods provided different insulin concentrations within the same samples. The CLIA results observed in this study were markedly lower than the results supplied by the ELISA and the RIA. Data analyses with Bland-Altman plots and Deming regression analyses indicated that there was no constant systemic error, meaning that one method measures consistently higher or lower than the other ones. The ELISA supplied higher insulin concentrations than the CLIA and the difference between both methods even increased with rising concentrations, indicating the presence of proportional error. The CLIA also provided lower results than the RIA for samples with concentrations under 100 µIU/mL, but when the concentration exceeded 100 µIU/mL, the CLIA supplied higher results for five samples compared to the RIA. These differences between insulin concentration results provided by the three assays may be caused by underlying antibodies used in the different assays. The ELISA and the RIA were based on antibodies directed against porcine insulin, which differs in amino acid 9 of the A chain (glycine in equine insulin and serine in porcine insulin), whereas human and equine insulin differ in amino acid 9 on the A chain and amino acid 30 of the B chain (alanine in equine insulin and threonine in human insulin) [22]. Differences at amino acid 9 on the A chain are not considered to be important for biological activity, but changes in amino acid 30 on the B chain were discussed to affect the 3D structure [23, 24]. Therefore, the markedly lower results provided by the CLIA may be caused by a loss of homology in this position and variation in 3D structure, which may impair sufficient antigen-antibody binding by steric interference. Furthermore, alterations in the antigenic epitope for the anti-human insulin antibody used in the CLIA may cause insufficient antigen-antibody binding. Variations in insulin concentrations, such as occurred in our study, have already been described by several authors for other assays and methods used frequently. Tinworth and colleagues [6] compared the equine-optimized, porcine-specific insulin ELISA and a modified version of the human-specific RIA[8] in their investigations. Both assays gave different insulin concentration results for the same samples as well, and underestimated the insulin concentration compared to liquid chromatography and high-resolution/high-accuracy mass spectrometry (LC-MS) [6]. The LC-MS provides absolute amounts of insulin and could be assessed as a gold standard to measure insulin concentrations in blood samples. Due to a lack of LC-MS in our study, potential over- or underestimation of the assays examined could not be evaluated. Banse and colleagues [7] compared the human-specific RIA[8] with another commercially available CLIA[9] which is often used for the measurement of equine insulin in both clinical practice and research. Both the human-specific RIA[8] and the CLIA[9] resulted in poor recovery rates within each assay and showed poor accordance. They, furthermore, examined the recovery of equine insulin standard within both assays. The recovery rate was poor in both assays, but the CLIA[9] investigated showed recovery rates of only up to 10 %. However, no equine insulin standard was available in our experiments; therefore, we could not examine the equine insulin standard recovery in the three immunoassays investigated in this study.

Clinical relevance of valid insulin measures

The clinical relevance of result variation becomes obvious when focusing on basal samples. According to the current state in equine veterinary practice, analysis of basal samples from fasted patients as a screening method for IR and ID accounts for a large proportion of insulin measurements in veterinary diagnostics. In veterinary practice, basal insulin concentrations of 20 μIU/mL were stated as cut-off values to distinguish between healthy and insulin-resistant or insulin-dysregulated horses in the consensus statements of the American College of Veterinary Internal Medicine [12]. On closer inspection, this cut-off value was suggested as being valid for sample analyses resting upon the Coat-A-Count insulin radioimmunoassay[8] (Siemens Medical Solutions Diagnostics, Los Angeles, CA), Immulite insulin solid-phase chemiluminescent assay[9] (Siemens Medical Solutions Diagnostics) and DSL-1600 insulin radioimmunoassay (Diagnostic Systems Laboratory Inc, Webster, TX) by the authors of the consensus statement. However, this cut-off value is often used without consideration of the underlying analysis method. Thus, assay-dependent differences in the insulin concentration levels measured can cause false classification of patients. Calculations of indices, which are based on basal insulin and glucose concentrations for estimation of IR, were used more commonly [25] and, therefore, could result in false diagnosis of IR. Exact differentiation between 15 and 30 μIU/mL may be very important for the interpretation of basal samples. In six of the seventeen animals investigated, analysis of basal, fasted blood samples by the three assays investigated led to different classification as insulin-sensitive or insulin-dysregulated depending on the data used. One of our samples illustrates this problem. The CLIA supplied an insulin concentration of 7.8 μIU/mL, whereupon this horse is classified as insulin-sensitive when considering the cut-off value of 20 μIU/mL recommended by the American College of Veterinary Internal Medicine. By contrast, ELISA and RIA supplied insulin concentrations of 50.73 μIU/mL and 46.53 μIU/mL, respectively, after which this horse is classified as insulin-resistant or insulin-dysregulated on the basis of basal, fasted blood analysis. Both commercial laboratories we commissioned with the analyses of our samples provided laboratory-specific reference values for insulin in fasted serum blood samples of horses. The reference range stated for the RIA was < 20 μIU/mL in fasted samples and was < 23.4 μIU/mL for the CLIA. These findings indicate that even consideration of laboratory and assay-specific reference ranges leads to different classification of this horse, depending on the assay used to determine the basal, fasted blood insulin concentration. As a consequence, independent of laboratory and assay-specific reference ranges or cut-off values, a general harmonization of equine insulin analyses is urgently required to allow accurate and safe diagnosis of hyperinsulinemia as one of the leading symptoms in horses suffering from EMS.

Conclusion

In conclusion, analyses of equine serum samples for insulin with three different immunoassay methods revealed three different results for the same sample. Adjustment of insulin measurements and methods is essential to allow the consistence of information of several studies and research approaches, enabling evidence-based criteria to compare insulin concentrations consistently between assays, studies and laboratories. Since harmonization is an interminable process, further assay comparison research studies would be helpful to allow the comparison of results and cut-off values determined with the use of one method to another. This study illustrates the importance of considering the insulin analysis method when results of different laboratories or studies are compared or interpreted for the diagnosis of insulin-related endocrine and metabolic pathologies.

Endnotes

[1]Equine Insulin ELISA, Mercodia AB, Uppsala, Sweden.

[2]Porcine Insulin RIA, Millipore, St. Charles, MO, USA.

[3]ADVIA Centaur Insulin Assay, Siemens Healthcare Diagnostics GmbH, Eschborn, Germany.

[4]Vacuette® Greiner Bio One, Frickenhausen, Germany.

[5]Animal Insulin Control, 362 Mercodia AB, Uppsala, Sweden.

[6]Diabetes Sample Buffer, Mercodia AB, Uppsala, Sweden.

[7]GraphPad Prism, Version 6.02 for Windows, GraphPad Inc. La Jolla, CA, USA.

[8]Coat-a-Count Insulin RIA, Siemens Medical Solutions Diagnostics, Los Angeles, CA, USA.

[9]Immulite Insulin Assay, Siemens Healthcare, Malvern, PA, USA.

Abbreviations

CV: Coefficient of variation; EMS: Equine metabolic syndrome; Fig: Figure; ID: Insulin dysregulation; IR: Insulin resistance; ELISA: Enzyme-linked immunosorbent assay; RIA: Radioimmunoassay; CLIA: Chemiluminescent immunoassay; RUD: Recovery upon dilution; LC-MS: Liquid chromatography and high-resolution/high-accuracy mass spectrometry; SD: Standard deviation; Tab: Table

Acknowledgements

The authors are grateful for financial support from Boehringer Ingelheim Vetmedica GmbH. Furthermore the authors thank Mrs. Kathrin Hansen for technical support in the laboratory and language editor Philip Saunders for the validation of the English language used in this article.

Funding

This research was financially supported by Boehringer-Ingelheim-Vetmedica GmbH.

Authors' contributions

TW, KH and KF designed the study. TW collected and analyzed data, wrote the manuscript and made figures. KH and KF helped to edit the manuscript. All authors read and approved the final manuscript.

Competing interests

The authors declare that they have no competing interests.

Consent for publication

Not applicable.

Author details

[1]Department of Physiology, University of Veterinary Medicine Hannover, Foundation, Bischofsholer Damm 15, 30173 Hannover, Germany. [2]Institute of Animal Science, Faculty of Agricultural Sciences, University of Hohenheim, Fruwirthstraße 35, 70593 Stuttgart, Germany. [3]Clinic for Horses, University of Veterinary Medicine Hannover, Foundation, Bünteweg 9, 30559 Hannover, Germany.

References

1. Schuver A, Frank N, Chameroy KA, Elliott SB. Assessment of insulin and glucose dynamics by using an oral sugar test in horses. J Equine Vet Sci. 2014;34:465–70.
2. Ralston SL. Insulin and glucose regulation. Vet Clin North Am Equine Pract. 2002;18:295–304.
3. Eiler H, Frank N, Andrews FM, Oliver JW, Fecteau KA. Physiologic assessment of blood glucose homeostasis via combined intravenous glucose and insulin testing in horses. Am J Vet Res. 2005;66:1598–604.
4. Rijnen KE, van der Kolk JH. Determination of reference range values indicative of glucose metabolism and insulin resistance by use of glucose clamp techniques in horses and ponies. Am J Vet Res. 2003;64:1260–4.
5. Frank N, Tadros EM. Insulin dysregulation. Equine Vet J. 2014;46:103–12.
6. Tinworth KD, Wynn PC, Boston RC, Harris PA, Sillence MN, Thevis M, Thomas A, Noble GK. Evaluation of commercially available assays for the measurement of equine insulin. Domest Anim Endocrinol. 2011;41:81–90.
7. Banse HE, McCann J, Yang F, Wagg C, McFarlane D. Comparison of two methods for measurement of equine insulin. J Vet Diagn Invest. 2014;26:527–30.
8. Öberg J, Bröjer J, Wattle O, Lilliehöök I. Evaluation of an equine-optimized enzyme-linked immunosorbent assay for serum insulin measurement and stability study of equine serum insulin. Comp Clin Path. 2011;21:1291–300.
9. Brojer JT, Nostell KE, Essen-Gustavsson B, Hedenstrom UO. Effect of repeated oral administration of glucose and leucine immediately after exercise on plasma insulin concentration and glycogen synthesis in horses. Am J Vet Res. 2012;73:867–74.
10. Brunner J, Wichert B, Burger D, von Peinen K, Liesegang A. A survey on the feeding of eventing horses during competition. J Anim Physiol Anim Nutr (Berl). 2012;96:878–84.
11. Frank N, Elliott SB, Brandt LE, Keisler DH. Physical characteristics, blood hormone concentrations, and plasma lipid concentrations in obese horses with insulin resistance. J Am Vet Med Assoc. 2006;228:1383–90.
12. Frank N, Geor RJ, Bailey SR, Durham AE, Johnson PJ, American College of Veterinary Internal M. Equine metabolic syndrome. J Vet Intern Med. 2010;24:467–75.
13. Borer-Weir KE, Bailey SR, Menzies-Gow NJ, Harris PA, Elliott J. Evaluation of a commercially available radioimmunoassay and species-specific ELISAs for measurement of high concentrations of insulin in equine serum. Amn J Vet Res. 2012;73:1596–602.
14. European Medicines Agency: Guideline on bioanalytical method validation. 2011. http://www.ema.europa.eu/docs/en_GB/document_library/Scientific_guideline/2011/08/WC500109686.pdf. Accessed 21 Jan 2016.
15. Livesey JH, Hodgkinson SC, Roud HR, Donald RA. Effect of time, temperature and freezing on the stability of immunoreactive LH, FSH, TSH, growth hormone, prolactin and insulin in plasma. Clin Biochem. 1980;13:151–5.

16. Mechanic L, Mendez A, Merrill L, Rogers J, Layton M, Todd D, Varanasi A, O'Brien B, Meyer Iii WA, Zhang M, et al. Planned variation in preanalytical conditions to evaluate biospecimen stability in the National Children's Study (NCS). Clin Chem Lab Med. 2013;51:2287–94.
17. Reimers TJ, McCann JP, Cowan RG, Concannon PW. Effects of storage, hemolysis, and freezing and thawing on concentrations of thyroxine, cortisol, and insulin in blood samples. Proc Soc Exp Biol Med. 1982;170:509–16.
18. Nostell KE, Essen-Gustavsson B, Brojer JT. Repeated post-exercise administration with a mixture of leucine and glucose alters the plasma amino acid profile in Standardbred trotters. Acta Vet Scand. 2012;54:7.
19. Brojer J, Lindase S, Hedenskog J, Alvarsson K, Nostell K. Repeatability of the combined glucose-insulin tolerance test and the effect of a stressor before testing in horses of 2 breeds. J Vet Intern Med. 2013;27:1543–50.
20. Volund A. Conversion of insulin units to SI units. Am J Clin Nutr. 1993;58:714–5.
21. Mercodia AB. Technical Note No: 34–0128. How to compare insulin values between mU/L and µg/L when using Mercodia Insulin ELISAs (Version 1.0). https://www.mercodia.com. Acessed 17 Feb 2015.
22. Ho EN, Wan TS, Wong AS, Lam KK, Stewart BD. Doping control analysis of insulin and its analogues in equine plasma by liquid chromatography-tandem mass spectrometry. J Chromatogr A. 2008;1201:183–90.
23. Kuuranne T, Thomas A, Leinonen A, Delahaut P, Bosseloir A, Schanzer W, Thevis M. Insulins in equine urine: qualitative analysis by immunoaffinity purification and liquid chromatography/tandem mass spectrometry for doping control purposes in horse-racing. Rapid Commun Mass Spectrom. 2008;22:355–62.
24. Conlon JM. Evolution of the insulin molecule: insights into structure-activity and phylogenetic relationships. Peptides. 2001;22:1183–93.
25. Treiber KH, Kronfeld DS, Hess TM, Boston RC, Harris PA. Use of proxies and reference quintiles obtained from minimal model analysis for determination of insulin sensitivity and pancreatic beta-cell responsiveness in horses. Am J Vet Res. 2005;66:2114–21.

4

FOXO1, PXK, PYCARD and *SAMD9L* are differentially expressed by fibroblast-like cells in equine synovial membrane compared to joint capsule

Line Nymann Thomsen[1], Preben Dybdahl Thomsen[2], Alison Downing[3], Richard Talbot[3] and Lise Charlotte Berg[1*]

Abstract

Background: The synovial membrane lines the luminal side of the joint capsule in synovial joints. It maintains joint homeostasis and plays a crucial role in equine joint pathology. When trauma or inflammation is induced in a joint, the synovial membrane influences progression of joint damage. Equine synovial membrane research is hampered by a lack of markers of fibroblast-like synoviocytes (FLS) to distinguish FLS from other fibroblast-like cells in musculoskeletal connective tissues. The aim of this study is to identify potential FLS markers of the equine synovial membrane using microarray to compare between gene expression in equine synovial membrane and the joint capsule in metacarpophalangeal joints.

Results: Microarray analysis of tissues from 6 horses resulted in 1167 up-regulated genes in synovial membrane compared with joint capsule. Pathway analysis resulted in 241 candidate genes. Of these, 15 genes were selected for further confirmation as genes potentially expressed by fibroblast-like synoviocytes. Four genes: *FOXO1, PXK, PYCARD* and *SAMD9L* were confirmed in 9 horses by qPCR as differentially expressed in synovial membrane compared to joint capsule.

Conclusions: In conclusion, *FOXO1*, *PXK*, *PYCARD* and *SAMD9L* were confirmed as differentially expressed in synovial membrane compared to joint capsule. These four genes are potential markers of fibroblast-like synoviocytes of the synovial membrane. As these genes are overexpressed in synovial membrane compared to joint capsule, these genes could shed light on synovial membrane physiology and its role in joint disease.

Keywords: Synovial membrane, Fibroblast-like synoviocytes, *FOXO1, PXK, PYCARD, SAMD9L*, Equine, Marker

Background

Horses are susceptible to joint disease and trauma leading to joint pathology. Joint pathology involves complex processes, and joint inflammation affects several tissues within the joint [1]. Regardless which intra-articular tissue type is first affected, the synovial membrane orchestrates and reinforces inflammatory responses of joints [2, 3]. Hence, the synovial membrane is key to enhance understanding of the pathophysiological processes within synovial joints.

In the healthy joint, the synovial membrane maintains joint homeostasis [4]. From its position as luminal lining of the joint capsule, the synovial membrane facilitates diffusion of plasma ultra-filtrate through synovial extracellular matrix and produces lubricating synovial additives [4]. Two cell types constitute the majority of the synovial membrane, macrophage-like synoviocytes (MLS) and fibroblast-like synoviocytes (FLS) [5]. MLS are macrophages resident to the joint [6, 7] which tend to distribute unevenly adjacent to the joint lumen [8] and to congregate at the top of synovial villi [6]. From this localisation within the synovial membrane, the MLS engulf foreign substances [6, 7]. MLS are characterized by expression of typical general resident macrophage-markers such as CD11b, CD14, CD68 and

* Correspondence: lcb@sund.ku.dk
[1]Department of Veterinary Clinical Sciences, University of Copenhagen, Hoejbakkegaards alle 5, 2630 Taastrup, Denmark
Full list of author information is available at the end of the article

CD206 [9–11]. The fibroblast-like synoviocytes (FLS) are modified fibroblasts with cytoplasmatic processes reaching to the joint lumen. FLS produce and secrete lubricating additives to the synovial fluid and produce and maintain the synovial extracellular matrix (ECM). FLS are the dominating cell-type in the synovial intima (80%) and the fibroblastic lineage results in characteristic expression of general fibroblast markers such as vimentin and prolyl hydroxylase [12]. Due to FLS's specific functions in the synovial lining [13], FLS differ from other fibroblasts, but there are only a few reports of selective markers of FLS differentiating them from other musculoskeletal fibroblasts. In humans, the hyaluronan precursor Uridine diphosphoglucose dehydrogenase (UDPGD) has been proposed as a specific marker of FLS [14–16]. In mice and humans, the cell contact mediating protein Cadherin-11, has been suggested as marker of FLS. Cadherin-11 is believed to mediate synovial lining integrity and to be implicated in synovial inflammation [3, 17]. None of these molecules have been investigated in horses and new equine FLS markers in the synovial membrane could elucidate the normal function of the synovial membrane and potentially contribute to unravel mechanisms behind the role of FLS in arthritic conditions. In addition, novel markers could improve the consistency and reproducibility of FLS based studies.

Thus, the aim of this study was to identify new markers of equine healthy FLS in the healthy joint. We compared the gene expression of the synovial membrane to the gene expression of the fibrous joint capsule using a microarray technique. This is a comparison of two tissues each containing different types of fibroblasts, the FLS of the synovial membrane and fibroblasts from the dense connective tissue of the joint capsule. This comparison was made to highlight differences in gene expression between the two types of cells and thus search out markers not previously explored. Based on the microarray data we used pathway analysis to select markers relevant to FLS, and we confirmed the selected target genes by qPCR to support general applicability of the markers.

Methods

Microarray experiment

Tissue samples

Synovial membrane and joint capsule tissue samples were collected from six horses of mixed breed (aged 15–25 years). The horses were euthanized at the Faculty of Health and Medical Sciences, University of Copenhagen according to regulations of Danish law for reasons unrelated to orthopaedic disorders in the metacarpophalangeal joints. The joint capsule was transected and the macroscopic appearance of the joint evaluated. Synovial membrane and joint capsule were excised from the entire fetlock joint to avoid site-specific variation within the joint. The synovial membrane was separated from the joint capsule by careful dissection. This method was verified by embedding dissected synovial membrane and joint capsule separately in paraffin and subsequently staining with Mayer's Hematoxylin[1] and Eosin[2] prior to histologic evaluation. The separated exised tissue from the six horses generated a yield of 0.6–4.0 g of synovial membrane and joint capsule from each joint, which was snap frozen in liquid nitrogen. To evaluate for signs of acute inflammation a full thickness tissue sample was excised from the proximo-dorsal part of the joint, fixed in 4% paraformaldehyde[3] in PBS,[4] and embedded in paraffin. This sample was later sectioned and stained with Mayer's Hematoxylin and Eosin and evaluated for histologic signs of acute inflammation. The histologic signs of acute inflammation evaluated were: Cellular infiltrates in general and in vascular surroundings and/or hyperplasia of the synovial membrane. Tissues from joints with macroscopic or cellular signs of inflammation were excluded from the study.

RNA extraction

For the microarray study, RNA was extracted using Trizol. Briefly, 1 ml Trizol[5] was added to the tissue and the tissue was homogenised in a Fastprep[6] homogeniser for 20 s at 4 m/s. Following 5 min incubation, 0.2 mL of 1-bromo-3-chloro propane[7] was added for phase separation and the tubes were shaken vigorously by hand for 15 s. Then the samples were incubated for 2 min, and centrifuged at 12,000 *g* for 15 min. 500 μL of the aqueous upper phase was carefully aspirated and transferred to new RNAse free 1.5 mL tubes. The RNA was precipitated by adding 1 μL of linear polyacrylamide,[8] gentle mixing and the addition of 500 μL isopropyl alcohol.[9] Following 10 min incubation the tubes were centrifuged at 14,000 *g* for 10 min. The supernatant was carefully removed and the RNA pellet washed with 1 mL 75% Ethanol, the tubes inverted and centrifuged for 5 min at 14,000 *g*. After a second wash in 75% ethanol the RNA pellet was left to air dry until any residual liquid had evaporated, approximately 5 to 15 min. The RNA was re-dissolved in 20 μL of MilliQ water. The yield of RNA was quantified using a NanoDrop[10] spectrophotometer (Labtech International Ltd., UK) and the quality assessed using an Agilent RNA 6000 nanochip on the Agilent Bioanalyser.[11]

Amplification of the RNA, aminoallyl incorporation and dye coupling

Total RNA was amplified using the MessageAmp™ II aRNA Amplification Kit,[12] and spiked control RNA's were added according to the Agilent Spike in Kit[13] with 500 ng RNA and 5 μl Spike mix added to each well and the volume adjusted to 10 μl by adding nuclease-free water.

Reverse transcription of RNA was initiated by adding 10 μl of a reverse transcription mastermix containing reverse transcriptase, NTP and T7 Olio d(T) primer and incubated according to the manufacturer's instructions. After 2 h at 42 °C the temperature was dropped to 16 ° C and second strand synthesis started using a mixture of fresh NTP with DNAse and Ribonuclease H and incubated for a further 2 h. Double stranded cDNA was purified on a spin column before aRNA synthesis.

The aRNA was synthesized by in vitro transcription. In vitro transcription was performed by adding 16 µl cDNA to a mixture consisting of 3 µl 5-(3-aminoallyl)-UTP,[14] 4 µl T7 ATP solution, 4 µl T7 CTP solution, 4 µl T7 GTP solution, 2 µl T7 UTP solution, 4 µl T7 10 Reaction Buffer and 4 µl Enzyme mix, according to manufacturer's instructions. The aRNA was purified according to the MessageAmp™ II aRNA Amplification Kit protocol. The yield of the aRNA was assessed using a NanoDrop. The uniformity of the size of the aRNA was assessed using a RNA 6000 nanochip on the Agilent Bioanalyser.

Cy3[15] dye was coupled to aminoallyl UTP (aaUTP) in the amplified aRNA, according to manufacturer's instructions. The yield and specific activity of the eluate containing the purified labelling reactions was quantified using a Nanodrop TM Spectrophotometer (Labtech International Ltd., UK).

Fragmentation and hybridisation of the samples to the Agilent Array

Twelve Agilent equine microarrays were used in the study. Each microarray was hybridised with a single labelled RNA sample. The Cy 3 labelled aRNA (10 µl) was mixed with 10 µl blocking Agent, 31.8 µl nuclease free water, and 2.2 µl fragmentation buffer,[16] and incubated at 60 °C for 30 min. The reaction was stopped by adding GE hybridization buffer. The samples were loaded onto the array and the sides were placed in rotisserie in a hybridization oven at 65 °C for 17 h.

Scanning of array and subsequent data analysis

After hybridization, the arrays were scanned on a 4200A axon scanner[17] using autoPMT and the appropriate GAL file.

Image files were imported using Agilent Feature extraction software[18] and Grids were manually fitted to the arrays according to the manual. The extracted intensities were imported into Partek Genomics Suite[19] where they were analysed using the gene expression workflow. The array data was normalised using a quantile normalisation where the software calculates the distribution model from the data files and normalize the probes according to the calculated distribution model. Analysis of variance was performed on the normalized data, with Bonferroni's correction applied.

The Agilent equine microarray was not fully annotated, but the annotation was improved by importing ENS numbers and LOC numbers from AgBase [18, 19]. A list of the accession numbers for each non-annotated probe was prepared and used to query the AgBase database using the online search tool.

The significantly up-regulated genes were imported into IPA [20]. Significantly overexpressed canonical pathways were explored.

Confimation of microarray data

Tissue samples

The microarray results were confirmed using qPCR on synovial membrane and joint capsule in nine horses (three of the horses from the microarray experiment and six additional horses of mixed breed (aged 2–22 years)). Equine synovial membrane and joint capsule was excised and processed as described previously, and presence of joint pathology was evaluated as previously described.

RNA extraction

RNA from the tissue samples was extracted using the Promega SV total RNA Isolation system,[20] according to manufacturer's instructions.

The yield of RNA was measured using a NanoDrop TM Spectrophotometer[21] and the quality assessed by using an Agilent Bioanalyser.[22]

Subsequently, the RNA was transcribed to cDNA by adding 5× MMLV RT buffer,[23] dNTP[24] (10 nM), random hexamer primer[25] (2 µg/µL), oligo(dt)[26] (0.5 µg/µL), RNasin RNase inhibitor[27] (40 U/µL) and MMLV reverse transcriptase enzyme[28] (200 U/ µL) to 800 ng RNA diluted in H_2O to a 25 µL reaction volume. The cDNA was transcribed at 25 °C for 10 min, 42 °C for 60 min, 95° C for 5 min. The cDNA was stored at −20 ° C and subsequently diluted to a concentration of 4 ng/µL for all samples.

Selection of reference genes

The following reference genes were tested for qPCR: 18S ribosomal RNA (18 S), glyceraldehyde-3-phosphate dehydrogenase (GAPDH), β-actin, mitochondrial ribosomal protein S7 (MRPS7), and 5-aminoimidazole-4-carboxamide ribonucleotide formyltransferase /IMP cyclohydrolase (ATIC). 18S was chosen as reference gene.

Primer design, candidate genes

The equine gene sequences corresponding to the 15 selected candidate genes were found in Ensembl [21]. If multiple transcript variants existed, primers were designed in conserved sequences. The transcript sequences (mRNA sequences) were inserted in the programme Primer 3 [22]. The chosen primer sequences were subsequently screened against the horse genome using BLAST ref. [23] to ensure primer specificity. Amplified gene products were confirmed for specificity through sequencing. Only primer sets with amplification efficiency between 1.8–2.2 were used. Primer sequences are listed in Table 1.

Table 1 Primers used to confirm the expression of candidate genes found in microarray experiment

Target	Accession	Forward primer	Reverse primer	Product size (base-pair)
ATXN1	[Genbank: XM_001496252]	CCCAAAAGCGAGAACTTCAG	TCCGTTTTCAAGTCCTCCAC	206
COL11A1	[Genbank: XM_001918115]	GGCAATTCCATTAACATGGTG	TTGATCCCAGGAATCGAAGT	151
COL28A1	[Genbank: XM_001494969]	AGACTCCGCTAGAGCTGCTG	AAGTCGTCCTTGCTGGAGAA	202
FSTL1	[Genbank: XM_001500510]	ACAAGAGGCCTGTGTGTGG	TGTCCGTCATAATCGACCTG	130
FOXO1	[Genbank: XM_005601179]	CCAGCCCAAACTACCAAAAA	AACACATTCTGGCCAAGGAC	241
GALNT3	[Genbank: XM_001496049]	CCTTGCTCTGTTGTTGGACA	TGCTGCATCTGTGTTTCTCC	153
GRN	[Genbank: XM_001489741]	TGTGAGGAGGGACTGAGGAC	TTGTTACGTGGCTTTCACCA	207
ITPR3	[Genbank: XM_005604053]	GCAGTTTGGGATGATGCAGT	CTGGGCTCTCGGTTCTTG	159
NF1	[Genbank: XM_005597955]	TGGCCCTGTACATGTTTCTG	TTGCTGACAGACGCAAATTC	172
PXK	[Genbank: XM_008519676]	CATTACCTCCACCTCCTCCA	GATCACAGGTTTCGGCTTTC	180
PYCARD	[Genbank: XM_001500509]	CCATCCTAGAGGCACTGGAA	CTCCGTACGCCTCCAGATAG	177
RGL2	[Genbank: XM_001497200]	GGATGGAGCTTCACACGATT	CAGGATGGCTACCCACATCT	235
SAMD9L	[Genbank: XM_005609186]	GAACCGGAAAACGTCTGTGT	GGGAGAAAGTCGGTGCATTA	171
SOS1	[Genbank: XM_005600032]	CACGAGAACCTGTGAGGACA	AAGTGCTTTGTCGAGGAGGA	249
TIMP3	[Genbank:NM_001081870]	CCTGCTACTACCTGCCTTGC	GGTCTGTGGCATTGATGATG	186

qPCR

The quantitative real time polymerase chain reaction (qPCR) was performed in 96 wells plates on Lightcycler 480 using SYBR Green I[29] detection. Each reaction consisted of 10 µl containing 2 µl cDNA, 1 µl forward primer, 1 µl reverse primer, 5 µl master mix and 1 µl H_2O. All samples were tested for genomic contamination by the use of an intron-spanning primer set unrelated to the study.

Gene expression of the target genes was calibrated between qPCR runs, and all samples were run in triplicate. The relative gene expression was calculated using Pfaffl's methods [24, 25] and normalised to the reference gene 18S. The relative normalised gene expression was analysed between synovial membrane and joint capsule. Mean normalised gene expression was analysed using Student's t test in SAS JMP software.[30] The assumption of normal distribution was not met, but the assumption of equal variances was met for 10 of 15 groups (Table 2).

Results

Microarray analysis

The microarray analysis of variance resulted in 2995 probes from 1907 genes significantly differentially expressed in synovial membrane compared to joint capsule ($p \le 0.05$). Selection of the genes up-regulated in synovial membrane rendered 1167 genes.

Ingenuity Pathway Analysis (IPA) [20] was used to select relevant up-regulated genes in synovial membrane as potential FLS candidate genes. The top biological functions explored were: 'Connective tissue disorders', 'Inflammatory disease' and 'Skeletal and Muscular disorders'. Within 'Connective tissue disorder', the functions annotations of *rheumatic disease, arthritis*, and *rheumatoid arthritis* were analysed. In addition, we explored the molecular and cellular function categories: 'Tissue morphology', 'Small molecule biochemistry', 'Cell morphology', and 'Cellular movement'.

A total of 241 genes were evaluated as potential candidate genes. Genes were included in the final candidate gene-list if they related to FLS function or cellular origin

Table 2 Real-time polymerase chain reaction results for the 15 selected candidate genes

Target gene	Synovial membrane	Joint capsule
ATXN1	0.40 ± 0.30[b]	0.43 ± 0.30
COL11A1	1.35 ± 1.81	0.43 ± 0.47
COL28A1	0.36 ± 0.47[b]	0.17 ± 0.28
FSTL1	7.75 ± 9.99[a]	0.27 ± 0.32
FOXO1	3.35 ± 2.97*	0.93 ± 1.15
GALNT3	0.99 ± 1.70	0.15 ± 0.17
GRN	7.05 ± 7.63[a]	1.44 ± 1.66
ITPR3	1.90 ± 1.11[b]	1.07 ± 0.83
NF1	2.96 ± 3.42	5.60 ± 5.96
PXK	7.49 ± 7.69*	0.49 ± 0.62
PYCARD	2.70 ± 2.43*	0.34 ± 0,29
RGL2A	2.06 ± 1.70[b]	1.37 ± 1,66
SAMD9L	5.39 ± 4.79*	0.45 ± 0.45
SOS1	1.83 ± 2.19[b]	0.77 ± 0.77
TIMP3	32.26 ± 46.34[a]	2.95 ± 4.06

Results are reported as expression of target gene relative to expression of 18S, the expression is furthermore normalised to a calibrator. The reported results are mean expression ± standard deviation. Difference between groups was tested with a Student's t-test. *Significant difference between Synovial membrane and Joint capsule, $p \le 0.05$. [a]Differential expression in synovial membrane compared to joint capsule p value ≤0.10. [b]Test of equal variances in the groups was not significant (mesenchymal cells). This led to a sub-classification of the genes into the following groups: genes potentially relating to FLS function, genes potentially relating to FLS cell morphology, genes relating to mesenchymal cells, or adverse genes (genes relevant to diseases involving mesenchymal cells).

This led to selection of 15 candidate genes. The fold change and the *p*-values of the 15 candidate genes in the microarray study are presented in Table 3.

Real-time PCR confirmation of candidate genes

The relative quantification analysis detected significantly different gene expression of *FOXO1*, *PXK*, *PYCARD* and *SAMD9L* at significance level $p \leq 0.05$ between synovial membrane and joint capsule (Table 2). In addition, the genes FSTL1, GRN and TIMP3 were expressed differently in synovial membrane compared to joint capsule at significance level $p \leq 0.10$ (Table 2).

Discussion

In this microarray study, we investigated new markers of the equine synovial membrane by comparing gene expression of synovial membrane to the gene expression of the joint capsule. Of 1167 genes up-regulated in the synovial membrane, we evaluated 241 genes with potential relevance to FLS to search for FLS markers. Genes were included in the final candidate gene-list, if the genes could be related to mesenchymal origin or FLS function. By careful selection of relevant candidate genes, the final candidate gene-list was reduced to 15 genes.

We reduced variations in probe set expression and gene expression between horses by normalisation between microarrays. Barrey et al., 2009 used data mining software to identify networks and associated genes in an equine microarray study [26]. In the present study relevant networks were visualised using Ingenuity Pathway Analysis (IPA), and the filtering process applied related to IPA results and a subjective selection of putative candidate genes. The subjective filtering process involved exclusion of genes with known expression in macrophages, endothelial cells or neurons, and genes with known ubiquitous gene expression. Genes with known mesenchymal expression were included in the candidate gene-list, even though this could include genes simultaneously expressed in adipocytes, connective tissue fibroblasts, chondrocytes and osteoblasts. This filtering process is likely to have excluded some potential novel markers of FLS, and it is highly relevant to further explore the gene list of up-regulated genes in synovial membrane.

Inter-horse variation and anatomical location may affect the results. Huang et al., 2008 compared ten different types of tissue to cartilage to identify novel markers of cartilage. Most of the samples in the study by Huang et al., 2008 were taken from the same horse [27]. Rinn et al., 2006 showed different gene expression in fibroblasts from various anatomical locations [28]. We aimed to minimize anatomical variation by excising samples from only one joint in a total of 12 horses. However, our candidate gene list may include genes differentially expressed merely as a result of the anatomical position of the cells instead of being specific to synovial membrane fibroblast-like cells.

Table 3 Microarray Fold change and p-values for the 15 selected candidate genes

Gene symbol	Function	Fold change	*p*-value ANOVA
ATXN1	Embryonal ECM remodelling	2.48	2.96E-03
COL11A1	Collagen in ECM	11.82	2.93E-04
COL28A1	Collagen in vascular ECM	21.29	1.63E-04
FSTL1	Unknown, embryonal mesenchymal location	4.10	2.59E-03
FOXO1	Transcription factor, function in osteoblast homeostasis	2.33	2,50E-05
GALNT3	Glycosylation, golgi complex	5.65	8.88E-07
GRN	Regulate cell growth	1.99	5.49E-06
ITPR3	Second messenger, possible role in exocrine secretion	3.09	6.64E-04
NF1	Regulate cell growth	3.27	4.35E-04
PXK	A functionel sorting nexin	4.81	1.72E-04
PYCARD	Function in inflammasomes	3.70	8.63E-05
RGL2A	Protein transport in cells	1.58	1.64E-05
SAMD9L	Protein-protein interaction	4.88	5.80E-04
SOS1	Promotes reorganization of actin cytoskeleton	2.90	1.20E-03
TIMP3	Metalloproteinase inhibitor	1.75	9.79E-04

ATXN1 ataxin 1, *COL11A1* collagen XI α1, *COL28A1* collagen XXVIII α1, *FOXO1* forkhead box O1, *FSTL1* follistatin-like 1, *GALNT3* UDP-N-acetyl-alpha-D-galactosamine:-polypeptide N-acetylgalactosaminyltransferase 3 (GalNac-T3), *GRN* granulin, *ITPR3* inositol 1,4,5-trisphosphate receptor, type 3, *NF1* neurofibromin 1, *PXK* PX domain containing serine/threonine kinase, *PYCARD* PYD and CARD domain containing, *RGL2* RGL2, ral guanine nucleotide dissociation stimulator-like 2, *SAMD9L* sterile alpha motif domain containing 9-like, *SOS1* son of sevenless homolog 1 (Drosophila), *TIMP3* TIMP metallopeptidase inhibitor 3

It has also been shown that age, gender, breed, and activity level can contribute to variation in levels of biomarkers in the joint or serum in horses with osteoarthritis [29–31]. In our study, we included mixed breed, gender, and age searching for candidate genes of general applicability. It was not the aim of this study to investigate the influence of breed, gender or age on the gene expression of the synovial membrane compared to the joint capsule which would call for a larger sample size.

The differential gene expression of *FOXO1, PXK, PYCARD* and *SAMD9L* was confirmed using qPCR. In addition, the genes: *FSTL1, GRN* and *TIMP3* showed a tendency towards higher gene expression in synovial membrane compared to joint capsule.

The genes presented here represent possible new markers of the synovial membrane. Knowledge of function of the four significantly differentially expressed genes in horses is limited. But in humans, the forkhead transcription factor *FOXO1* has turned out to be a regulator of glucose expenditure in cells and plays a role in regulation of glucose homeostasis in osteoblasts [32]. This suggests that *FOXO1* may regulate glucose homeostasis in FLS. Human *FOXO1* is also expressed in activated macrophages in lung tissue [33]. It could therefore be expressed in activated macrophages in joints, but a study in human rheumatoid arthritis synovial membrane showed that *FOXO1* expression occur mainly in FLS and only occasionally in MLS, supporting the use of *FOXO1* as a marker of FLS [34].

PXK encodes a phox homology domain, which is suggested as a sorting nexin localised in endosomes in cells and possibly interacting with actin. These functions suggest that *PXK* is involved in sorting processes in the cells and possibly in receptor trafficking [35].

PYCARD encodes for a protein in inflammasomes, a multiprotein complex which contributes to initiating the inflammatory process by activation of pro-caspase 1, which further leads to production of inflammatory cytokines [36]. Evidence indicates that the protein encoded by *PYCARD* is expressed in both FLS and MLS [37], but not at a high level in joint capsule fibroblasts according to the present study. Further studies are warranted to clarify if the expression of *PYCARD* is found in both FLS and MLS of the synovial membrane.

Expression of the gene *SAMD9L* has been located to a variety of tissues, but the exact function of the gene transcript is unknown. A recent study has investigated proliferation depressing effects of *SAMD9L*, and *SAMD9L* expression was reduced in tumours [38]. Thus, if *SAMD9L* is specifically expressed by FLS in synovial membrane and the gene has a proliferation depressing effect, the expression of the gene in FLS from joints with inflammatory arthritis or rheumatoid arthritis should be investigated. It is worth noting that neither of the human FLS markers suggested in the literature, UDPGD nor cadherin 11, appeared on the list of significantly upregulated genes in our microarray study.

Conclusions

In conclusion, the genes *FOXO1, PXK, PYCARD* and *SAMD9L* were confirmed as differentially expressed between synovial membrane and joint capsule. We suggest inclusion of the genes *FOXO1, PXK, PYCARD* and *SAMD9L* as potential markers of FLS in future studies of the equine synovial membrane.

Endnotes

[1]VWR & Bie & Berntsen A/S, Herlev, Denmark
[2]Merck, Darmstadt, Germany
[3]VWR & Bie & Berntsen A/S, Herlev, Denmark
[4]University of Copenhagen, Copenhagen, Denmark
[5]Life Technologies Ltd., Paisley UK.
[6]Fisher Scientific UK Ltd.,Loughborough UK
[7]Sigma Aldrich Ltd. Dorset UK
[8]Life Technologies Ltd., Paisley UK
[9]Fisher Scientific UK Ltd.,Loughborough UK
[10]Labtech International Ltd., Uckfield, UK
[11]Agilent Technologies UK Limited, Stockport UK
[12]Life Technologies Ltd., Paisley UK
[13]Agilent Technologies UK Limited, Stockport UK
[14]Life Technologies Ltd., Paisley UK
[15]GE Healthcare Life Sciences, Little Chalfont, UK
[16]Agilent Technologies UK Limited, Stockport UK
[17]Molecular Devices (UK), Wokingham, UK
[18]Molecular Devices (UK), Wokingham, UK
[19]Partek Incorporated. Saint Louis, USA
[20]Promega Biotech AS, Nacka, Sweden
[21]Thermo Scientific, Wilmington, US
[22]Agilent Technologies Denmark Aps, Hørsholm, Denmark
[23]Promega Biotech, Nacka, Sweden
[24]Promega Biotech, Nacka, Sweden
[25]Promega Biotech, Nacka, Sweden
[26]Promega Biotech, Nacka, Sweden
[27]Promega Biotech, Nacka, Sweden
[28]Promega Biotech, Nacka, Sweden
[29]Roche, Hvidovre, Denmark
[30]SAS Institute Inc., North Carolina, USA

Abbreviations

ECM: Extracellular matrix; FLS: Fibroblast-like synoviocytes; IPA: Ingenuity pathway analysis; MLS: Macrophage-like synoviocytes; PBS: Phosphate- buffered saline solution; qPCR: Quantitative polymerase chain reaction; UDPGD: Uridine diphosphoglucose dehydrogenase

Acknowledgements

We would like to thank Lennart E. Carlsen and Vibeke B. Hansen for help with the horses, and laboratory technician Anne F. Petersen for skilful help with the qPCR experiments.

Funding
The study was funded by University of Copenhagen and the following funds: Hesteafgiftsfonden, Torben og Alice Frimodts fond, Foreningen Kustos af 1881 and Vetfond.

Authors' contributions
LNT designed the study, collected all samples, analysed microarray data, performed qPCR experiments, analysed qPCR results and drafted the manuscript. PDT designed the study, helped interpret microarray results and qPCR results. AD and RT designed the microarray study, performed the microarray study and analysed initial results. LCB designed the study, helped analysing and interpreting microarray results and qPCR results. All authors revised the manuscript and approved the final manuscript.

Competing interests
The authors declare that they have no competing interests.

Consent for publication
Not applicable.

Author details
[1]Department of Veterinary Clinical Sciences, University of Copenhagen, Hoejbakkegaards alle 5, 2630 Taastrup, Denmark. [2]Department of Veterinary and Animal Sciences, University of Copenhagen, Groennegaardsvej 7, 1870 Frederiksberg C, Denmark. [3]Edinburgh Genomics, Ashworth Laboratories, University of Edinburgh, Edinburgh EH9 3FL, UK.

References
1. Goodrich LR, Nixon AJ. Medical treatment of osteoarthritis in the horse - a review. Vet J. 2006;171:51–69.
2. Bondeson J, Wainwright SD, Lauder S, Amos N, Hughes CE. The role of synovial macrophages and macrophage-produced cytokines in driving aggrecanases, matrix metalloproteinases, and other destructive and inflammatory responses in osteoarthritis. Arthritis Res Ther. 2006; 8:R187.
3. Chang SK, Gu Z, Brenner MB. Fibroblast-like synoviocytes in inflammatory arthritis pathology: the emerging role of cadherin-11. Immunol Rev. 2010; 233:256–66.
4. Levick JR, McDonald JN. Fluid movement across synovium in healthy joints: role of synovial fluid macromolecules. Ann Rheum Dis. 1995;54:417–23.
5. Graabaek PM. Ultrastructural Evidence for 2 Distinct Types of Synoviocytes in Rat Synovial-Membrane. J Ultrastruct Res. 1982;78:321–39.
6. Iwanaga T, Shikichi M, Kitamura H, Yanase H, Nozawa-Inoue K. Morphology and functional roles of synoviocytes in the joint. Arch Histol Cytol. 2000;63: 17–31.
7. Key J. The mechanisms involved in the removal of colloidal and particulate carbon from joint cavities. J Bone Joint Surg. 1926;8:666–83.
8. Graabaek PM. Characteristics of the 2 Types of Synoviocytes in Rat Synovial-Membrane - An Ultrastructural-Study. Lab Investig. 1984;50:690–702.
9. Athanasou NA, Quinn J. Immunocytochemical analysis of human synovial lining cells: phenotypic relation to other marrow derived cells. Ann Rheum Dis. 1991;50:311–5.
10. Bartok B, Firestein GS. Fibroblast-like synoviocytes: key effector cells in rheumatoid arthritis. Immunol Rev. 2010;233:233–55.
11. Kunisch E, Fuhrmann R, Roth A, Winter R, Lungershausen W, Kinne RW. Macrophage specificity of three anti-CD68 monoclonal antibodies (KP1, EBM11, and PGM1) widely used for immunohistochemistry and flow cytometry. Ann Rheum Dis. 2004;63:774–84.
12. Muller-Ladner U, Ospelt C, Gay S, Distler O, Pap T. Cells of the synovium in rheumatoid arthritis. Synovial fibroblasts. Arthritis Res Ther. 2007;9:223.
13. Kiener HP, Watts GF, Cui Y, Wright J, Thornhill TS, Skold M, et al. Synovial fibroblasts self-direct multicellular lining architecture and synthetic function in three-dimensional organ culture. Arthritis Rheum. 2010;62:742–52.
14. Clarkin CE, Allen S, Kuiper NJ, Wheeler BT, Wheeler-Jones CP, Pitsillides AA. Regulation of UDP-glucose dehydrogenase is sufficient to modulate hyaluronan production and release, control sulfated GAG synthesis, and promote chondrogenesis. J Cell Physiol. 2011;226:749–61.
15. Edwards JC. The nature and origins of synovium: experimental approaches to the study of synoviocyte differentiation. J Anat. 1994; 184(Pt 3):493–501.
16. Wilkinson LS, Pitsillides AA, Worrall JG, Edwards JC. Light microscopic characterization of the fibroblast-like synovial intimal cell (synoviocyte). Arthritis Rheum. 1992;35:1179–84.
17. Valencia X, Higgins JM, Kiener HP, Lee DM, Podrebarac TA, Dascher CC, et al. Cadherin-11 provides specific cellular adhesion between fibroblast-like synoviocytes. J Exp Med. 2004;200:1673–9.
18. Agbase.2012. http://www.agbase.msstate.edu. Accessed 29 June 2012.
19. McCarthy F, Gresham C, Buza T, Chouvarine P, Pillai L, Kumar R, et al. AgBase: supporting functional modeling in agricultural organisms. Nucleic Acids Res. 2011;39:D497–506.
20. Ingenuity Pathway Analysis (IPA).29–6-2012. www.ingenuity.com/products/ipa. Accessed 3 July 2012.
21. Ensembl. 2012. http://Oct2012.archive.ensembl.org.1-10-2012. Accessed 1 Oct 2012.
22. Rozen S, Skaletsky H. Primer3 on the WWW for general users and for biologist programmers. Methods Mol Biol. 2000;132:365–86.
23. NCBI BLAST .2012. http://blast.ncbi.nlm.nih.gov/Blast.cgi. Accessed 1 Oct 2012.
24. Livak KJ, Schmittgen TD. Analysis of relative gene expression data using real-time quantitative PCR and the 2(−Delta Delta C(T)) Method. Methods. 2001;25:402–8.
25. Pfaffl MW. A new mathematical model for relative quantification in real-time RT-PCR. Nucleic Acids Res. 2001;29:e45.
26. Barrey E, Mucher E, Jeansoule N, Larcher T, Guigand L, Herszberg B, et al. Gene expression profiling in equine polysaccharide storage myopathy revealed inflammation, glycogenesis inhibition, hypoxia and mitochondrial dysfunctions. BMC Vet Res. 2009;5:29.
27. Huang L, Zhu W, Saunders CP, MacLeod JN, Zhou M, Stromberg AJ, et al. A novel application of quantile regression for identification of biomarkers exemplified by equine cartilage microarray data. BMC Bioinformatics. 2008;9:300.
28. Rinn JL, Bondre C, Gladstone HB, Brown PO, Chang HY. Anatomic demarcation by positional variation in fibroblast gene expression programs. PLoS Genet. 2006;2:e119.
29. Frisbie DD, Al-Sobayil F, Billinghurst RC, Kawcak CE, McIlwraith CW. Changes in synovial fluid and serum biomarkers with exercise and early osteoarthritis in horses. Osteoarthr Cartil. 2008;16:1196–204.
30. Brommer H, van Weeren PR, Brama PA, Barneveld A. Quantification and age-related distribution of articular cartilage degeneration in the equine fetlock joint. Equine Vet J. 2003;35:697–701.
31. Todhunter RJ, Fubini SL, Freeman KP, and Lust G. Concentrations of keratan sulfate in plasma and synovial fluid from clinically normal horses and horses with joint disease. J Am Vet Med Assoc. 1-2-1997;210:369-374.
32. Kode A, Mosialou I, Silva BC, Josha S, Ferron M, Rached MT, et al. FoxO1 protein cooperates with ATF4 protein in osteoblasts to control glucose homeostasis. J Biol Chem. 2012;287:8757–66.
33. Chung S, Lee TJ, Reader BF, Kim JY, Lee YG, Park GY, et al. FoxO1 regulates allergic asthmatic inflammation through regulating polarization of the macrophage inflammatory phenotype. Oncotarget. 2016;7(14):17532–46.
34. Ludikhuize J, de LD GD, Smeets TJ, Vinkenoog M, Sanders ME, et al. Inhibition of forkhead box class O family member transcription factors in rheumatoid synovial tissue. Arthritis Rheum. 2007;56:2180–91.
35. Takeuchi H, Takeuchi T, Gao J, Cantley LC, Hirata M. Characterization of PXK as a protein involved in epidermal growth factor receptor trafficking. Mol Cell Biol. 2010;30:1689–702.

36. Taniguchi S, Sagara J. Regulatory molecules involved in inflammasome formation with special reference to a key mediator protein, ASC. Semin Immunopathol. 2007;29:231–8.
37. Rosengren S, Hoffman HM, Bugbee W, Boyle DL. Expression and regulation of cryopyrin and related proteins in rheumatoid arthritis synovium. Ann Rheum Dis. 2005;64:708–14.
38. Gallant-Behm CL, Ramsey MR, Bensard DL, Nojek I, Tran J, Liu M, et al. Delta Np63 alpha represses anti-proliferative genes via H2A.Z deposition. Genes Dev. 2012;26:2325–36.

5

In vitro susceptibility of *Borrelia burgdorferi* isolates to three antibiotics commonly used for treating equine Lyme disease

Sanjie Caol[1,2], Thomas Divers[3*], Mark Crisman[4] and Yung-Fu Chang[1]

Abstract

Background: Lyme disease in humans is predominantly treated with tetracycline, macrolides or beta lactam antibiotics that have low minimum inhibitory concentrations (MIC) against *Borrelia burgdorferi*. Horses with Lyme disease may require long-term treatment making frequent intravenous or intramuscular treatment difficult and when administered orally those drugs may have either a high incidence of side effects or have poor bioavailability. The aim of the present study was to determine the in vitro susceptibility of three *B. burgdorferi* isolates to three antibiotics of different classes that are commonly used in practice for treating *Borrelia* infections in horses.

Results: Broth microdilution assays were used to determine minimum inhibitory concentration of three antibiotics (ceftiofur sodium, minocycline and metronidazole), for three *Borrelia burgdorferi* isolates. Barbour-Stoner-Kelly (BSK K + R) medium with a final inoculum of 10^6 *Borrelia* cells/mL and incubation periods of 72 h were used in the determination of MICs. Observed MICs indicated that all isolates had similar susceptibility to each drug but susceptibility to the tested antimicrobial agents varied; ceftiofur sodium (MIC = 0.08 μg/ml), minocycline hydrochloride (MIC = 0.8 μg/ml) and metronidazole (MIC = 50 μg/ml).

Conclusions: The MIC against *B. burgorferi* varied among the three antibiotics with ceftiofur having the lowest MIC and metronidazole the highest MIC. The MIC values observed for ceftiofur in the study fall within the range of reported serum and tissue concentrations for the drug metabolite following ceftiofur sodium administration as crystalline-free acid. Minocycline and metronidazole treatments, as currently used in equine practice, could fall short of attaining MIC concentrations for *B. burgdorferi*.

Keywords: Horse, *Borrelia burgdorferi*, Lyme, Ceftiofur, Metronidazole, Doxycycline

Background

Lyme borreliosis, the most common tick-transmitted disease in the northern hemisphere, is caused by Gram-negative spirochetes of the *Borrelia burgdorferi* sensu *lato* complex [1]. In North America, *B. burgdorferi* sensu stricto is the cause of Lyme disease although several strains of that organism are found [2]. Adult horses in some endemic areas of the U.S. have seroprevalence rates of 33% or greater indicating a high incidence of *B. burgdorferi* exposure [3–7]. Lyme disease in horses is documented by numerous case reports [8–15], but the proportion of seropositive horses with clinical Lyme disease is unknown. Lyme disease in humans is treated predominantly with tetracyclines or beta lactam antimicrobials with low minimum inhibitory concentrations (MIC) against *Borrelia burgdorferi* [16]. Successful treatment of chronic infections is believed to require longer treatment duration than for early infections [17]. Early infections are rarely recognized in the horse due to absence of erythema migrans. Antibiotic treatment, mostly minocycline or doxycycline given per os, for four or more weeks' duration is common in seropositive horses with signs thought to be associated with Lyme disease [18].

As with most other bacterial diseases, susceptibility of the causative agent to the antibiotics used for therapy is an important prerequisite for antibiotic selection and successful treatment. Multiple in vitro studies have indicated susceptibility of *B. burgdorferi* to several antibiotics

* Correspondence: tjd8@cornell.edu
[3]Department of Clinical Sciences College of Veterinary Medicine, Cornell University, Ithaca, NY 14853, USA
Full list of author information is available at the end of the article

including amoxicillin, azithromycin, several cephalosporins such as ceftriaxone and cefuroxime, and most of the tetracycline group of antibiotics [19–28]. Resistance has been demonstrated to trimethoprim sulfas and flouroquinolones, two commonly used oral antibiotics in horses [25, 29]. Differences in the antibiotic susceptibilities of different strains and forms (motile or cysts) of *B. burgdorferi* sensu stricto have been reported [25, 30, 31]. Most in vitro testing is performed on viable, motile *B. burgdorferi,* as they are readily available for testing and assumed to be the *Borrelia* stage most commonly causing disease [1]. A persistent (cystic, non-growing) form of *Borrelia* has been associated with, but is not proven to cause, chronic Lyme disease. This form of *Borrelia* is known to have a very high resistance pattern to the antibiotics commonly used for acute infections but has a moderate sensitivity to metronidazole [30–32].

In adult horses, use of many of the antibiotics recommended for treatment of Lyme disease in humans (tetracyclines, beta lactams and macrolides) is associated with a high incidence of diarrhea (beta lactams and macrolides) when administered orally and all three classes of antibiotics have highly variable and frequently low oral bioavailability in the horse [33–38]. In addition, parenteral administration of tetracycline or beta lactam drugs that are known to be effective in vitro against *Borrelia spp.* can cause injection site reactions following repeated intramuscular administration. Prolonged intravenous administration of tetracycline can cause renal failure and thrombophlebitis [39].

The aim of the present study was to determine the in vitro susceptibility of three *B. burgdorferi* isolates to three antibiotics that might be of practical use for treating *Borrelia* infections in horses.

Methods

The susceptibility of clinical isolates of *B. burgdorferi* to three different classes of antibiotics was evaluated by measurement of their minimum inhibitory concentrations (MICs). The three antibiotics evaluated in the study were antibiotics commonly used in equine practice and administered either orally (minocycline and metronidazole) or intramuscularly (ceftiofur) in treating equine Lyme disease.

Borrelia burgdorferi isolates

Three *B. burgdorferi* sensu stricto isolates obtained from the collection of Dr. Yung-Fu Chang at the Department of Population Medicine and Diagnostic Sciences, College of Veterinary Medicine, Cornell University were tested for antibiotic susceptibility. Isolates were obtained from skin biopsies of ponies that had been experimentally infected by attaching adult ticks (*Ixodes scapularis*), collected in a forested area of Westchester County, New York.

Antimicrobial agents

The antibiotics tested were ceftiofur sodium (Sigma-Aldrich, USA), minocycline hydrochloride (MP Biomedicals, LLC), and metronidazole (Sigma-Aldrich, USA). All antibiotics were obtained as United States Pharmacopeia reference standard powders and were prepared according to the manufacturer's recommendations. The concentration ranges used for testing were: ceftiofur sodium 0.04–2.56 μg/ml; minocycline hydrochloride 0.20–12.80 μg/ml; and metronidazole 12.50–800.00 μg/ml. The ranges were chosen according to data in the literature [24, 30].

Culturing *B. burgdorferi*

Borrelia burgdorferi was cultured in Barbour-Stoner-Kelly (BSK K + R) medium [40]. The cultures were incubated at 34 °C with 5% CO_2 and maintained in sterile 15 mL Corning tubes (cat.# 89,093–186 from VWR, Radnor, PA USA) with 10 mL BSK K + R medium for about 3–5 days.

In vitro testing of the antibacterial agents

Broth microdilution procedures were used in the determination of minimum inhibitory concentrations (MICs) [28, 40, 42]. For this procedure, MICs were determined using sterile 48-wells Tissue Culture Plates (VWR International, LLC, USA). The last column served as the negative control and contained BSK K + R medium only. In the wells of all other columns, 900 μL of *B. burgdorferi* culture was added at a final density of 10^6 cells/mL, determined in a Neubauer counting chamber (Brand, Wertheim, Germany). The second column, which served as the positive control, contained no antimicrobial agent. From the third column onward, 100 μL portions of two fold decreasing concentrations of antibiotics were added. Each antimicrobial agent was tested in triplicate assays, to demonstrate reproducibility of the assay, before being sealed with adhesive plastic and incubated at 34 °C in aerobic conditions for 72 h [28, 30]. Following incubation, all wells were examined by the same observer (SC) for visible growth and motility of isolates by dark-field microscopy. The lowest antibiotic concentration at which isolates demonstrated no visible growth and motility was determined for each antibiotic [30]. The MIC was determined three times for each isolate and provided as descriptive data in Table 1. The final MIC was defined as the lowest concentration of an antimicrobial that inhibits the visible growth and motility of a microorganism after 72 h of incubation [30] and reported as the mode value for all isolates (Table 1).

Results

The in vitro susceptibility of three isolates of *B. burgdorferi* to three antibiotics was tested. All of the isolates were

Table 1 Minimum inhibitory concentrations (MICs) of three antibiotics for individual *Borrelia burgdorferi* isolates

Isolate		MIC[a] (μg/ml)		
		Ceftiofur	Minocycline	Metronidazole
	Experiment	MIC	MIC	MIC
6342	1	0.08	0.8	25
	2	0.08	0.4	50
	3	0.04	0.8	50
TLSH-916	1	0.08	0.8	50
	2	0.04	0.8	25
	3	0.08	0.4	25
21,343/3	1	0.04	0.8	50
	2	0.08	0.8	50
	3	0.08	0.4	25
Mode Value		0.08	0.8	50

[a]MIC, are reported for individual isolates and assays [1–3]; the mode value is reported as the final MIC

susceptible to all tested antibiotics at some concentration. Table 1 shows the MICs values of each antibiotic tested against the individual *B. burgdorferi* isolates. According to MIC breakpoints[a] (Table 1), the lowest drug concentrations to inhibit all the strains was 0.08 μg/ml for ceftiofur sodium, 0.80 μg/ml for minocycline hydrochloride, and 50 μg/ml for metronidazole. Complete inhibition of motility was noted at the MIC concentration for each antibiotic. Ultimately, the *B. burgdorferi* strains tested were most susceptible in vitro to ceftiofur and least susceptible to metronidazole.

Discussion

There are multiple reports on in vitro susceptibility of isolates of *B. burgdorferi* to macrolides, tetracyclines and beta lactam antibiotics [19–28]. However, susceptibility to ceftiofur sodium has not been previously reported and susceptibility to metronidazole has only been reported twice with conflicting findings [27, 31]. Ceftiofur sodium is a third-generation cephalosporin approved for use in horses as both a sodium salt formulation with a rapid absorption and elimination time, and in a crystalline formulation intended for IM administration with a slowed absorption and greater area under the curve (AUC). Ceftiofur is rapidly metabolized to desfuroylceftiofur in the horse but the in vitro activity of desfuroylceftiofur is almost identical to that of ceftiofur for Gram-negative bacteria and similar for most Gram-positive bacteria tested [43]. Desfuroylceftiofur was not available for testing in this study and ceftiofur was used instead. In vitro activity testing against *B. burgdorferi* with either desfuroylceftiofur or ceftiofur has not previously been reported. In vitro testing of ceftiofur against *B. burgdorferi* seemed pertinent because cephalosporins have variable in vitro activity against *B. burgdorferi* with ceftriaxone, cefotaxime, cefdinir, and cefixime having a high activity while others such as cefetamet-pivoxil, ceftibuten, and cefpodoxime-proxetil were ineffective in vitro [22]. We were particularly interested in determining the MIC of ceftiofur because the crystalline form of the drug is approved to be administered IM in horses on days one, four and then weekly, and will maintain serum ceftiofur and desfuroylceftiofur combined concentrations >0.22 μg/ml throughout the treatment period and for at least six days following the end of treatment [44–46]. This serum concentration would be well above the MIC of ceftiofur against *B. burgdorferi* reported here. The most important determinant of efficacy with a time dependent antimicrobial such as ceftiofur is the length of time that concentrations exceed the MIC [47]. It would be important to know tissue concentrations of an antibiotic when considering treatment of *Borrelia* infections and if those concentrations are above the MIC. In the only publication reporting tissue concentrations of ceftiofur following IM administration of ceftiofur crystalline free acid, tissue levels of its metabolite in the uterus were maintained between 0.1–0.2 μg/g, which is above the MIC of *Borrelia* found in our in vitro report [44]. Even more important would be to know the in vivo response to treatment and in the only experimental equine *B. burgdorferi* antibiotic study performed to date, the aqueous solution of sodium ceftiofur at 2.2 mg/kg IM daily for 28 days eliminated *B. burgdorferi* from two of four experimentally infected ponies [48]. Although serum trough levels of sodium ceftiofur, administered at this dose and interval would be predicted to have remained above the MIC for *B. burgdorferi*, trough levels in tissue would have likely been near or below the MIC and this could be one possible explanation why two of the four ponies remained infected [48, 49]. A similar long-acting antibiotic, cefovecin, has recently been shown to decrease the number of dogs with joint lesions and to induce a marked reduction in serum antibody in an in vivo experimental *B. burgdorferi* study [50].

Minocycline and doxycycline have been shown in numerous studies to have similar MICs to *B. burgdorferi* with most MIC results ranging between 0.12–0.63 μg/ml, which are similar to our findings [22]. Minocycline and doxycycline are known to be highly efficacious in treating early onset Lyme disease in humans [51–53]. Bioavailability of the two drugs after per os (PO) administration is significantly lower in horses than humans (20–30% versus 95–100%) and duration of infection prior to beginning treatment is likely longer in horses, therefore, difference in efficacy in treating *B. burgdorferi* between the two species might be expected [33–35, 54]. In the horse, minocycline has better bioavailability than doxycycline, and at the currently recommended dosage of 4 mg/kg PO every 12 h(q12h) provides a peak serum concentration of approximately 0.67 µg/ml which is below the *B. burgdorferi* MIC (0.8µg/ml) found in this study but higher than the MIC found in several other reports [22, 34]. Highest reported trough synovial fluid concentrations of minocycline in horses were 0.33 µg/ml, below the MIC in the current study [34]. Minocycline concentration in the CSF of normal horses dosed at 4 mg/kg q 12 h was 0.39 µg/ml and in aqueous fluid was 0.11 µg/ml ± 0.04 in needle-disrupted blood aqueous barrier suggesting minocycline may have marginal efficacy for treating neuroborreliosis and low efficacy for treating Lyme uveitis [34]. Doxycycline administered at 10 mg/kg is reported to result in a peak serum concentration between 0.32–0.97 µg/ml with a lower percentage distribution into CSF and aqueous fluid than minocycline [33, 55]. In the only experimental equine *B. burgdorferi* antibiotic study, doxycycline administered at 10 mg/kg PO q24h for 28 days was able to eliminate *B. burgdorferi* in only one of four treated ponies [48]. It should be pointed out that the dose of doxycycline used in that 2005 study was less than the current commonly used dose of 10 mg/kg q 12 h. These pharmacokinetic studies and the current and previously reported MIC data may help explain why horses treated, even long term, for *B. burgdorferi* infection with either minocycline or doxycycline often have only modest or no decrease in *B. burgdorferi* serologic titers [18]. There does appear to be some accumulation of doxycycline in synovial fluid due to delayed elimination in horses and both minocycline and doxycycline are commonly reported to reduce stiffness and lameness in field cases of Lyme disease, but this might be a result of their known anti-inflammatory effects on synovium and cartilage [56–58]. In the pony experimental infection and treatment study, oxytetracycline (5 mg/kg administered once daily IV) was the only drug used that eliminated *B. burgdorferi* from all treated ponies and based upon the pharmacokinetic study by Teske 1973, trough levels at that dose would have remained above the MIC_ for *B. burgdorferi* [48, 59].

Published studies evaluating in vitro sensitivity of *B. burgdorferi* to metronidazole had significant differences; Sapi (2011) reported an MIC of 0.3 µg/ml, while Brorson (1999) found motile *Borrelia* to be minimally affected by metronidazole even at very high concentrations (516 µg/ml) [30, 31]. Reported differences in in vitro *B. burgdorferi* sensitivity to metronidazole and other antimicrobials may be due to differing strains, methods of broth dilution, incubation periods, and end-point spirochete inoculation concentrations [22, 41]. For instance, our study measured inhibitory concentration as opposed to minimum bactericidal concentration (MBC) effects, which determines the killing of all organisms in the test inoculum. MBC values for *Borrelia* are generally three or more times greater than MIC for tetracyclines, cephalosporins, and metronidazole [22, 28]. In vitro test result differences against different morphologic forms (motile spirochetes or round bodies) of *B. burgdorferi* have also been reported. Sapi (2011) showed *B. burgdorferi* can develop increasing antibiotic tolerance as morphology changes from typical spirochetal form in log phase growth to variant round body "cyst" forms in a stationary phase [30]. Much of the controversy that surrounds Lyme disease in humans pertains to chronic Lyme disease and concern by some that the round morphologic variants of *B. burgdorferi* may play a role in the pathogenesis of chronic Lyme disease [32]. These antibiotic resistant cyst forms are known to occur in vitro more frequently under antibiotic pressure from drugs commonly used to treat Lyme disease [30]. Metronidazole is one of the few antimicrobials that has been shown in one study to be moderately effective in vitro against the round form and at concentrations practically achievable in the patient [30]. The importance of the cyst forms and, therefore, the value of metronidazole treatment is questionable as a recent systematic review of the literature suggested there is no clinical literature to justify specific treatment of *B. burgdorferi* morphologic variants [60]. We know that some equine practitioners in the U.S. are using metronidazole as a treatment for *Borrelia* infections and we were therefore interested in testing the MIC of metronidazole against the free-living spirochete [30, 32]. The in vitro finding of the current study would at least suggest that ceftiofur and minocycline would be preferred over metronidazole for inhibition of the motile *B. burgdorferi* with ceftiofur having the lowest MIC of the three drugs tested.

One important limitation of this study is that measurements of MIC made in vitro cannot be directly applied to in vivo situations. In addition to in vitro testing, differences in pharmacokinetics and dynamics of each drug must be considered along with the immune responses of the patient. Regardless, in vitro testing provides a guide to antimicrobial selection of an antimicrobial with good sensitivity pattern against *Borrelia*. A second limitation to the study was that MIC and not MBC was determined in this study and higher concentration of the drugs would likely be needed to kill the bacteria [42]. Another potential limitation of the study is that the main metabolite of ceftiofur, desfuroylceftiofur, was not available and could not be used in the in vitro study. Although ceftiofur and desfuroylceftiofur, have been shown to have nearly identical activity against all previously tested gram negative bacteria, there is no proof the same would occur with *Borrelia* [43]. Lastly, although we only tested *B. burgdorferi* sensu stricto and not the other genospecies of *B. burgdorferi* sensu *lato*, *B. afzelii* and *B. garinii* which are common in Europe, a previous publication determined that differences in antibiotic (amoxicillin, ceftriaxone, and doxycycline) sensitivity between the three genospecies did not seem sufficiently pronounced to be of fundamental clinical relevance [42].

In summary, this study provides information that might help guide equine practitioners' decisions on antibiotic treatment of suspected Lyme disease. Based upon current dosing recommendations and pharmacokinetic studies in horses, ceftiofur sodium administered as crystalline-free acid could maintain serum and some tissue drug metabolite concentrations in horses above the ceftiofur MIC for *B. burgdorferi*. It is currently unknown if adequate ceftiofur concentrations might be found in tissues most commonly infected by *B. burgdorferi* and if the treatment would actually eliminate the organism. Minocycline, as currently used in equine practice, could maintain serum concentrations near the MIC for *B. burgdorferi* but might not be expected to consistently provide adequate concentration in synovial fluid, aqueous humor or CSF. Metronidazole would only attain MIC against motile *B. burgdorferi* at peak serum concentrations following standard equine dosing. In-vivo treatment studies using either field cases or experimentally infected horses will be required to investigate the efficacy of the antibiotics in treating equine *B. burgdorferi* infections.

Conclusions

The results of this study provide information to assist practitioners in the therapeutic decision process for treatment of *B. burgdorferi* in horses. Minocycline might provide serum concentrations near or above the MIC for *B. burgdorferi* but may not provide adequate concentration in synovial fluid or CSF. Based upon current dosing recommendations, ceftiofur crystalline-free acid could maintain serum and some tissue concentrations in horses above the MIC for *B. burgdorferi*. Further in-vivo studies will be required to fully elucidate the efficacy of these and other antibiotics in treating equine Lyme borreliosis.

Abbreviations

AUC: Area under the curve; CSF: Cerebrospinal fluid; IM: Intramuscular; IV: Intravenous; MIC: Minimum inhibitory concentrations; PO: Per os; q12h: Every 12 h

Funding

This study was partly supported by Zoetis.

Authors' contributions

Study design: SC, TD, MC, YFC. MIC testing: SC. Provision of antimicrobials: YFC, MC. Writing of manuscript: TD, YFC, SC, MC. All authors read and approved the final manuscript.

Competing interests

Dr. Crisman is employed by Zoetis.

Author details

[1]Department of Population Medicine and Diagnostic Sciences, College of Veterinary Medicine, Cornell University, Ithaca, NY 14853, USA. [2]Currently Research Center of Swine Disease, College of Veterinary Medicine, Sichuan Agricultural University, Chengdu 611130, China. [3]Department of Clinical Sciences College of Veterinary Medicine, Cornell University, Ithaca, NY 14853, USA. [4]Virginia-Maryland College of Veterinary Medicine, Blacksburg, VA 24061, USA.

References

1. Stanek G, Wormser GP, Gray JS, Strle F. Lyme borreliosis. Lancet. 2012;379:461–73.
2. Ogden NH, Feil EJ, Leighton PA, Lindsay LR, Margos G, Mechai S, Michel P, Moriarty TJ. Evolutionary aspects of emerging Lyme disease in Canada. Appl Environ Microbiol. 2015;81:7350–9.
3. Burbelo PD, Bren KE, Ching KH, Coleman A, Yang X, Kariiu T, Iadarola MJ, Pal U. Antibody profiling of *Borrelia burgdorferi* infection in horses. Clin Vaccine Immunol. 2011;18:1562–7.
4. Magnarelli LA, Ido JW, Van Andel AE, Wu C, Padula SJ, Fikrig E. Serologic confirmation of *Ehrlichia equi* and *Borrelia burgdorferi* infections in horses from the Northeastern United States. J Am Vet Med Assoc. 2000;217:1045–50.
5. Funk RA, Pleasant RS, Witonsky SG, Reeder DS, Werre SR, Hodgson DR. Seroprevalence of *Borrelia burgdorferi* in horses presented for Coggins testing in Southwest Virginia and change in positive test results approximately one year later. J Vet Intern Med. 2016;30(4):1300–4.
6. Durrani AZ, Goyal SM, Kamal N. Retrospective study of seroprevalence of *Borrelia burgdorferi* antibodies in horses in Minnesota. J Equine Vet Sci. 2011;31:427–9.
7. Metcalf KB, Lilley CS, Revenaugh MS, Glaser AL, Metcalf ES. The prevalence of antibodies against *Borrelia burgdorferi* found in horses residing in the Northwestern United States. J Equine Vet Sci. 2008;28:587–9.
8. Priest HL, Irby NL, Schlafer DH, Divers TJ, Wagner B, Glaser AL, Chang YF, Smith MC. Diagnosis of *Borrelia*-associated uveitis in two horses. Vet Ophthalmol. 2012;15:398–405.
9. James FM, Engiles JB, Beech J. Meningitis, cranial neuritis, and radiculoneuritis associated with *Borrelia burgdorferi* infection in a horse. J Am Vet Med Assoc. 2010;237:1180–5.
10. Imai DM, Barr BC, Daft B, Bertone JJ, Feng S, Hodzic E, Johnston JM, Olsen KJ, Barthold SW. Lyme neuroborreliosis in two horses. Vet Pathol. 2011;48:1151–7.
11. Sears KP, Divers TJ, Neff RT, Miller WH Jr, McDonough SP. A case of *Borrelia*-associated cutaneous pseudolymphoma in a horse. Vet Dermatol. 2012;23:153–6.
12. Passamonti F, Veronesi F, Cappelli K, Capomaccio S, Reginato A, Miglio A, Vardi DM, Stefanetti V, Coletti M, Bazzica C, Pepe M. Polysynovitis in a horse due to Borrelia burgdorferi sensu lato infection - case study. Ann Agric Environ Med. 2015;22:247–50.
13. Burgess EC, Mattison M. Encephalitis associated with *Borrelia burgdorferi* infection in a horse. J Am Vet Med Assoc. 1987;191:1457–8.
14. Burgess EC, Gillette D, Pickett JP. Arthritis and panuveitis as manifestations of *Borrelia burgdorferi* infection in a Wisconsin pony. J Am Vet Med Assoc. 1986;189:1340–2.
15. Hahn CN, Mayhew IG, Whitwell KE, Smith KC, Carey D, Carter SD. A possible case of Lyme borreliosis in a horse in the UK. Equine Vet J. 1996;28:84–8.
16. Sanchez JL. Clinical manifestations and treatment of Lyme disease. Clin Lab Med. 2015;35:765–8.
17. Cameron DJ, Johnson LB, Maloney EL. Evidence assessments and guidelines recommendations in Lyme disease: the clinical management of known tick bites, erythema migrans rashes and persistent disease. Expert Rev Anti-Infect Ther. 2014;12:1103–35.

18. Divers TJ, Grice AL, Mohammed HO, Glaser AL, Wagner B. Changes in *Borrelia burgdorferi* ELISA antibody over time in both antibiotic treated and untreated horses. Acta Vet Hung. 2012;60:421–9.
19. Johnson RC, Kodner CB, Jurkovich PJ, Collins JJ. Comparative in vitro and in vivo susceptibilities of the Lyme disease spirochete *Borrelia burgdorferi* to cefuroxime and other antimicrobial agents. Amtimicrob Agents Chemother. 1990;34:2133–6.
20. Ates L, Hanssen-Hübner C, Norris DE, Richter D, Kraiczy P, Hunfeld KP. Comparison of in vitro activities of tigecycline, doxycycline, and tetracycline against the spirochete *Borrelia burgdorferi*. Ticks Tick Borne Dis. 2010;1:30–4.
21. Johnson RC, Kodner C, Russell M, Girard D. In-vitro and in-vivo susceptibility of *Borrelia burgdorferi* to azithromycin. J Antimicrob Chemother. 1990;25:33–8.
22. Hunfeld KP, Brade V. Antimicrobial susceptibility of *Borrelia burgdorferi* sensu lato: what we know, what we don't know, and what we need to know. Wien Klin Wochenschr. 2006;22:659–68.
23. Johnson SE, Klein GC, Schmid GP, Feeley JC. Susceptibility of the Lyme disease spirochete to seven antimicrobial agents. Yale J Biol Med. 1984;57:549–53.
24. Hunfeld KP, Kraiczy P, Winchelhaus TA, Schäfer V, Brade V. Colorimetric in vitro susceptibility testing of penicillins, cephalosporins, macrolides, streptogramins, tetracyclines, and aminoglycosides against *Borrelia burgdorferi* isolates. Int J Antimicrob Agents. 2000;15:11–7.
25. Baradaran-Dilmaghani R, Stanek G. In vitro susceptibility of thirty *Borrelia* strains from various sources against eight antimicrobial chemotherapeutics. Infection. 1996;24:60–3.
26. Yang X, Nguyven A, Qiu D, Luft BJ. In vitro activity of tigecycline against multiple strains of *Borrelia burgdoferi*. J Antimicrob Chemother. 2009;63:709–12.
27. Johnson RC, Kodner DC, Russell M. In vitro and in vivo susceptibility of the Lyme disease spirochete, *Borrelia burgdoferi*, to four antimicrobial agents. Antimicrob Agents Chemother. 1987;31:164–7.
28. Dever LL, Jorgensen JH, Barbour AG. In vitro antimicrobial susceptibility testing of *Borrelia burgdorferi*: a microdilution MIC method and time-kill studies. J Clin Microbiol. 1992;30:2692–7.
29. Kim D, Kordick D, Divers T, Chang YF. In vitro susceptibilities of *Leptospira* spp. and *Borrelia burgdorferi* isolates to amoxicillin, tilmicosin, and enrofloxacin. J Vet Sci. 2006;7:355–9.
30. Sapi E, Kaur N, Ananwu S, Luecke DF, Luecke DF, Datar A, Patel S, Rossi M, Stricker RB. Evaluation of in-vitro antibiotic susceptibility of different morphological forms of *Borrelia burgdoferi*. Infect Drug Resist. 2011;4:97–113.
31. Brorson O, Brorson SH. An in vitro study of the susceptibility of mobile and cystic forms of *Borrelia burgdorferi* to metronidazole. APMIS. 1999;107:566–76.
32. Feng J, Wang T, Shi W, Zhang S, Sullivan D, Auwaerter PG, Zhang Y. Identification of novel activity against *Borrelia burgdorferi* persisters using an FDA approved drug library. Emerg Microbes Infect. 2014;3:e49.
33. Bryant JE, Brown MP, Gronwall RR, Merritt KA. Study of intragastric administration of doxycycline: pharmacokinetics including body fluid, endometrial and minimum inhibitory concentrations. Equine Vet J. 2000;32:233–8.
34. Schnabel LV, Papich MG, Divers TJ, Altier C, Aprea MS, McCarrel TM, Fortier LA. Pharmacokinetics and distribution of minocycline in mature horses after oral administration of multiple doses and comparison with minimum inhibitory concentrations. Equine Vet J. 2012;44:453–8.
35. Davies JL, Salmon JH, Papich MG. Pharmacokinetics and tissue distribution of doxycycline after oral administration of single and multiple doses in horses. Am J Vet Res. 2006;67:310–6.
36. Leclere M, Magdesian KG, Cole CA, Szabo NJ, Ruby RE, Rhodes DM, Edman J, Vale A, Wilson WD, Tell LA. Pharmacokinetics and preliminary safety evaluation of azithromycin in adult horses. J Vet Pharmacol Ther. 2012;35:541–9.
37. Wilson WD, Spensley MS, Baggot JD, Hietala SK. Pharmacokinetics and estimated bioavailability of amoxicillin in mares after intravenous, intramuscular, and oral administration. Am J Vet Res. 1988;49:1688–94.
38. Ensink JM, Klein WR, Mevius DJ, Klarenbeek A, Vulto AG. Bioavailability of oral penicillins in the horse: a comparison of pivampicillin and amoxicillin. J Vet Pharmacol Ther. 1992;15:221–30.
39. de Castillo JFE. Tetracyclines. In: Giguère S, Prescott JF, Dowling PM, editors. Antimicrobial therapy in veterinary medicine. 5th ed. Hoboken: John Wiley Blackwell; 2013. p. 262.
40. Barbour AG. Isolation and cultivation of Lyme disease spirochetes. Yale J Biol Med. 1984;57:521–5.
41. Boerner J, Failing K, Wittenbrink MM. In vitro antimicrobial susceptibility of *Borrelia burgdorferi*: influence of test conditions on minimal inhibitory concentration (MIC) values. Zenetralbl Bakteriol. 1995;283:49–60.
42. Sicklinger M, Wienecke R, Nubert U. In vitro susceptibility testing of four antibiotics against *Borrelia burgdorferi*: a comparison of results for the three genospecies *Borrelia afzelii, Borrelia garinii,* and *Borrelia burgdoferi sensu stricto*. J Clin Microbiol. 2003;41:1791–3.
43. Salmon SA, Watts JL, Yancey RJ Jr. In vitro activity of ceftiofur and its primary metabolite, desfuroylceftiofur, against organisms of veterinary importance. J Vet Diagn Investig. 1996;8:332–6.
44. Scofield D, Black J, Wittenburg L, Gustafson D, Ferris R, Hatzel J, Traub-Dargatz J, McCue P. Endometrial tissue and blood plasma concentration of ceftiofur and metabolites following intramuscular administration of ceftiofur crystalline free acid to mares. Equine Vet J. 2014;46:606–10.
45. Collard WT, Cox SR, Lesman SP, Grover GS, Boucher JF, Hallberg JW, Robinson JA, Brown SA. Pharmacokinetics of ceftiofur crystalline-free acid sterile suspension in the equine. J Vet Pharmacol Ther. 2011;34:476–81.
46. Fultz L, Giguère S, Berghaus LJ, Davis JL. Plasma and pulmonary pharmacokinetics of desfuroylceftiofur acetamide after weekly administration of ceftiofur crystalline free acid to adult horses. Equine Vet J. 2014;46:252–5.
47. Auckenthaler R. Pharmacokinetics and pharmacodynamics of oral beta-lactam antibiotics as a two-dimensional approach to their efficacy. J Antimicrob Chemother. 2002;50:13–7.
48. Chang YF, Ku YW, Chang CF, Chang CD, McDonough SP, Divers T, Pough M, Torres A. Antibiotic treatment of experimentally *Borrelia burgdoferi*-infected ponies. Vet Microbiol. 2005;107:285–94.
49. Witte TS, Bergwerff AA, Scherpenisse P, Drillich M, Heuwieser W. Ceftiofur derivates in serum and endometrial tissue after intramuscular administration in healthy mares. Theriogenol. 2010;74:466–72.
50. Wagner B, Johnson J, Garcia-Tapia D, Honsberger N, King V, Strietzel C, Hardham JM, Heinz TJ, Marconi RT, Meeus PFM. Comparison of effectiveness of cefovecin, doxycycline, and amoxicillin for the treatment of experimentally induced early Lyme borreliosis in dogs. BMC Vet Res. 2015;11:163–70.
51. Arvikar SL, Steere AC. Diagnosis and treatment of Lyme arthritis. Infect Dis Clin N Am. 2015;29:269–80.
52. Carris NW, Pardo J, Montero J, Shaeer KM. Minocycline as a substitute for doxycycline in targeted scenarios: a systematic review. Open Forum Infect Dis. 2015;2:178–89.
53. Dersch R, Freitag MH, Schmidt S, Sommer H, Rauer S, Meerpohl JJ. Efficacy and safety of pharmacological treatments for acute Lyme neuroborreliosis - a systemic review. Eur J Neurol. 2015;22:1249–59.
54. Salvin S, Houin G. Clinical pharmacokinetics of doxycycline and minocycline. Clin Pharmacokinet. 1988;15:355–66.
55. Gilmour MA, Clarke CR, MacAllister CG, Dedeo JM, DI C, Morton DJ. Pugh M. Ocular penetration of oral doxycycline in the horse. Vet Ophthalmol. 2005;8:331–5.
56. Schnabel LV, Papich MG, Watts AE, et al. Orally administered doxycycline accumulates in synovial fluid compared to plasma. Equine Vet J. 2010;42:208–12.
57. Fortier LA, Motta T, Greenwald RA, Divers TJ, Mayr KG. Synoviocytes are more sensitive than cartilage to the effect of minocycline and doxycycline on IL-1alpha and MMP-13-induced catabolic gene responses. J Orthop Res. 2010;28:522–8.
58. Bernardino AL, Kaukshal D, Philipp MT. The antibiotics doxycycline and minocycline inhibit the inflammatory responses to the Lyme disease spirochete *Borrelia burgdorferi*. J Infect Dis. 2009;199:1379–88.
59. Teske RH, Rollins LD, Condon RJ, Carter GG. Serum oxytetracycline concentrations after intravenous and intramuscular administration in horses. J Am Vet Med Assoc. 1973;162:119–20.
60. Lantos PM, Auwaerter PG, Wormser GP. A systematic review of *Borrelia burgdorferi* morphologic variants does not support a role in chronic Lyme disease. Clin Infect Dis. 2014;58:663–71.

6

Clinicopathological findings in horses with a bi- or tripartite navicular bone

Ellen J. van der Zaag[1], Erik A. W. S. Weerts[2], Antoon J. M. van den Belt[3] and Willem Back[4,5*]

Abstract

Background: Navicular bone partition is a rare condition reported in horses, which is during the evaluation of a lameness or prepurchase examination often misinterpreted for a parasagittal fracture. In this report, the clinicopathological findings of three cases of navicular bone partition are evaluated. The possible pathomechanisms underlying the condition are hypothesised, focusing on a potential origin of foetal vascular disturbance. This study is furthermore aiming at a clearer and earlier recognition of navicular bone partition, since this condition would finally predispose for a clinical lameness with a poor prognosis.

Case presentations: Case 1 was a 10-year-old Belgian Warmblood gelding with a Grade 3/5 chronic, recurrent left-forelimb lameness that had persisted for 4 months. Perineural palmar digital nerve block of the distal foot abolished the lameness. Radiographic examination revealed a bipartite navicular bone in the left forelimb. Unfortunately, the animal was lost to follow-up. Case 2 was a 7-year-old Quarter Horse stallion with a Grade 3/5 recurrent right forelimb lameness that had persisted for 2 years. The lameness switched to the contralateral left forelimb with a palmar digital nerve block. Radiographic examination identified a tripartite navicular bone in both forelimbs. Pathological examination additionally revealed chronic degenerative changes of the cartilage and subchondral bone with marked cystic changes. Case 3 was a 5-year-old Dutch Warmblood gelding with a Grade 3/5 recurrent left hindlimb lameness that had persisted for 6 months. Owing to the uncooperative behaviour of the horse, only a combined peroneal and tibial nerve block could be performed, which abolished the lameness. Radiographic examination revealed a bipartite navicular bone in the left hindlimb. Pathological examination showed a navicular bipartition in the left hindlimb, with microscopic changes comparable to those evident in Case 2; additionally, cartilage indentations were also found in the navicular bones of the right front- and hindlimb at a similar location as the partition site in the left hindlimb.

Conclusions: It is speculated that a navicular bone partition has a congenital origin and is caused by vascular disturbance during foetal development. This may lead to aberrant endochondral ossification or the formation of multiple ossification centres resulting in navicular bone partitioning. In the adult horse, chronic repetitive biomechanical challenges at the partition sites may induce local degenerative changes with subchondral cyst formation and thus would cause a gradually developing chronic lameness with a poor prognosis.

Keywords: Bi-tripartite navicular bone, Bi-tripartite sesamoids, Partitioning, Navicular fracture, Navicular pathology

Background

Bi- or tripartite development of the navicular bone is a rare condition in horses that is often a coincidental finding during the evaluation of lameness or at a prepurchase examination [3, 9, 15, 22]. Early recognition of the condition and distinction from its main differential diagnosis, a parasagittal fracture, is essential. Misinterpretation of radiographs can lead to disputes between the vendor, buyer, veterinarian and/or insurance company. The clinical appearance of navicular bone partition usually is a mild lameness, but affected horses may initially exhibit normal athletic function [9]. Its radiographic findings are characterized by rounded bone edges of the adjacent pieces of navicular bone, a wide radiolucent region at the partition site and often cystic development of adjacent subchondral bone, which are in contrast with

* Correspondence: W.Back@uu.nl
[4]Department of Equine Sciences, Faculty of Veterinary Medicine, Utrecht University, Yalelaan 112-114, NL-3584 CM Utrecht, The Netherlands
[5]Department of Surgery and Anaesthesia of Domestic Animals, Faculty of Veterinary Medicine, Ghent University, Salisburylaan 133, B-9820 Merelbeke, Belgium
Full list of author information is available at the end of the article

the usual characteristics of a fracture [4, 5, 8, 9, 15, 18]. The veterinary literature on navicular bone partition is limited and pathological examination of the partite navicular bone has rarely been performed [18]. In the human literature, the condition has been reported more frequently, but to date it is merely considered a relatively uncommon anatomical variation [19]. Possible pathomechanisms underlying this condition mentioned, include the formation of multiple ossification centres and/or a disturbance in the blood supply, leading to incomplete ossification of the cartilaginous bone model [10, 11, 19].

In this article, the clinical, radiological and pathological examination of 3 horses with navicular bone partition will be presented and possible underlying pathogenetic mechanisms are discussed with a review of the existing relevant literature.

Case presentation

Between 2008 and 2011, two adult Warmblood horses and one Quarter Horse were referred to the Equine Clinic of the Faculty of Veterinary Medicine at Utrecht University, The Netherlands, for the evaluation of a chronic, recurrent lameness of the left forelimb (LF), right forelimb (RF) and left hindlimb (LH), respectively. All horses were clinically evaluated i.e. the lameness was scored on a hard ground surface at trot on a straight line and on the lunge according to a modified AAEP lameness scale (Grade 0–5, separately for walk and trot). Palpation and percussion of the lame- and contralateral limb was performed and a flexion test of the distal limbs was scored as positive or negative. Local perineural analgesia was used to block the lameness and evaluated after 5–10 min. In two horses (Case 1 and 2) the lameness was blocked after a perineural analgesia of the distal foot using a palmar digital nerve (PDN) block. The third horse (Case 3) became sound after a peroneal and tibial nerve block. Bilateral radiographic evaluation of the navicular bone consisted of a dorsoproximal-palmarodistal oblique view, a lateromedial view and, if possible, a palmaro 45 ° proximal-palmarodistal oblique view. All three horses were diagnosed with a bi- or tripartite navicular bone in the lame limb; in one horse (Case 2) the condition was bilateral. Two horses (Cases 2 and 3) were euthanized because of the chronicity of the lameness and their poor prognosis, allowing a subsequent pathological examination (of both front feet in Case 2 and all four feet in Case 3). Case 1 was unavailable for follow-up.

Case 1

A 10-year-old Belgian Warmblood gelding presented with recurrent LF lameness of gradual onset that had persisted for approximately 4 months. Clinical examination revealed a Grade 3/5 LF lameness on a straight line that became more severe on the left hand lunge. Palpation of the palmar aspect of the foot and percussion of the frog identified no abnormalities. The horse exhibited a positive response to a flexion test of the left distal forelimb and no response at the right distal forelimb. The lameness was abolished by a PDN block of the LF limb.

Radiographic examination of the LF in dorsopalmar view revealed bipartition of the navicular bone at the lateral wing with rounded adjacent edges. There was no clear formation of osteophytes and the distal border of the navicular bone had some small lucent zones. The lateral trabecular part of the navicular bone, however, showed several well-defined, radiolucent cystic areas adjacent to the radiolucent line (Fig. 1a). In the RF, in a similar dorsopalmar view of the navicular bone, a few radiolucent zones were visible on the distal border but there were no indications of partitioning or cystic development (Fig. 1b).

After the horse was diagnosed with a unilateral, bipartite navicular bone of the LF, the owner initially decided to further evaluate the use of the horse for hunting using

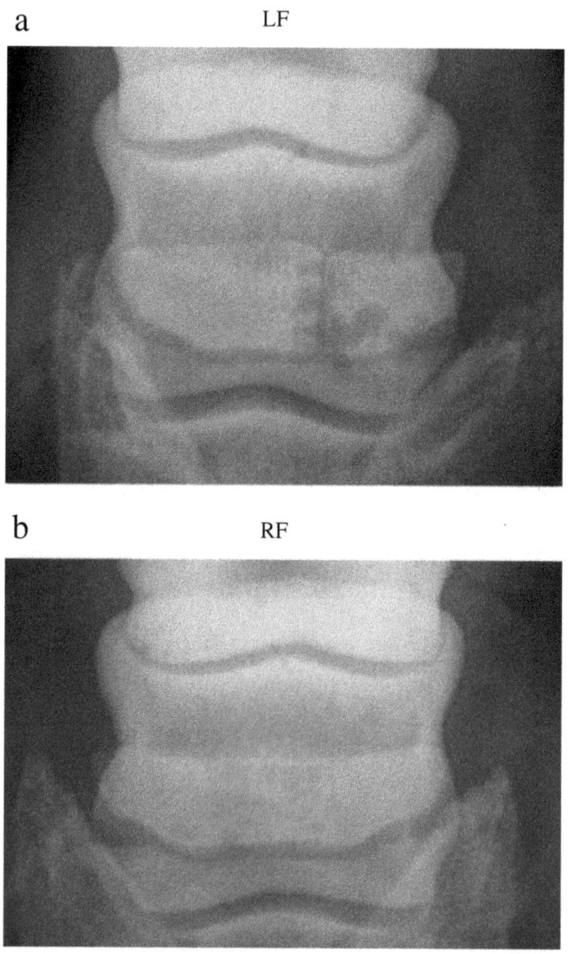

Fig. 1 Case 1. Radiographic examination of a horse, dorsoproximal-palmarodistal oblique view. **a** Bipartite navicular bone conformation in the left forelimb. **b** Normal navicular bone conformation in the right forelimb

orthopaedic shoeing, but finally the horse was lost to follow-up.

Case 2

A 7-year-old Quarter Horse stallion presented with a recurrent RF lameness of gradual onset that had persisted for over 2 years. Clinical examination revealed a Grade 3/5 RF lameness comparable lame on both a straight line and on the right hand lunge. Palpation of the palmar pastern region of the RF elicited a mild pain response as-well as percussion of the frog. The horse exhibited a (slightly) positive response to a flexion test of the right forelimb and no response to the left forelimb. Using a PDN block of the distal RF the lameness switched to the LF limb.

Radiographic examination of both distal forelimbs in dorsopalmar view revealed that both navicular bones had extensive disseminated irregular radiolucent areas in the lateral and medial wings. The radiolucent lines, located at more or less one third of their length in the navicular bones, showed rounded corners at the proximal and distal edges. Both forefeet were diagnosed with a tripartite navicular bone (Fig. 2a, b). Due to the long history of lameness and poor prognosis, the horse was euthanized and both distal forelimbs became available for pathological examination.

Pathological examination of the navicular bones of the RF and LF revealed similar abnormalities. Macroscopically, both bones appeared to be composed of three similar-sized pieces of bone, connected by small amounts of irregular slightly protruding pale yellow-brown to red fibrillar tissue, resulting in marked irregularity of the articular surface of the navicular bone to the middle phalanx at these connection sites. The cartilage of the middle third part of the bone lying in between these sites showed marked smooth indentation. On the transverse cut surface in the middle of the navicular bone, the connection sites consisted of fibrous tissue. Directly adjacent to these connection sites, the subchondral bone contained several variably-sized and sharply-demarcated grey-brown to white cystic areas. The synovial lining of the coffin joint and the adherent deep digital flexor tendon were slightly irregular, with yellow and red discoloration. The articular surfaces of the middle and distal phalanx of both coffin joints appeared smooth. Both coffin joints contained normal amounts of slightly viscous and slightly yellow synovial fluid (Fig. 3a). The transverse cut surface of the LF navicular bone was examined histopathomorphologically after fixation in 4 % formalin, decalcification, paraffin-embedding, sectioning at 4 μm, and Haematoxylin and Eosin (H&E) staining (Fig. 3b). The bone showed a continuous smooth lining of articular cartilage, deviating into the bone at the partition sites and culminating in extensive accumulations of fibrocartilaginous tissue in the centre of these sites. The adjacent articular cartilage at the coffin joint surface showed multifocal, irregular thickening and a tendency for mild chondrocyte clustering (chondrone formation). The adjacent subchondral bone appeared markedly sclerotic. The cystic lesions (lateral side of the middle third piece and the medial side of the lateral third piece) consisted of concentrically-layered accumulations of partly densely-arranged, moderately-cellular and highly-vascularized connective tissue and partly loosely-arranged myxoid-like connective tissue; no clear synovial epithelial lining was evident in the cysts. There were no signs of active inflammation or osteonecrosis. The macroscopically-described fibrillar tissue protruding at the partitions' connection site surfaces could histologically not be traced back (Fig. 4).

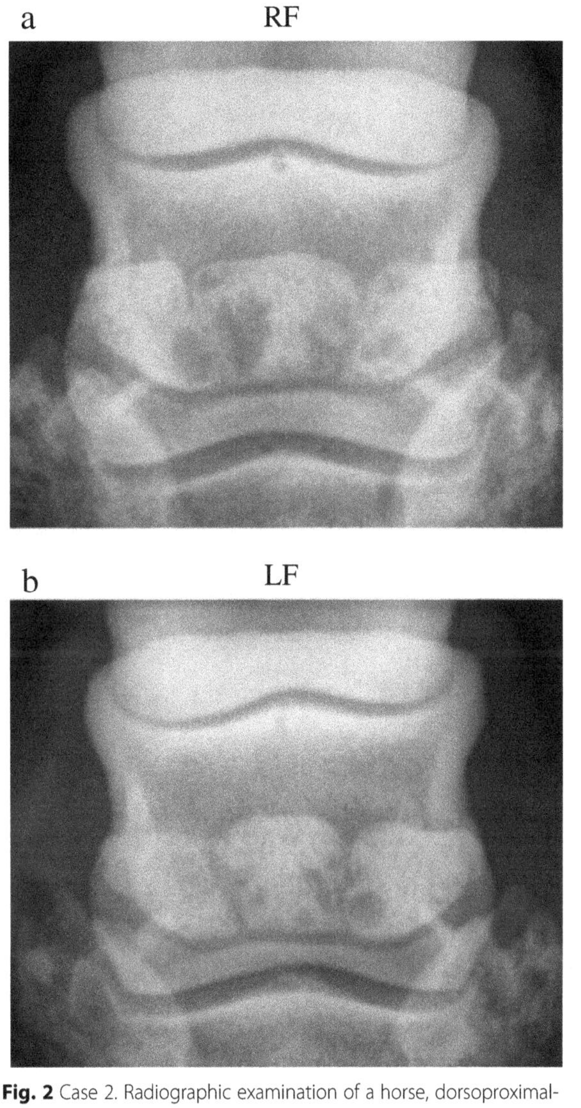

Fig. 2 Case 2. Radiographic examination of a horse, dorsoproximal-palmarodistal oblique view with a tripartite navicular bone conformation in both forelimbs. **a** Right forelimb. **b** Left forelimb

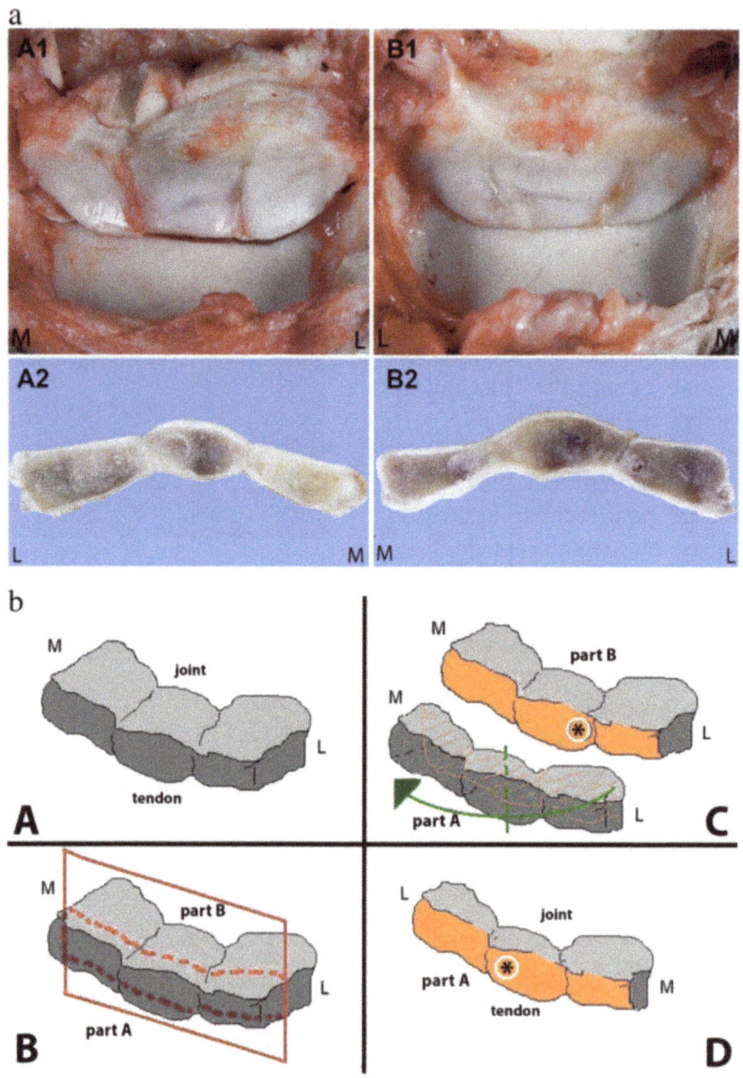

Fig. 3 **a** Case 2. Macroscopic pathological findings of the LF (**A**) and RF (**B**) navicular bones (*panels 1*: distal interphalangeal joint articular surface, *panels 2*: transverse cut surface, formalin-fixed). Both navicular bones consist of three approximately same-sized bony pieces connected by fibrillary tissue. On the transverse cut surface of both bones, round variable-sized well-demarcated cystic structures can be appreciated near the connection sites. The articular surface is markedly irregular. L?=?lateral side, M?=?medial side. **b** Case 2. Transverse section of the navicular bones was made as shown in panel B (*red lines*), resulting in a front part A and a back part B. The Fig. 3: A2 and B2 are the front parts A of the bones after making the transverse sections and rotated 180°, as shown in panels C and D. The cystic change as situated in Fig. 3a: A2 is shown here as a *white circle (asterisk)*; L?=?lateral; M?=?medial

Case 3

A 5-year-old Dutch Warmblood gelding presented with recurrent LH lameness of gradual onset that had persisted for more than 6 months. Clinical examination revealed a Grade 3/5 LH lameness on a straight line and on the left hand lunge. Palpation of the palmar aspect of the foot and percussion of the frog identified no abnormalities. Owing to its dangerous, uncooperative behaviour, a flexion test and PDN block of the distal hindlimb could not be performed. Finally, a peroneal and tibial nerve block was performed on the lame LH, by which the lameness of the horse became abolished. Routine radiographs were made of the left hock and the left distal limb. After the horse was diagnosed with a LH bipartite navicular bone the contralateral navicular bone was also radiographically evaluated.

Radiographic examination of the LH revealed a wide, radiolucent line in the lateral wing of the navicular bone at one-third of its total length in a parasagittal plane, with rounded margins of the two osseous pieces and cystic changes in the adjacent subchondral bone, together indicative of bipartition; there were no clear findings consistent with degenerative joint disease (Fig. 5a). Further radiographic evaluation of the LH (fetlock and

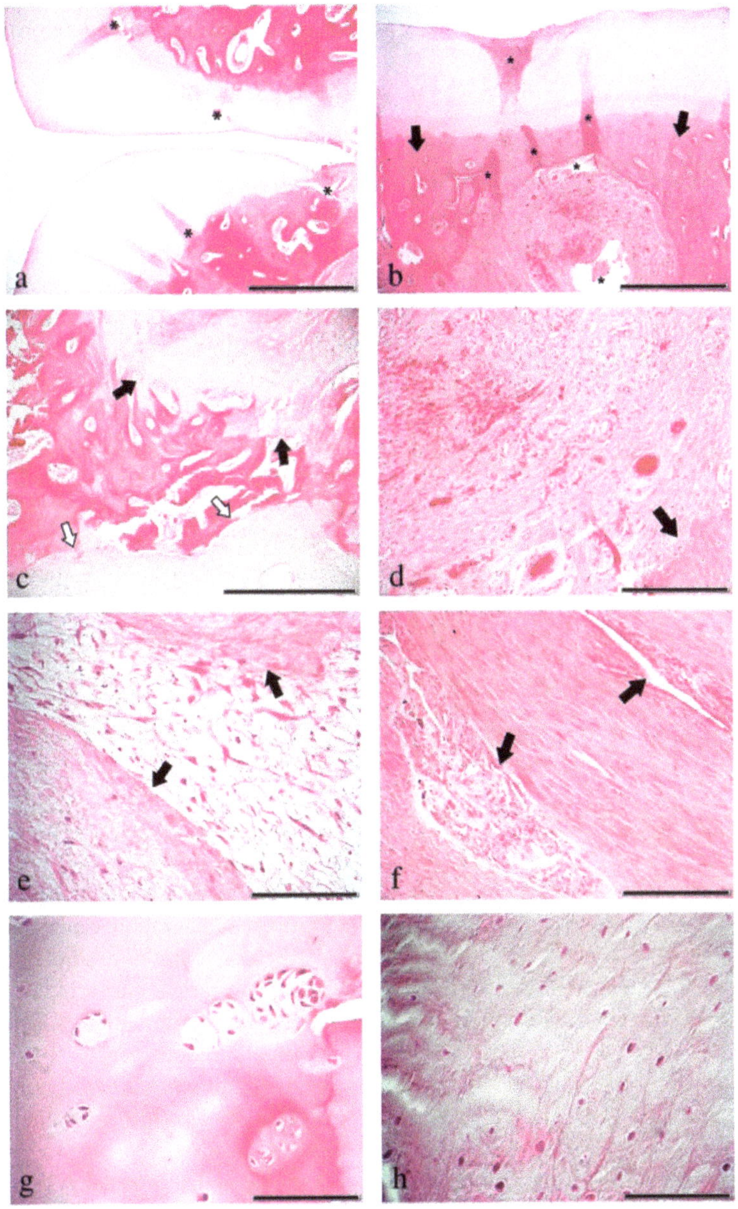

Fig. 4 Combined histopathomorphological findings of Case 2 and 3, Haematoxylin and Eosin (H&E). **A** continuous smooth lining of cartilage, deviating into the bone at the partition's connection site (bar = 1 mm). **B** cystic change of the subchondral bone with marked sclerosis of the adjacent bone (*arrows*) (bar = 1 mm). **C** retained cartilage cores (*black arrows*) at a connection site of one of the partitions near the outer concentric layer of a subchondral bone cyst (*white arrows*) (bar = 500 μm). **D** densely-packed highly vascular aspect of cysts, with adjacent sclerotic bone (*arrow*) (bar = 200 μm). **E** myxoid-like aspect of the cysts, surrounded by more densely-packed fibrous tissue (*arrows*) (bar = 100 μm). **F** island of tissue in the fibrous cystic wall resembling synovial membrane (*arrows*) (bar = 200 μm). **G** chondrone formation in the articular cartilage (bar = 100 μm). **H** fibrocartilaginous tissue at the partition's connection sites (bar = 100 μm). **B**, **C**, **D**, **E**, **G** and **H** Case 2, LF navicular bone. **A** and **F**: Case 3, LH navicular bone; * = tissue processing artefacts

talocrural joint) and right hindlimb (RH) navicular bone (Fig. 5b) revealed no abnormalities. Due to the duration of the lameness, the depressed character change of the horse possibly due to the persistent pain, and the given poor prognosis, the horse was euthanized and a pathological examination was performed on all four feet.

Macroscopically, the LH navicular bone seemed to consist of two partly joined pieces with a clear sagittal indentation at the lateral side at one-third of its length, possessing smooth cartilaginous edges and causing marked focal irregularity of the articular surface to the middle phalanx. Contrasting with the partition sites in Case 2, the articular surface at this location did not contain any protruding irregular fibrillar tissue. On the transverse cut surface, adjacent to the edges of both bone parts, round variably-sized and sharply-demarcated

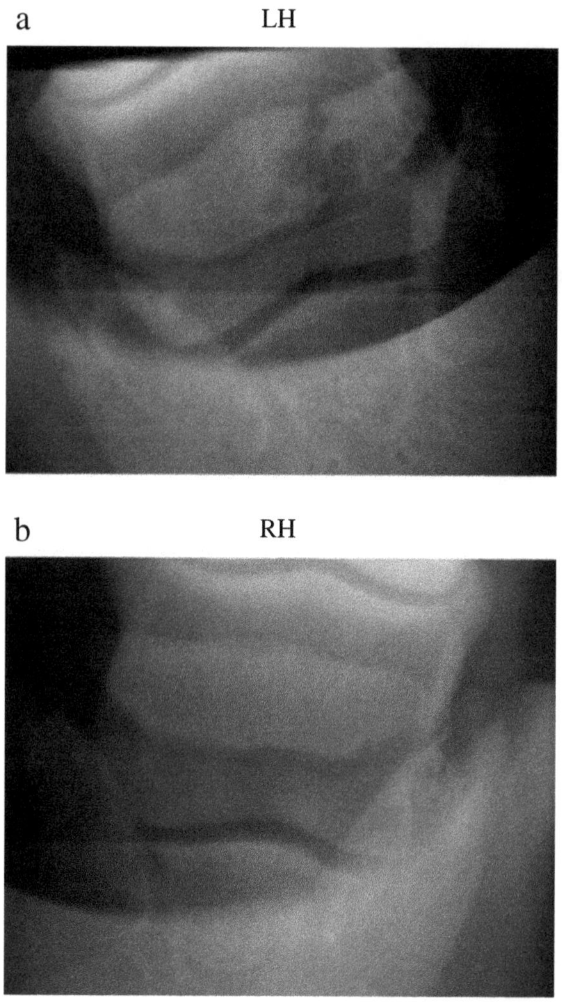

Fig. 5 Case 3. Radiographic examination of a horse, dorsoproximal-palmarodistal oblique view. **a** Bipartite navicular bone confirmation in the left hindlimb. **b** Normal navicular bone confirmation in the right hindlimb

white to grey-brown cystic structures were visible in the subchondral and trabecular bone. The RH and RF navicular bones showed only mild indentations in the lateral and the medial wings respectively of the articular cartilage at approximately one-third of the total length. The navicular bone of the LF appeared unremarkable. The articular surfaces of the middle and distal phalanx of all coffin joints were smooth. As described for Case 2, all coffin joints contained normal amounts of slightly viscous and slightly yellow synovial fluid (Fig. 6).

Histopathomorphologic examination of the LH navicular bone (fixation, decalcification, paraffin-embedding, slide thickness and staining identical to Case 2 revealed changes similar, though milder, to those in Case 2. There was mild irregular thickening of the otherwise continuous smooth lining of the cartilage, which deviated, similarly to Case 2, into the bone at the partition site. At this location, the navicular bone showed a similar accumulation of fibrocartilaginous tissue, few retained cartilage cores and sclerosis of the adjacent trabecular bone. Cystic lesions were of the same composition as in Case 2 and contained focally in the outer concentric layers near the connection site small tissue pieces resembling synovial lining (Fig. 4).

Discussion

In both human medical and veterinary literature, navicular (distal sesamoid) bone partition is considered to be congenital [2–4, 11, 15, 21, 22]. In human medical literature, the most frequently suggested pathogenesis underlying sesamoid partition would involve the formation of multiple ossification centres within a single cartilage bone model, which fail to fuse [10, 11, 19], or the formation of extra aberrant ossification centres in addition to the normally present single cartilage bone model [19]. Regarding these hypothetic pathogenetic mechanisms, repetitive vascular micro-trauma [11] has been cited as an important aetiological factor, as it interferes with the ossification process by impairing blood, and therefore oxygen and nutrient, supply.

Typical features of a partitioned navicular bone, distinguishing it from parasagittal fractures, provided by this study are:

- a chronic, persistent lameness, often with a gradual onset;
- a correlation between relatively mild lameness and overt radiographic findings in the same limb(s);
- the possibility of multiple limbs involved;
- no substantial peri-articular remodelling;
- a uniform distribution of partition sites at approximately one-third of the length of the navicular bone; and
- a continuous, smooth lining of articular cartilage that deviates into the bone at the partition site.

Aetiologically, especially the last two characteristics, being rather distinctive anatomical features, seem to plead for abnormal congenital development of the involved navicular bones. The uniform distribution of the partition sites seen in all cases, notably in the case of tripartition when two partitions exist in the same bone evenly-spread along its total length simultaneously, is difficult to consider as being the result of a coincidental traumatic event leading to fractures. In addition, even if such simultaneous evenly-spread parallel fractures would be able to occur, clinical presentation would likely be very different with marked acute lameness dominating. Moreover, regarding the normal features of fracture repair and the normal regenerative properties of bone and cartilage, it is hardly thinkable that an uncomplicated smooth cartilaginous lining as seen at the partition sites in the described cases would develop at a site of a

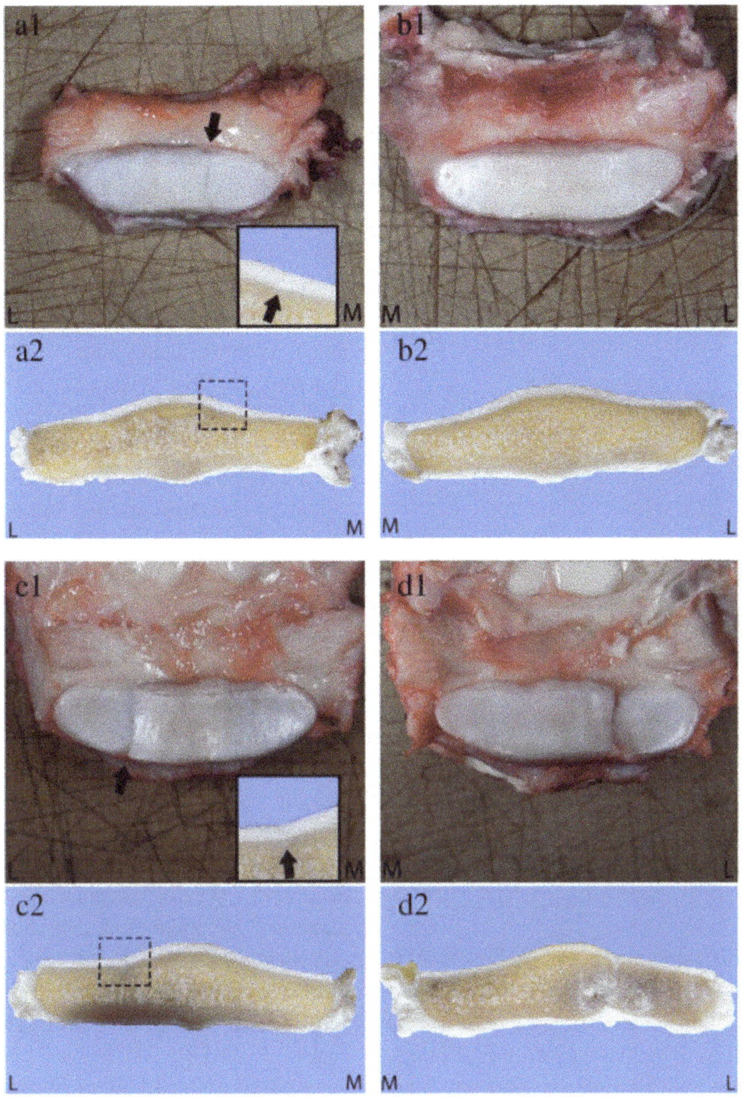

Fig. 6 Case 3. Macroscopic pathological findings of the RF (**A**), LF (**B**), RH (**C**) and LH (**D**) navicular bones (*panels 1*: coffin joint articular surface, *panels 2*: transverse cut surface, formalin-fixed). The LH navicular bone shows marked sagittal indentation at its articular surface at the lateral side at one third of its length. On the transverse cut surface, cystic structures resembling those described in Case 2 can be appreciated at both sides of the connection site. The RF and RH navicular bones show only mild superficial sagittal indentation of the articular cartilage (picture insets), without changes of the subchondral bone; L = lateral side, M = medial side

fracture. Whether the partitioning of distal or proximal sesamoid bones is a hereditary defect, is not that clear yet and beyond the scope of the findings presented in this study. In veterinary literature a 'multiple ossification centre hypothesis' is suggested as a possible pathomechanism underlying partitioning [3, 4, 8, 9]. Some authors otherwise regard this sesamoid bone partitioning as a mild form of polydactyly [2, 21]. If navicular bone partitioning is an expression of polydactyly, however, one might expect further partitioning of also of the more proximal bones and surrounding structures (e.g. tendons) distal to the involved navicular bone. The authors of this paper found no evidence of this in their case series and can therefore would rather not support polydactyly as the underlying mechanism of navicular bone partition.

During fetal development, the blood supply to the navicular bone takes two routes from day 125 of gestation [17]. One supply is temporarily situated in the superficial layer of the fibrocartilage; the other supply is the one becoming permanent in the adult horse [16], in which arteries enter the navicular bone from four different directions (Fig. 7a). The temporary secondary circulation gradually disappears after 270 days of gestation when the first ossification centres appear. A disruption of either of these vascular networks might trigger aberrant ossification, although the temporary vessels are less likely to be of influence as

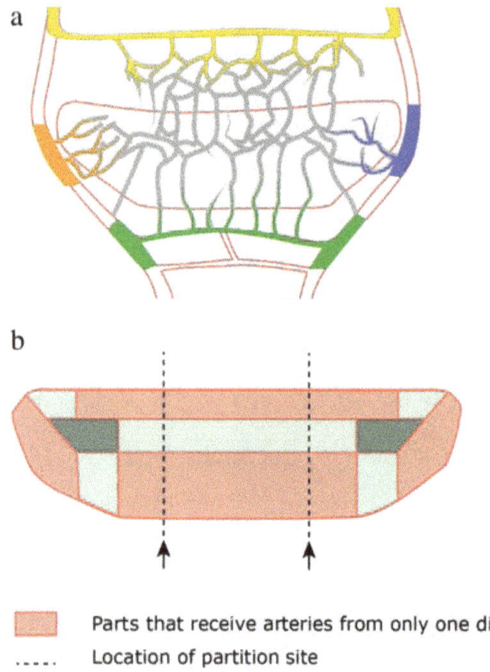

Fig. 7 a Schematic drawings of the arterial supply of the navicular bone (modified with permission from Rijkenhuizen et al. [16]). Arteries entering from four different directions illustrated in colours. **b** Diagram of the suggested arterial anatomy of the navicular bone (modified with permission from Rijkenhuizen et al. [16]). The red coloured areas indicate the parts of the navicular bone that receive arteries from only one direction. The light green coloured areas receive arteries from two directions and the dark green coloured areas from three directions

they do not penetrate the navicular bone itself. Vascular micro-trauma or intra-articular blood vessel compression due to e.g. unusual positioning of the foetus in utero may lead to hypoxia and result in defective endochondral ossification, eventually causing partitioning of the navicular bone. In a similar way, regarding the pathogenesis of osteochondrosis during bone and joint growth, failure of the blood supply is considered responsible for the defective transformation of cartilage into bone [13, 14, 23]. In a study of foals [13, 14], early lesions of osteochondrosis in the distal tibia occurred secondary to failure of the blood supply of the physeal plate, leading to ischaemic necrosis and locally preventing normal ossification. Moreover, osteochondrotic lesions are formed in a limited time window, namely the period when the epiphyseal cartilage is supplied by vulnerable blood vessels [23]. Our hypothesis is based on the same idea that disturbance at a specific moment during gestation, when angiogenesis in the cartilage navicular bone model occurs [17], can initiate the partitioning process.

Partitions of the navicular bone seem to be consistently located at approximately one-third of the length of the bone in the parasagittal plane in the lateral and/or medial wing. Certain parts of the navicular bone receive arteries from only one side, whereas other parts receive arteries from more than one direction [16, 17]. The parts supplied by only one artery are likely more susceptible to ischemia due to vascular disturbance, which during foetal development might predispose these areas for partitioning (Fig. 7b).

The histological changes found at partition sites can be briefly summarized as features of degeneration. A suitable explanation for these changes is that the partition sites are subjected to abnormal pressure through micro-movement of the bone segments during life [6, 12], again (as with the hypothesized pathogenesis of the partitioning process in utero) compromising local circulation and leading to ischaemic necrosis. Within the rather non-specific spectrum of degenerative lesions of bone, subchondral cysts are chronic lesions that likely develop from necrotic cartilaginous fragments retained in the subchondral bone, focal necrosis of the subchondral bone itself or as a result of synovial fluid being forced into the subchondral bone via defects in the articular cartilage [1]. This is seen as a sequel of osteochondrosis [1], in the course of navicular disease [7] and also after trauma to the articular surface [1, 20]. It is unclear whether the cysts in the presented cases have been formed via necrotic retained cartilaginous fragments (which are a finding in Case 3) or from the subchondral bone itself. Nevertheless, because of their known gradual development during various chronic joint disease, subchondral cysts are more likely to co-occur with partitioning and its subsequent degenerative features than with an acutely occurring fracture; therefore, they might be of differential diagnostic significance during radiological examination.

In our case series there was a clear age difference (5-7-10 years) in occurrence of clinical and radiological symptoms, which is consistent with other reported cases [2, 3, 18]. All horses had similar workload and were regularly ridden; only one horse had a history of sliding in the pasture. Apparently, the onset of specific clinical and radiological features is not directly age-related, but indirectly related to the severity of the joint pressure imbalances and therefore subsequently the amount of time needed for the bone parts to develop the described degenerative features presumably as a result of these imbalances.

In summary, when we compare our case findings, the final lameness in the partitioned navicular bone seems to be mainly linked to bone degeneration with cystic changes. Case 1 showed cyst-like abnormalities during radiographic examination in only the lame limb; in this case, lameness was abolished after a PDN block. In Case 2, the involved horse showed cystic changes in the navicular bones of both front legs; the horse became lame in the contralateral limb after PDN block. In Case 3, after the lame limb was blocked (unfortunately the diagnosis could not definitively be confirmed by a PDN block), lameness did not switch to the contralateral limb, which is in accordance with the radiological and pathological appearance of the bone, showing no overt abnormalities.

Conclusions

Navicular bone partitioning likely has a congenital origin and can possibly be explained as the result of articular vascular disturbance in utero, causing bone abnormalities during foetal bone formation by endochondral ossification. Its clinical appearance is a relatively mild lameness of gradual onset. The partitioning of the bone is already overt at birth and moreover due to the biomechanical challenges at the partioning site during life this induces a progressing focal (cystic) degeneration of the navicular bone causing a gradually developing lameness. The cysts and other characteristic radiographic findings, which include the presence of two or three clearly separate pieces at approximately one-third of the length of the bone with rounded margins and minimal peri-articular remodelling, are important features distinguishing navicular bone partitioning from its main differential diagnosis, the parasagittal fracture.

The horses described in this study already showed clinical signs consistent with the abnormal radiographs and pathological findings. They seem to have exhibited the chronic stage of the disease, mainly reflecting secondary degenerative processes. Detection of the partitioning process earlier in life may result in a better understanding of the progression of this condition, thus leading to a better insight in the disease. Earlier recognition by the veterinarian might prevent occurrence of prepurchase issues and furthermore save time and money otherwise becoming invested in horses suffering from an abnormality with finally a poor prognosis.

Consent for publication

In all admitted cases their owners gave their written informed consent for publication by means of signing our official client acceptance form.

Abbreviations

LF: left forelimb; LH: left hindlimb; PDN: palmar digital nerve; RF: right forelimb; RH: right hindlimb.

Competing interests

The authors declare that they have no competing interests.

Authors' contributions

EZ coordinated the case report and reviewed the literature; WB and AJB contributed to the diagnostic work-up; EW conducted post-mortem examinations; all authors contributed to writing the manuscript and read and approved the final manuscript.

Acknowledgements

Professor Astrid B.M. Rijkenhuizen is acknowledged for her suggestions and critically reading the manuscript.

Funding

There was a kind funding in time from Veterinary Clinic De Delta, The Netherlands for E.J. van der Zaag, from the Department of Pathobiology, division Pathology, Faculty of Veterinary Medicine, Utrecht University, The Netherlands for E.A.W.S.Weerts, from the Department of Clinical Sciences of Companion Animals, Faculty of Veterinary Medicine, Utrecht University, The Netherlands for A.J.M. van den Belt, and finally the Department of Equine Sciences, Faculty of Veterinary Medicine, Utrecht University, The Netherlands, and the Department of Surgery and Anaesthesia of Domestic Animals, Faculty of Veterinary Medicine, Ghent University, Belgium, both for W. Back.

Author details

[1]Veterinary Clinic De Delta, Foppenpolder 1, NL-3155 EA Maasland, The Netherlands. [2]Department of Pathobiology, Division Pathology, Faculty of Veterinary Medicine, Utrecht University, Yalelaan 1, NL-3584 CL Utrecht, The Netherlands. [3]Department of Clinical Sciences of Companion Animals, Faculty of Veterinary Medicine, Utrecht University, Yalelaan 108, NL-3584 CM Utrecht, The Netherlands. [4]Department of Equine Sciences, Faculty of Veterinary Medicine, Utrecht University, Yalelaan 112-114, NL-3584 CM Utrecht, The Netherlands. [5]Department of Surgery and Anaesthesia of Domestic Animals, Faculty of Veterinary Medicine, Ghent University, Salisburylaan 133, B-9820 Merelbeke, Belgium.

References

1. Baxter GM. Subchondral cystic lesions in horses. In: McIlwraith CW, Trotter GW, editors. Joint disease in the horse. 1st ed. Collorado: W.B. Saunders Company; 1996. p. 384–97.
2. Benninger MI, Deiss E, Ueltschi G. Bipartite distal phalanx and navicular bone in an Andalusian stallion. Vet Rad Ultras. 2005;46:69–71.
3. Beherens E, Yllesca MV. An unusual lameness: a case report. Equine Vet Sci. 1990;10:53–4.
4. Biggi M, Dyson S. Unusual navicular bone fractures in hindlimbs: clinical and imaging findings. J Equine Vet Science. 2013;33:1082–7.
5. Colles CM. Navicular bone fractures in the horse. Equine Vet Educ. 2011;23:255–61.
6. Dik KJ, Van den Belt AJ, Van den Broek J. Relationships of age and shape of the navicular bone to the development of navicular disease: a radiological study. Equine Vet J. 2001;33:172–5.
7. Dyson SJ, Murray R, Blunden T, Schramme M. Current concepts of navicular disease. Equine Vet Educ. 2006;18:45–56.
8. Dyson SJ. Fracture of the navicular bone and congenital bipartite navicular bone. In: Ross MW, Dyson SJ, editors. Diagnoses and management of lameness in the horse. 2nd ed. St. Louis: Elsevier; 2010. p. 243–344.
9. Dyson SJ. Radiological interpretation of the navicular bone. Equine Vet Educ. 2008;20:268–80.
10. Frankel JD, Harrington J. Symptomatic bipartite sesamoids. J of Foot Surg. 1990;29:318–23.
11. Gerber M, Roberto PD. Sesamoids. In: Hetherington VJ, editor. Textbook of hallux valgus and forefoot surgery. 1st ed. Ohio College of Podiatric Medicine (OCPM). Cleveland: Kent State University (KSU); 2000. p. 153–162. https://www.kent.edu/cpm/textbook-hallux-valgus-forefoot-surgery.
12. Komosa M, Purzyc H, Wojnar M, Frąckowiak H, Kobryńczuk F. Navicular syndrome in sport horses as a result of the disorder of biological bone tissue turnover rhythm: a review. Biol Rhythm Res. 2013;44:339–51.
13. Olstad K, Ytrehus B, Ekman S, Carlson CS, Dolvik NI. Early lesions of osteochondrosis in the distal tibia of foals. J Orth Res. 2007;25:1094–105.
14. Olstad K, Ytrehus B, Ekman S, Carlson CS, Dolvik NI. Epiphyseal cartilage canal blood supply to the tarsus of foals and relationship to osteochondrosis. Equine Vet J. 2008;40:30–9.
15. Reeves MJ, Yovich JV, Turner AS. Miscellaneous conditions of the equine foot. Vet Clin North Am Equine Pract. 1989;5:221–42.

16. Rijkenhuizen ABM, Németh F, Dik KJ, Goedegebuure SA. The arterial supply of the navicular bone in the normal horse. Equine Vet J. 1989;21:399–404.
17. Rijkenhuizen ABM, Németh F, Dik KJ, Goedegebuure SA. Development of the navicular bone in foetal and young horses, including the arterial supply. Equine Vet J. 1989;21:405–12.
18. Schröer NJ, Niederacher V, Horstmann W, Hermanns W. Fallbericht über strahlbeinveränderungen bei einem 7-jährigen bayrischen Warmblutpferd-klinik, röntgen, computertomographie und histologische untersuchung. Pferdeheilkunde. 2000;16:474–8.
19. Shawdon A, Kiss ZS, Fuller P. The bipartite tarsal navicular bone: radiographic and computed tomography findings. Austr Rad. 1995;39:192–4.
20. Sloet van Oldenruitenborgh-Oosterbaan MM, Barneveld A, van den Belt AJM. De veterinaire keuring van het paard. Leeuwarden: Uitgeverij Libre BV; 2007.
21. Smith DRK, Leach DH, Bell RJ. Phalangeal and navicular bone hypoplasia and hoof malformation in the hind limbs of a foal. Can Vet J. 1986;27:28–34.
22. Thompson KN, Rooney JR. Bipartite proximal sesamoid bones in young thoroughbred horses. Vet Rad & Ultr. 1994;35:368–70.
23. Ytrehus B, Carlson CS, Ekman S. Etiology and pathogenesis of osteochondrosis. Vet Path. 2007;44:429–48.

7

Measurement variability of right atrial and ventricular monophasic action potential and refractory period measurements in the standing non-sedated horse

Dominique De Clercq*, Barbara Broux, Lisse Vera, Annelies Decloedt and Gunther van Loon

Abstract

Background: In human and veterinary medicine, monophasic action potential (MAP) analysis and determination of local refractory periods by contact electrode technique gives valuable information about local cardiac electrophysiological properties. It is used to investigate dysrhythmias and the impact of drugs on the myocardium. Precise measurement of total MAP duration is difficult, therefore the MAP duration is usually determined at a repolarization level of 90% (APD90). Until now, no studies are published about the feasibility of this technique in the standing non-sedated horse. In 6 healthy Warmblood horses, on two different days, an 8F quadripolar contact catheter was passed through a jugular introducer sheath and placed under ultrasound guidance at the level of the intervenous tubercle or right atrial free wall (RA), and in the right ventricular apex (RV) to record the MAP. The MAP amplitude and APD90 were measured at a resting sinus rhythm (heart rate of 30-42 bpm) and at pacing cycle lengths (PCL) of 1000 and 600 ms. The effective refractory period (ERP) was determined at PCL of 1000 and 600 ms.

Results: The overall mean (±SD) APD90 (rest), APD90 (1000) and APD90 (600) were 263 ± 39 ms, 262 ± 41 ms, 236 ± 47 ms for the RA and 467 ± 23 ms, 412 ± 38 ms, 322 ± 29 ms for the RV. The mean ERP1000 and ERP600 were 273 ± 24 ms and 256 ± 22 ms for the RA and 386 ± 40 ms and 293 ± 30 ms for the RV. The measurement variability for the amplitude, APD90 and ERP measurements in the RA ranged between 36 and 44, 9–22 and 7–8%, respectively. The measurement variability for the amplitude, APD90 and ERP measurements in the RV ranged between 49 and 66, 6–7 and 10–12%, respectively.

Conclusions: RA and RV MAP duration and ERP can be obtained by a contact electrode in standing non-sedated horses. The measurement variability varies with catheter location.

Keywords: Electrophysiology, Echocardiography, Monophasic action potential, Equine

Background

Electrocardiography (ECG) is the most commonly used technique for recording electrical activity of the myocardium. However, it does not provide precise information regarding the dispersion of repolarization or afterdepolarizations since an ECG represents the summation of electrical activity of many myocardial cells from a relatively large region of the heart [1]. Local activity can be recorded by an intracellular action potential (IAP). However, IAP recordings can only be made in an individual cardiac cell and thus can only be performed in vitro [2]. Monophasic action potential (MAP) has a smaller magnitude than the IAP but it is proven that MAP recordings can give highly accurate information about local atrial or ventricular activation time and the entire time course of myocardial repolarization [3–11]. The local electrical activity of the in situ beating heart can be assessed by monophasic action potentials. This waveform can be obtained by slight pressure of a contact electrode on the heart muscle and is a result of membrane potential discharges (depolarisation) and slowly recharges

* Correspondence: Dominique.DeClercq@UGent.be
Department of Large Animal Internal Medicine, Ghent University, Salisburylaan 133, B-9820 Merelbeke, Belgium

(repolarisation). MAPs are useful for investigation of the electrophysiology of different dysrhythmias (i.e. effects of cycle length changes and the role of afterdepolarizations in the genesis of triggered dysrhythmias) and the impact of drugs (i.e. the effect of antiarrhythmic drugs on the action potential duration in the in situ heart) on the heart of different species [3–11]. The recording of MAP signals in awake dogs without any interference of anaesthetic drugs is possible. With this procedure the effect of anaesthetics drugs on the autonomic nervous system, which has an important impact on cardiac electrophysiology, can be avoided [12, 13]. The underlying principles concerning MAP recordings are described elsewhere [1, 14, 15].

Until now, very little information about the possibility of making MAP recordings in standing horses is available. This paper describes the technique and the day-to-day variability of right cardiac MAP recording and refractory period measurement in the standing non-sedated horse.

Methods

The experimental protocol was approved by the Ethical Committee of the Faculty of Veterinary Medicine and the Faculty of Bioscience Engineering at Ghent University (case number EC 2015/85, date of approval 9th of September, 2015). The horses used in this study are Faculty owned horses.

The study population consisted of 6 healthy warmblood horses (3 geldings, 3 mares) aged (mean ± standard deviation) 10 ± 5 years, with a body weight of 531 ± 70 kg and a height at the withers of 160 ± 6 cm. Cardiovascular or other diseases were excluded based on a physical examination, a complete blood exam, an echocardiographic examination, a 30-min resting ECG and a 20-min standardized exercise ECG. All horses underwent the procedure two times on two different days with at least 2 weeks interval. In each horse an 8.5F introducer sheath (Introflex, Edwards Lifesciences, Dilbeek, Belgium) was placed using the Seldinger technique in the lower third of the jugular vein. An 8F quadripolar contact catheter with separated recording and pacing electrodes (EasyMap MAP, Medfact, Lörrach, Germany) was used [3, 5] (Fig. 1). The catheter was inserted via the introducer sheath and placed under ultrasound guidance at the level of the intervenous tubercle (in 9/12 procedures) or at the level of the right atrial free wall (RA) (in 3/12 procedures) and in the right ventricular apex (RV) (in all procedures =12/12) (Fig. 2). The catheter tip was pressed softly against the endocardium. Exact pressure determination against the wall was not possible with this type of catheter. MAPs were recorded after placement of the catheter in a position that provided continuous recordings with a stable amplitude of > 2 mV in the RA and > 3 mV in the RV, a stable baseline, a sharp positive phase and an upward convex plateau phase [14]. MAP signal and a surface ECG (base-apex) were recorded

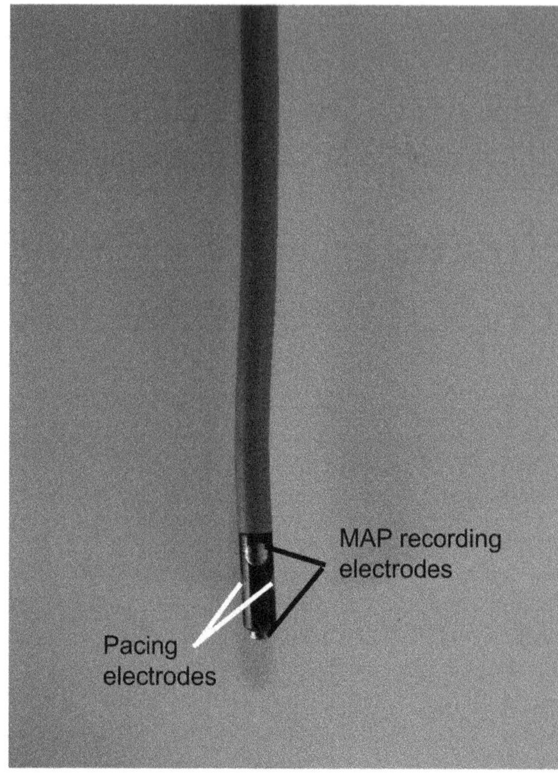

Fig. 1 Illustration of the tip of an 8F quadripolar contact catheter with separated recording and pacing electrodes

simultaneously using a multichannel recorder (PowerLab 8/35, ADInstruments, Oxford, United Kingdom) with amplifier (Bio Amp and Quad Bridge Amp, ADInstruments, Oxford, United Kingdom) and digitized on a computer. Measurements were performed off-line using the ECG analysis (used for determination of QT interval) and cardiac action potential peak analysis module of a specialized semi-automatic data analysis software program (LabChart 8, ADInstruments, Oxford, United Kingdom). The minimum peak height detection was set at 1.5 mV and an automatic resting membrane potential detection was used. The MAP signals were amplified using a low-pass filter with settings between 20 and 50 Hz because it reduces noise which results in a smoother signal. Each MAP curve was checked separately and only MAPs where the automatic resting membrane potential was automatically set at the level of true baseline were used. In every horse, 15 cardiac cycles were analysed per chamber and per day. The APD90 was automatically measured by the software as the duration between the upstroke and 90% repolarization. The amplitude was measured as the height from baseline to the crest of the plateau phase. First, measurements were made in sinus rhythm at rest (non-paced) at a cardiac rate between 30 and 42 beats per minute (bpm). Subsequently, to have a more stable rhythm and to follow the impact of heart frequency on action potential characteristics, the

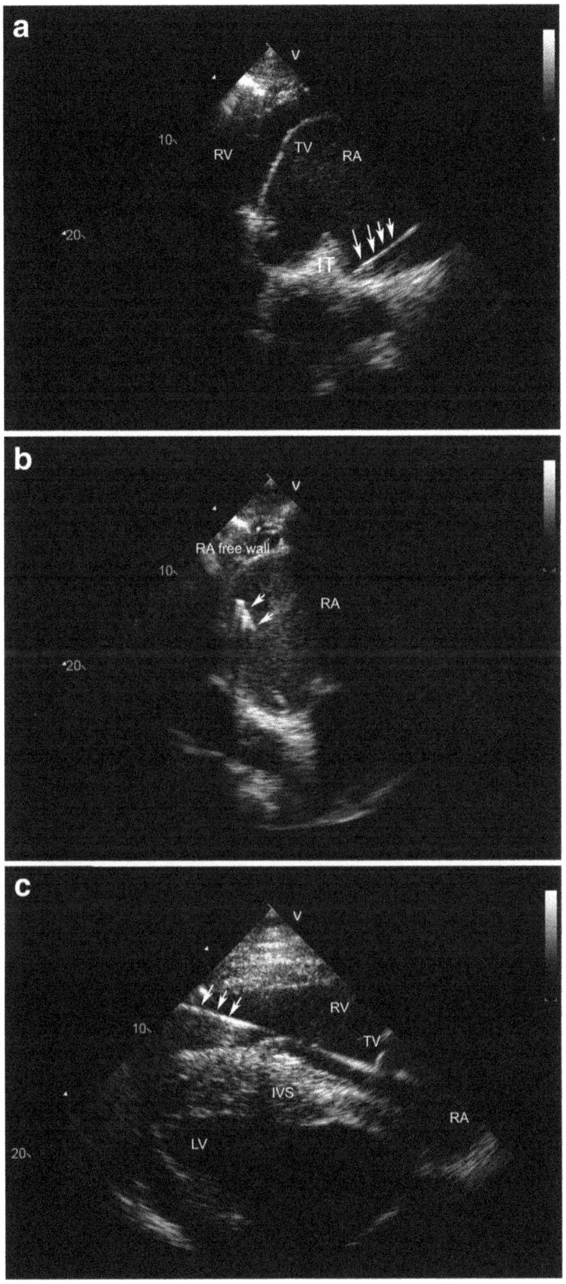

Fig. 2 Ultrasound images of the contact MAP electrode (arrow) positioned at the level of the intervenous tubercle (IT) in the right atrium (**a**), at the level of the right atrial free wall (**b**) and in the right ventricular apex (**c**). RA: right atrium, TV: tricuspid valve, RV: right ventricle, LV: left ventricle, IVS: interventricular septum

pacing electrodes of the MAP catheter were connected to a pacemaker programmer (Medtronic, Jette, Belgium) and pacing was performed at a pacing cycle length of 1000 and 600 ms to record MAPs at each pacing rate. Finally, atrial and ventricular effective refractory period (ERP) were determined as previously described [16, 17]. In brief, after pacing for 2 min with a fixed pacing interval (S1-S1), an extra stimulus (S2), at two times threshold amplitude, was introduced with a coupling interval (S1-S2) below the expected refractory period. The coupling interval was increased in steps of 8 ms until capture of the extra stimulus occurred (the atrial or ventricular S2 was followed by a P wave or QRS complex on the surface ECG, respectively). The longest S1-S2 interval without capture was taken as the effective refractory period. ERP was always determined 5 times to obtain a mean value. After the procedure, all catheters were gently removed under ultrasound guidance.

Data analysis

Statistical analyses were performed using dedicated software (SPSS Statistics 22.0, SPSS Inc., Chicago, Illinois). Mean and standard deviation (SD) were calculated from pooled measurements of all examinations for each horse. Measurement variability was obtained by comparing the results of the measurements from day 1 and day 2 in a one-way repeated measures analysis of variance with the horse as the unit of repeated measure. The numerical values for the reported coefficients of variation (CV) were calculated by dividing the square root of the mean square error (MSE) by the grand mean, multiplied by 100. The relationship between QT interval and APD90 was evaluated using a linear regression model with QT interval, day and horse as predictors. A value of $P < 0.05$ was considered significant.

Results

Right atrial (Fig. 3a) and right ventricular (Fig. 3b) MAP recordings of sufficient quality could be obtained and analysed in non-sedated standing horses. However, horse movements, occasionally resulted in loss of contact requiring MAP catheter repositioning. The total recording time per horse and per day ranged between 35 and 165 min. Catheter positioning was occasionally associated with self-limiting atrial or ventricular depolarizations which terminated after slight catheter tip movements. During atrial ERP measurements, a short episodes of self-limiting episode of atrial tachycardia was occasionally found and 1 horse showed a 20-min paroxysm of atrial fibrillation. Ventricular ERP measurements were never associated with tachyarrhythmias. No complications were observed after the procedure. Typical right atrial and ventricular MAP recordings are shown in Fig. 3. All the results of the two different days are shown in Table 1 and Fig. 4. In 1 horse a difference of up to 100 ms was observed during VERP1000 measurement. The QT interval was a significant predictor of APD90 in sinus rhythm (beta 0.604, $P < 0.001$) and at a pacing cycle length of 1000 ms (beta -0.177, $P = 0.018$), but not at a pacing cycle length of 600 ms.

Discussion

The present paper describes the technique to determine right atrial and right ventricular repolarization and

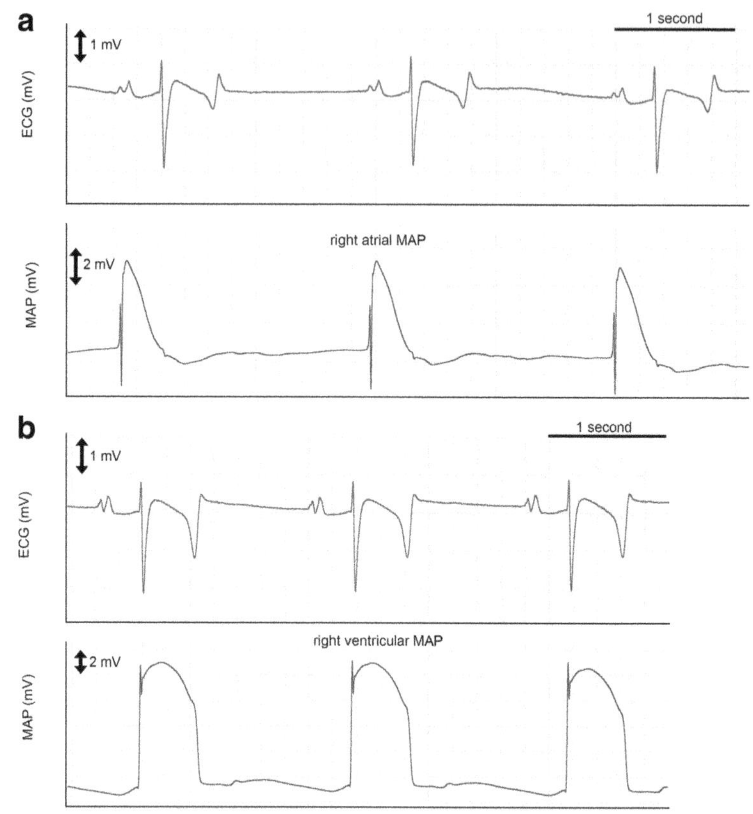

Fig. 3 Surface electrocardiogram (ECG) with corresponding right atrial (**a**) and right ventricular (**b**) monophasic action potential (MAP) at resting sinus heart rhythm in the standing non-sedated horse

refractoriness in the standing non-sedated horse. In humans, dogs, cats and pigs, MAP recordings play an important role in the investigation of myocardial electrophysiology under physiological, pathological and pharmacological conditions and have a low level of risk [18–21]. However, complications such as sustained dysrhythmias, movement artefacts or pericardial effusion have been reported [21]. In our study, complications were limited to some short episodes of ectopic rhythms and 1 short episode of atrial fibrillation during ERP measurement. The ERP values in our study were comparable with previously reported results [17]. The coefficient of variation of the ERP measurements ranged between 7 and 12% but in one horse the difference in VERP1000 measurement was > 100 ms. As observed in awake dogs, neck and body movements of the horse occasionally resulted in dislocation of

Table 1 Mean ± SD of monophasic action potential amplitude and repolarization level of 90% (APD90), and effective refractory periods (ERP) measured at rest and at a pacing cycle length of 1000 and 600 ms in the right atrium and right ventricle

	Right atrium				Right ventricle			
	Mean ± SD				Mean ± SD			
	Day 1	Day 2	Overall	Cv %	Day 1	Day 2	Overall	CV %
amplitude (rest) (mV)	6.9 ± 1.9	4.4 ± 1.9	5.7 ± 2.3	35.6	9.1 ± 4.3	7.2 ± 3.3	8.1 ± 3.8	49.2
amplitude (1000) (mV)	6.0 ± 2.0	4.0 ± 1.0	5.0 ± 2.0	44.0	7.8 ± 2.6	8.8 ± 5.4	8.3 ± 4.0	58.9
amplitude (600) (mV)	6.0 ± 1.0	6.0 ± 2.0	6.0 ± 2.0	37.3	7.7 ± 2.5	9.7 ± .8	8.7 ± 4.4	65.5
APD90 (rest) (ms)	282 ± 41	244 ± 29	263 ± 39	9.1	464 ± 30	469 ± 15	467 ± 23	6.2
APD90 (1000) (ms)	281 ± 47	243 ± 23	262 ± 41	12.3	399 ± 38	443 ± 24	421 ± 38	7.3
APD90 (600) (ms)	242 ± 17	231 ± 67	236 ± 47	21.6	325 ± 26	319 ± 34	322 ± 29	5.6
ERP1000 (ms)	269 ± 18	277 ± 31	273 ± 24	7.9	401 ± 34	370 ± 43	386 ± 40	11.8
ERP600 (ms)	251 ± 14	261 ± 28	256 ± 22	6.5	295 ± 38	291 ± 23	293 ± 30	9.6

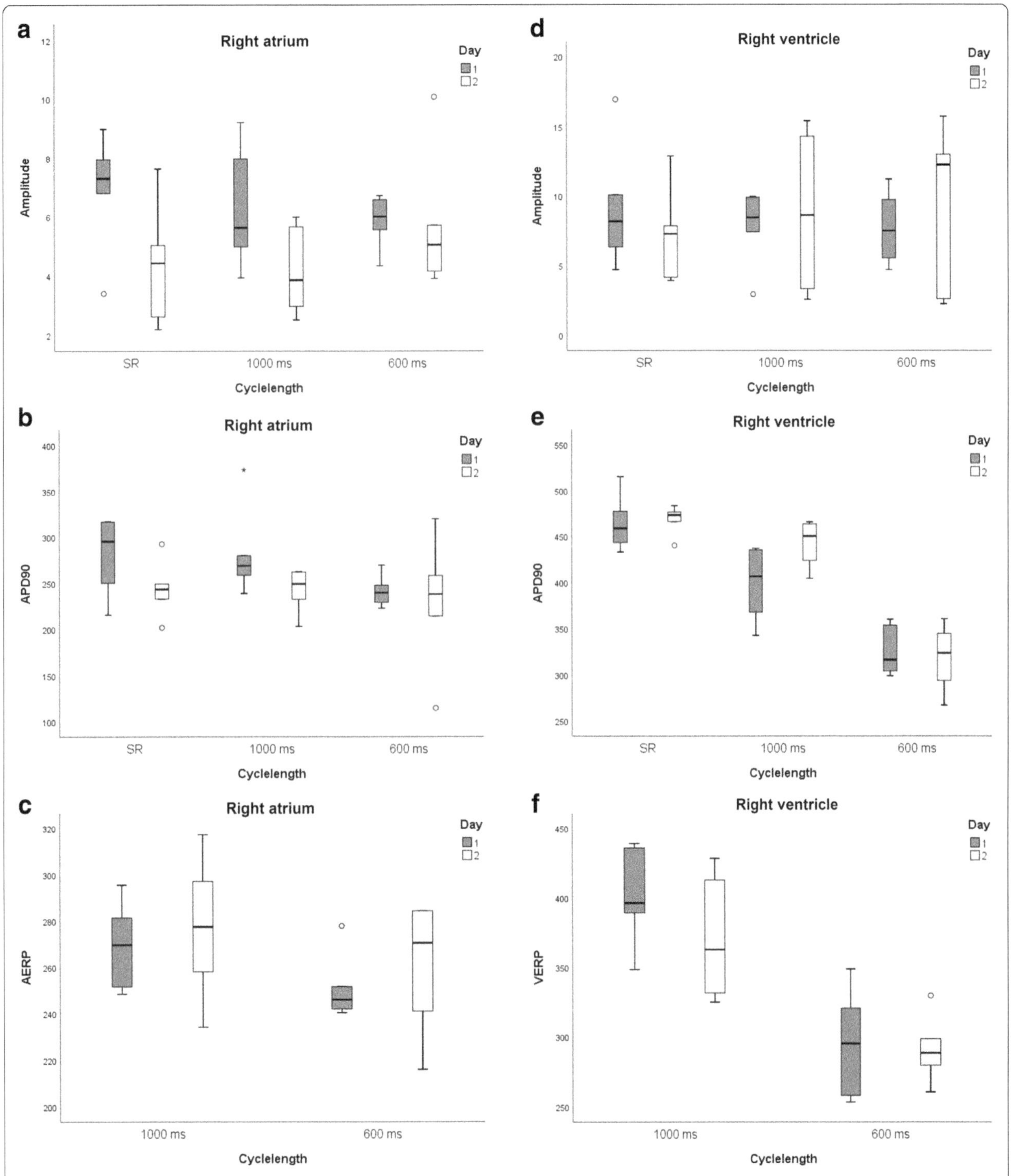

Fig. 4 Graphical illustration of monophasic action potential amplitude (**a** and **d**) and repolarization level of 90% (APD90) (**b** and **e**), and effective refractory periods (ERP) (**c** and **f**) measured at rest (sinus rhythm: SR) and at a pacing cycle length of 1000 and 600 ms in the right atrium (RA) (**a**, **b** and **c**) and right ventricle (RV) (**d**, **e** and **f**) in two different days. The median and spread of the value is indicated by a boxplot, with the box span indicating the middle half of the observations, the line in the box marking the median, and the whiskers indicating the range of observations. Extreme values were plotted individually at the end of the whiskers

the catheter and loss of tip pressure increasing recording time and measurement variability [12]. Due to a more fixed catheter position in the apex, dislodgement of the catheter tip was less frequently observed in ventricular MAP recordings. Probably, some spatial repolarization differences might also contribute to the measurement variability [12, 18]. MAP duration and amplitude can be variable due to angle differences of the catheter tip relative to the myocardium, the catheter location and varying endocardial contact pressures in the beating heart [14, 22]. Prolonged good quality MAP recordings from the same endocardial site (up to 1 h), as described for awake dogs, could not be achieved in our horses [13]. Mild sedation or the use of another type of catheter could have facilitated the procedure in our study. In humans and sedated dogs, the MAP amplitude of the atrium and ventricle ranges between 5 and 50 mV due to a variability in contact pressure but also due to tissue type and species [12, 14, 22]. The difference in amplitude between the atrium and the ventricle has partly been explained by the different thickness of the endocardial tissue beneath the tip of the catheter [12]. The reason for the species- and tissue-related difference in MAP amplitude remains unexplained [21]. Intracellular recordings have amplitudes of approximately 120 mV regardless species or tissue dependence [22]. The MAP amplitude in our horses ranged between 2 and 10 mV for the atrium and between 3 and 16 mV for the ventricle, which could be related to endocardial properties. The large size of the equine heart seems to be a large disadvantage and also has probably contributed to a less stable catheter position. The most frequently used filters with MAP measurements are the low-pass filter and the high-pass filter. The manuals of the used program advice to use a low-pass filter since it lets low frequencies pass and stops high frequencies. This results in a reduction of noise, it gives a smoother MAP signal. The use of a high pass filter lets high frequencies pass and remove any steady component or slow fluctuations from the signal. This setting is used to stabilize baseline. In this study, a low pass filter was used to obtain smoother MAP signals. In addition, each MAP signal was checked manually to evaluate if the automatic resting membrane potential was correctly set at the level of baseline.

Study limitations

Despite ultrasonographic guidance, dislocation of the catheter and changes in contact pressure by the beating heart and horse movements contributed to the variability. Further research is necessary to identify whether repeating the measurements in a higher number of horses, harder pressure on the catheter tip, measurement under general anaesthesia or using another type of catheter (with i.e. a spring-steel stylet that is inserted into the myocardium) results in a lower variability in horses. Also the influence of spatial differences within right atrium and right ventricle, and between the left and right heart should be investigated. In humans, access to the left heart is possible transseptal or via the arteria femoralis [23–25]. In horses, access to the left heart might be possible via the carotid artery or transseptal.

The present study describes the feasibility and measurement variability of MAP recordings and refractory period measurements with a quadripolar contact catheter in the standing non-sedated horse. Further research is necessary to evaluate if this MAP recording technique might be useful to investigate equine dysrhythmias and the impact of drugs on the myocardium.

Conclusion

RA and RV MAP duration and ERP can be obtained by a contact electrode in standing non-sedated horses. The measurement variability varies with catheter location.

Abbreviations

APD90: Action potential at a repolarization of 90%; bpm: Beats per minute; CV: Coefficient of variation; ECG: Electrocardiography; ERP: Effective refractory period; Hz: Hertz; MAP: Monophasic action potential; ms: Milliseconds; MSE: Mean square error; mV: Millivolt; PLC: Pacing cycle length; RA: Right atrium; RV: Right ventricle; SD: Standard deviation

Acknowledgements

None

Funding

No funding was received for this study.

Authors' contributions

DDC and GvL contributed to the study design. DDC performed the measurements. DDC and AD contributed to statistical data analysis and interpretation. DDC, BB and LV contributed to the practical execution of the study. All authors read and critically revised and approved the final manuscript.

Consent for publication

Not Applicable.

Competing interests

The authors declare that they have no competing interests.

References

1. Yang SG, Kittnar O. New insights into application of cardiac monophasic action potential. Physiol Res. 2010;59:645–50.
2. Behrens S, Franz MR. In vivo registration of monophasic action potentials - new possible applications in clinical electrophysiology. Z Kardiol. 1996;85: 231–42.
3. Franz MR. Long-term recording of monophasic action potentials from human endocardium. Am J Cardiol. 1983;51:1629–34.
4. Franz MR, Burkhoff D, Spurgeon H, Weisfeldt ML, Lakatta EG. In vitro validation of a new cardiac catheter technique for recording monophasic action potentials. Eur Heart J. 1986;7:34.

5. Ino T, Karagueuzian HS, Hong K, Meesmann M, Mandel WJ, Peter T. Relation of monophasic action potential recorded with contact electrode to underlying transmembrane action potential properties in isolated cardiac tissues: a systematic microelectrode validation study. Cardiovasc Res. 1988; 22:255–64.
6. Franz MR, Chin MC, Sharkey HR, Griffin JC, Scheinman MM. A new single catheter technique for simultaneous measurement of action potential duration and refractory period in vivo. J Am Coll Cardiol. 1990;16:878–86.
7. Franz MR. Method and theory of monophasic action potential recording. Prog Cardiovasc Dis. 1991;33:347–68.
8. Franz MR. Bridging the gap between basic and clinical electrophysiology: what can be learned from monophasic action potential recordings? J Cardiovasc Electrophysiol. 1994;5:699–710.
9. Eckardt L, Haverkamp W, Borggrefe M, Breithardt G. Experimental models of torsade de pointes. Cardiovasc Res. 1998;39:178–93.
10. Kirchhof P, Eckardt L, Mönnig G, Johna P, Loh E, Schulze-Bahr G, Breithardt M, Borggrefe W, Haverkamp J. A patient with "atrial torsades de pointes". J Cardiovasc Electrophysiol. 2000;11:806–11.
11. Moore HJ, Franz MR. Monophasic action potential recordings in humans. J Cardiovasc Electrophysiol. 2007;18:787–90.
12. Eckardt L, Meissner A, Kirchhof P, Weber T, Borggrefe M, Breithart G, Van Aken H, Haverkamp W. In vivo recording of monophasic action potentials in awake dogs – new applications for experimental electrophysiology. Basic Res Cardiol. 2001;96:169–74.
13. Eckardt L, Meissner A, Kirchhof P, Weber T, Milberg P, Breithardt G, Haverkamp W. In vivo recording of monophasic action potentials in awake dogs. Methods. 2003;30:109–14.
14. Franz MR. Current status of monophasic action potential recording: theories, measurements and interpretations. Cardiovasc Res. 1999;41:25–40.
15. Knollmann BC, Tranquillo J, Sirenko SG, Hendriquez C, Franz MR. Microelectrode study of the genesis of the monophasic action potential by contact electrode. Technique J Cardiovasc Electrophysiol. 2002;12:1246–52.
16. van Loon G, Fonteyne W, Rottiers H, Tavernier R, Jordaens L, D'Hont L, Colpaert R, De Clercq T, Deprez P. Dual-chamber pacemaker implantation via the cephalic vein in healthy horses. J Vet Intern Med. 2001;15:564–71.
17. De Clercq D, van Loon G, Tavernier R, Duchateau L, Deprez P. Atrial and ventricular electrical and contractile remodeling and reverse remodeling owing to short-term pacing-induced atrial fibrillation in horses. J Vet Intern Med. 2008;22:1353–9.
18. Franz MR, Bargheer K, Rafflenbeul W, Haverich A, Lichtlen PR. Monophasic action potential mapping in human subjects with normal electrocardiograms: direct evidence for the genesis of the T wave. Circulation. 1987;75:379–86.
19. Morgan JM, Cunningham D, Rowland E. Dispersion of monophasic action potential duration: demonstrable in humans after premature ventricular extra stimulation but not in steady state. J Am Coll Cardiol. 1992;19:1244–53.
20. Leirner AA, Cestari IA. Monophasic action potential. New uses for an old technique. Arq Bras Cardiol. 1999;72:237–42.
21. Tsuburaya R, Yasuda S, Ito Y, Shiroto T, Gao JI, Ito K, Shimokawa H. Eicosapentaenoic acid reduces ischemic ventricular fibrillation via altering monophasic action potential in pigs. J Mol Cell Cardiol. 2011;3:329–36.
22. Lerman BB, Burkhoff D, Yue DT, Franz MR, Sagawa K. Mechanoelectrical feedback: independent role of preload and contractility in modulation of canine ventricular excitability. J Clin Invest. 1985;76:1843–50.
23. Ross J, Braunwald E, Morrow AG. Left heart catheterization by the transseptal route: a description of the technique and its applications. Circ. 1960;22:927-34.
24. Early MJ. How to perform a trasseptal puncture. Heart. 2009;95:85-92.
25. Justino H, Petit CJ. Percutaneous common carotid artery access for pediatric interventional cardiac catheterization. Circ Cardiovasc Interv. 2016;9:1-9.

8

Searching for ivermectin resistance in a Strongylidae population of horses stabled in Poland

A. Zak[1*], N. Siwinska[1], M. Slowikowska[1], H. Borowicz[1], K. Kubiak[1], J. Hildebrand[2], M. Popiolek[2] and A. Niedzwiedz[1]

Abstract

Background: There are no available studies describing the possible resistance of strongyles to ivermectin in horses in Poland. One hundred seventy three horses from nine stud farms from South-Western Poland were studied. The effectiveness of ivermectin was studied on the 14th day after ivermectin administration using the fecal egg count reduction test, and a long-term observation of the egg reappearance period was carried out. The fecal study was carried out using a modified McMaster method, which typically detects 20 eggs per gram of stool. The results were subjected to statistical analysis that enabled quantification of the eggs in the stool samples.

Results: The study revealed high efficacy of ivermectin on the 14th day after administration without a shortening of the egg reappearance period.

Conclusion: The results indicate that strongyles resistance to ivermectin in Poland is not a serious problem.

Keywords: Strongylidae, Horses, Ivermectin resistance, Ivermectin effectiveness

Highlights

- Our results indicate that the efficacy of ivermectin measured on the 14th day after treatment is very high in horses stabled in Poland.
- There was no shortening of the egg reappearance period after ivermectin administration.
- Our results indicate that there is no ivermectin resistance in the strongylidae population in horses bred in South-Western Poland

Background

Strongyles are considered to be major gastrointestinal parasites in horses [1, 2]. Pasture grazed adult horses are usually infected with strongyles [1, 3, 4]. The clinical symptoms associated with a strongyle infection are often non-specific. They include a weight loss, poor quality horse hair, diarrhoea and recurring colic [1, 5]. The migration of the fourth-stage small strongyle larvae (L4) and their encystation in the intestinal wall may lead to "larval cyathostomosis". This, in turn, causes protein-losing enteropathy and diarrhoea [1, 6]. Anthelmintic drugs from three groups - the benzimidazoles (e.g. fenbendazole and oxibendazole), the tetrahydropyrimidine pyrantel and the macrocyclic lactones (ivermectin and moxidectin) are used in horses to treat strongyle infections. Long-term use of anthelmintic drugs may trigger parasite resistance in a given population. In the case of small strongyles (cyathostomins), benzimidazole resistance was reported in 14 countries while pyrantel resistance occurred in 12 countries [7]. Unlike benzimidazole and pyrantel, the efficacy of macrocyclic lactones is reported to be 99% [8]. However, there have been isolated cases of ivermectin resistance based on the fecal egg count reduction test (FECRT) and/or a shortened egg reapperance period (ERP) what is the first indicator of strongyle resistance to ivermectin [4, 8–15]. To date, the ivermectin resistance of strongyles in Poland has not been studied. The aim of the study was the assessment of ivermectin resistance of strongylidae in horses bred in Poland, considering routine horse deworming twice a year using ivermectin, without coproscopy.

* Correspondence: agnieszka.zak@up.wroc.pl
[1]Department of Internal Medicine and Clinic of Diseases of Horses, Dogs and Cats, Faculty of Veterinary Medicine, Wroclaw University of Environmental and Life Sciences, pl. Grunwaldzki 47, 50-366 Wroclaw, Poland

Methods

Study design

The study was carried out in the first half of 2016 in nine stables in Lesser Poland, Greater Poland and Lower Silesian Voivoideships. In total, 173 horses of both genders and various breeds, with a mean age of 11.5 years (range: 1.5–32) were studied. The first part of the study, which included an evaluation of the efficacy of ivermectin on the 14th day after its administration, using the FECRT, was carried out on all 173 horses. In the second part of the study, the egg reappearance period (ERP) was assessed in 42 horses, from two stables, selected based on an EPG value exceeding 20. All the horses were routinely dewormed twice a year in accordance with current standards using single-component pastes in spring and two-component pastes in autumn. In both groups, the horses were naturally infected with strongyles. The study included pasture grazing horses which were dewormed more than 8 weeks prior to the commencement of the study [16, 17]. The examination at T0 consisted of collecting fresh stool samples from the litter bedding, evaluating the number of eggs per gram of stool during the fecal examination and determining the contamination potential of the horses. Deworming was carried out using an equine single-component oral paste (Ecomectin, Eco Animal Health) containing 18.7 mg/g of ivermectin administered at a dosage of 200 μg/kg. The body weight of the horses was recorded using a "girth tape". Deworming was carried out by a qualified veterinary surgeon. No dose errors were reported. Stool samples were collected 14 days after deworming from all the horses. In addition, stool samples were collected on a weekly basis from 42 horses selected for the ERP, until at least 10% of the amount of eggs recorded at T0 was reached. The fecal examination followed a modified McMaster centrifugation – enhanced method using a sugar-salt flotation solution with a specific gravity of 1.3 g/ml [18, 19]. The detection limit of the method was 20 EPG [18].

Statistical analysis

At T0, the contamination potential of the horses older than 3 years was determined using the 2013 AAEP Parasite Control Guidelines. In each yard, the horses were classified into groups based on the varying degree of strongyle egg shedding: low contaminators – EPG <200, moderate contaminators – EPG range: 201–500 and high contaminators - EPG >500 [4]. The chi square independence test was used to compare contamination potentials in different age and sex groups. Statistical significance was set at $P < 0.05$.

The primary outcome measure in this study was the analysis of the length of the ERP. The ERP was defined as the time between the treatment and the first breaching of a 90% efficacy threshold [20].

The secondary outcome measure was the percentage reduction in the arithmetic mean of the strongyle FEC on day 14 (after) relative to day 0 (before): $(FEC_{D0} - FEC_{D14})/FEC_{D0}$ with the 95% confidence interval (CI). Anthelmintic efficacy was defined as a reduction in FEC ≥95% with the lower 95% CI >90% was considered as efficacious [21, 22]. Anthelmintic resistance was present if (1) the percentage reduction in FEC was ≤95% and (2) the lower 95% CI was <90%. Suspected anthelmintic resistance was present if (1) the percentage reduction in FEC was ≤95% or (2) the lower 95% CI was ≤90%.

In addition, the results of the efficacy of ivermectin on the 14th day after its administration underwent statistical analysis using a specifically designed "eggCounts" package in R (using Bayesian hierarchical models for faecal egg count data) and a web interface, (http://www.math.uzh.ch/as/index.php?id=eggCounts) enabling the elimination of the Poisson errors in the counting technique. [23]. A correction factor = 20 and zero-inflation were used in the calculations.

Results

All the animals were dosed correctly, and no abnormal health events linked to the treatment were recorded in this study. The mean EPG in the horses in this study was 131.56 (CI = 95.58–167.55) and ranged from 0 to 2300. The contamination potential of the horses older than 3 years was classified as 78.48% - low contaminators, 15.69% moderate contaminators and 5.83% high contaminators. There was no statistically significant correlation between the sex, the age and the contamination potential in the studied horses ($p = 0.05$). No *Parascaris equorum* eggs were detected in the feces.

The analysis of the efficacy of ivermectin on the 14th day after its administration indicated that it was effective in 99.9% (CI = 95%). Detailed results are presented in Table 1. The results are presented in the form of graphs in Fig. 1.

The ERP ranged from seven to 16 weeks from the time of ivermectin administration. The ERP was not shortened, and the complete results are presented in Table 2.

Discussion

The modified centrifugation – enhanced McMaster method according to Roepstoff and Nansen (1998) is considered the most accurate McMaster fecal egg counting technique in horses [19]. However, it is not possible to avoid errors associated with the choice of the fecal examination method.

Table 1 Summay statistics for each variable - generated using an "eggCounts" package in R (http://www.math.uzh.ch/as/index.php?id=eggCounts)

	2.5%	50%	97.5%
fecr	0.997	0.999	1
meanEPG.untreated	123.606	131.308	140.243
meanEPG.treated	0.003	0.078	0.433

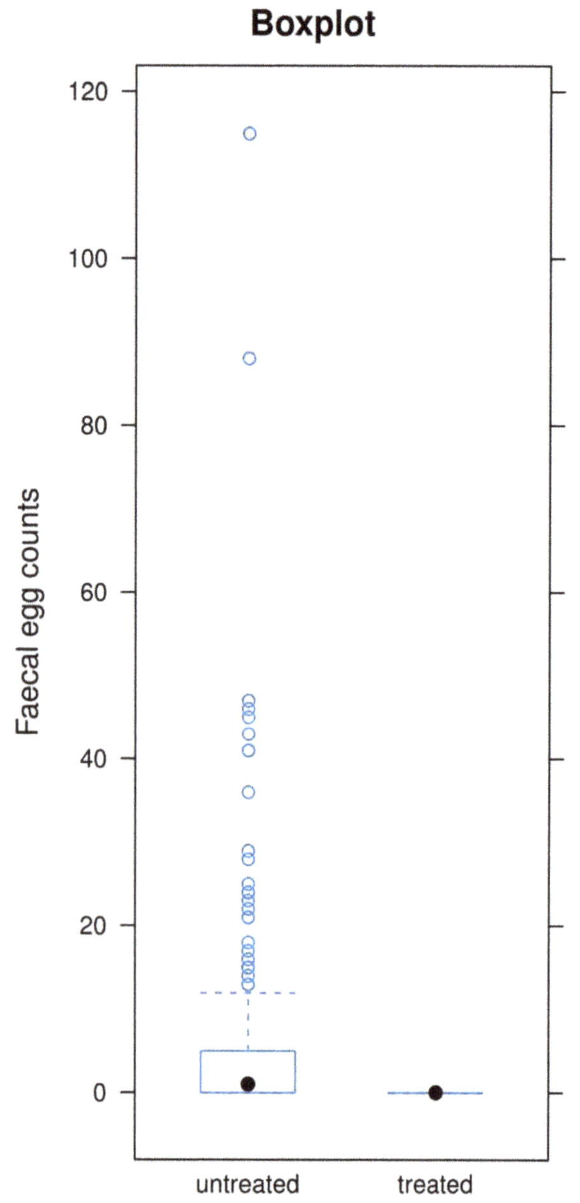

Fig. 1 The graphical display contains information regarding the amount of EPG prior to and after treatment of all the horses

Hence, a specifically designed "eggCounts" package in R and a user friendly web interface were used to eliminate errors associated with the inaccuracy of the test method as well as the uneven and random distribution of the eggs in the stool sample [23]. That allowed an accurate interpretation of the results, which were more reliable. Our results indicate a high efficacy of ivermectin on the 14th day after its administration. Other reports confirm our findings and state an ivermectin efficacy of almost 100% on the 14th day after administration, or describe single cases of partial or total resistance [8–15]. The first indicator of strongyle resistance to ivermectin is a shortened ERP [4, 24]. It is difficult to compare the results of studies assessing the ERP, due to a lack of a uniform definition of the ERP. The EPG after treatment, which is considered a criterion determining the ERP, varies. Some investigators calculate the ERP as the first week in which eggs appear in the stool [25–28]. Other authors consider the EPG as 100 [16], 200 [29], <80% of the FECR [30] or <90% of the FECR [31–33]. In our study, ERP was considered to be the time a < 90% FECR was obtained. Unshortened ERP and a high efficacy of ivermectin on the 14th day after its administration rule out strongyle ivermectin resistance in horses bred in Poland. In order to slow down the development of resistance, it seems necessary to introduce a selective therapy in horses that have an egg per gram count exceeding 500, proposed by the 2013 AAEP Parasite Control Guidelines, or the deworming of horses that either have a FEC exceeding 200 EPG or that shed *Parascaris equorum* eggs, which is used in Denmark [20]. This study was prompted by reports of incorrect use of horse deworming programmes in Poland. Although this data is not available in literature, reports of incorrect deworming were received from several horse breeders and owners. The use of off label ivermectin in horses orally, intended for intravenous or intramuscular use in pigs and ruminants, may cause ivermectin-resistance. There are no reports of the pharmacokinetic and pharmacodynamic properties of intravenous or intramuscular ivermectin administered orally. Ivermectin was most probably used off label for economical reasons. Currently, great emphasis is placed on deworming procedures in Poland. Veterinary professionals and horse owners are encouraged to use selective deworming schemes in order to limit future ivermectin resistance.

Conclusions

We found no strongyle ivermectin-resistance in horses bred in Poland based on the high efficacy of ivermectin on the 14th day after its administration and no shortening of the ERP. Despite these results, it is necessary to undertake measures that will limit future strongyle resistance, such as the use of correct deworming methods and selective deworming.

Table 2 Results of ERP study – normal range for ivermectin is 42–56 days

Farm	number of horses	mean EPG	EPG range	ERP (days) – number of horses									
				D49	D56	D63	D70	D77	D84	D91	D98	D105	>D105
1.	14	350	20–1760	0	1	1	1	3	2	0	1	3	3
2.	28	92.14	20–900	1	5	8	3	6	1	1	1	0	2

Abbreviations

EPG: Egg Per Gram; ERP: Egg Reappearance Period; FEC: Fecal Egg Counts; FECR: Fecal Egg Count Reduction; FECRT: Fecal Egg Count Reduction Test

Acknowledgements

This research was supported by Calier Polska, Eco Animal Health. The results of this study has been presented in preliminary form at the oral report session at 15th Congress of Polish Society of Veterinary Sciences, Lublin, Poland, 22-24 September 2016.

Funding

This research did not receive any specific grant from funding agencies in the public or not-for-profit sectors. This research was supported by Calier Polska, Eco Animal Health by donating deworming past for research.

Authors' contributors

AŻ participated in the design of the study, in the fecal egg count test and was a major contributor in writing the manuscript. NS, MS, HB participated in the design of the study, in the collection of the samples, in the fecal egg count test and in analysis and interpretation of data. JH and MP participated in the design of the study and performed the statistical analysis. AN and KK conceived of the study, and participated in its design and coordination and helped to draft the manuscript. All authors were involved in continuous discussion regarding manuscript. All authors read and approved the final manuscript.

Consent for publication

Not applicable.

Competing interests

The authors state that there are no conflicts of interest. None of the aforementioned companies had any involvement, direct or indirect, in the development of the manuscript and the works reflects the opinions solely of the authors.

Author details

[1]Department of Internal Medicine and Clinic of Diseases of Horses, Dogs and Cats, Faculty of Veterinary Medicine, Wroclaw University of Environmental and Life Sciences, pl. Grunwaldzki 47, 50-366 Wroclaw, Poland. [2]Department of Parasitology, Institute of Genetics and Microbiology, University of Wroclaw, ul. S. Przybyszewskiego 63/77, 51-148 Wroclaw, Poland.

References

1. Love S, Murphy D, Mellor D. Pathogenecity of cyathostome infection. Vet Parasitol. 1999;85:113–21.
2. Kaplan RM, Vidyashankar AN. An inconvenient truth: global worming and anthelmintic resistance. Vet Parasitol. 2012;186:70–8.
3. Kaplan RM. Anthelmintic resistance in nematodes of horses. Vet Res. 2002; 33:491–507.
4. Nielsen MK, Mittel L, Grice A, Erskine M, Graves E, Vaala W, Tully RC, French DD, Bowman R, Kaplan RM. AAEP Parasite Control Guidelines. 2013. https://aaep.org/sites/default/files/Guidelines/AAEPParasiteControlGuidelines_0.pdf.
5. Mair TS, Sutton DG, Love S. Caecocaecal and caecocolic intussusceptions associated with larval cyathostominosis in four young horses. Equine Vet J. 2000;32:77–80.
6. Corning S. Equine cyathostomins: a review of biology, clinical significance and therapy. Parasit Vectors. 2009;2:1–6.
7. Peregrine AS, Molento MB, Kaplan RM, Nielsen MK. Anthelmintic resistance in important parasites of horses: does it really matter? Vet Parasitol. 2014; 201:1–8.
8. Boersema JH, Eysker M, Mass J, van der Aar WM. Comparison of the reappearance of strongyle eggs in foals, yearlings, and adult horses after treatment with ivermectin or pyrantel. Vet Quart. 1996;18:7–9.
9. Canever RJ, Braga PRC, Boeckh A, Grycajuck M, Bier D, Molento MB. Lack of Cyathostomin sp. reduction after anthelmintic treatment in horses in Brazil. Vet. Parasitology. 2013;194:35–9.
10. Geurden T, van Doorn D, Claerebout E, Kooyman F, De Keersmaecker S, Vercruysse J, Besognet B, Vanimisetti B, Di Regalbono AF, Beraldo P, Di Cesare A, Traversa D. Decerased strongyle egg re-appearance period after treatment with ivermectin and moxidectin in horses in Belgium, Italy and The Netherlands. Vet Parasitol. 2004;204:291–6.
11. Kaplan RM, Klei TR, Lyons ET, Lester G, Courtney CH, French DD, Tolliver SC, Vidyashankar AN, Zhao Y. Prevalence of anthelmintic resistant cyathostomes on horse farms. J Am Vet Med Assoc. 2004;225:903–10.
12. Lester HE, Spanton J, Stratford CH, Bartley DJ, Morgan ER, Hodgkinson JE, Coumbe K, Mair T, Swan B, Lemon G, Cookson R, Matthews JB. Anthelmintic efficacy against cyathostomins in horses in southern England. Vet Parasitol. 2013;197:189–96.
13. Stratford CH, Lester HE, Pickles KJ, Mcgorum BC, Matthews JB. An investigation of anthelmintic efficacy against strongyles on equine yards in Scotland. Equine Vet J. 2014;46:17–24.
14. Traversa D., Von Samson-Himmelstjerna G., Demeler J., Milillo P., Schürmann S., Barnes H., Otranto D., Perrucci S., Regalbono AF., Beraldo P., Boeckh A., Cobb R. Anthelmintic resistance in cyathostomin populations from horse yards in Italy, United Kingdom and Germany. Parasites & Vectors. 2009;supl.2:1–7.
15. Traversa D, Castagna G, von Samson-Himmelstjerna G, Meloni S, Bartolini R, Geurden T, Pearce MC, Woringer E, Besognet B, Milillo P, D'Espois D. Efficacy of major anthelmintics against horse cyathostomins in France. Vet Parasitol. 2014;188:294–300.
16. Van Doorn DCK, Eysker M, Kooyman FNJ, Wagenaar JA, Ploeger HW. Searching for ivermectin resistance in Dutch horses. Vet Parasitol. 2012;185:355–8.
17. Demeulenaere D, Vercruyse J, Dorny P, Clarebout E. Comparative studies of ivermectin and moxidectin in the control of naturally acquired cyathostome infections in horses. Vet Rec. 1997;15:383–6.
18. Roepstorff A., Nansen P. Epidemiology, diagnosis and control of helminth parasites of swine. 1998, FAO Animal health manual, Rome,
19. Vadlejch J, Petrtýl M, Zaichenko I, Čadková Z, Jankovská I, Langrová I, Moravec M. Which McMaster egg counting technique is the most reliable. Parasitol Res. 2011;109:1387–94.
20. Larsen ML, Ritz C, Petersen SL, Nielsen MK. Determination of ivermectin efficacy against cyathostomins and Parascaris equorum on horse farm using selective therapy. Vet J. 2011;188:44–7.
21. Coles GC, Bauer C, Borgsteede FH, Geerts S, Klei TR, Taylor MA, Waller PJ. World Association for the Advancement of Veterinary Parasitology (W.A.A.V.P.) methods for the detection of anthelmintic resistance in nematodes of veterinary importance. Vet.Parasitol. 1992;44:35–44.
22. von Samson – Himmelstjerna G. Anthelmintic resistance in equine parasites – detection, potencial clinical relevance and implications for control. Vet. Parasitol. 2012;185:2–8.

23. Torgerson PR, Paul M, Furrer R. Evaluating faecal egg count reduction using a specifically designed package `eggCounts` in R and a user friendly web interface. Int J Parasitol. 2014;44(5):299–303.
24. Coles GC, Jackson F, Pomroy WE, Prichard RK, von Samson – Himmelstjerna G, Silvestre a, Taylor MA, Vercruysse J. The detection of anthelmintic resistance in nematodes of veterinary importance. Vet. Parasitol. 2006;136:167–85.
25. Dudeney A, Campbell C, Coles G. Macrocyclic lactone resistance in cyathostomins. Vet Rec. 2008;163:163–4.
26. Little D, Flowers JR, Hammerberg BH, Gardner SY. Management of drug-resistant cyathostominosis on a breeding farm in central North Carolina. Equine Vet J. 2013;35:246–51.
27. Lyons ET, Tolliver SC, Ionita M, Lewellen A, Collins SS. Field studies indicating reduced activity of ivermectin on small strongyles in horses on a farm in Central Kentucky. Parasitol Res. 2008;103:209–15.
28. Molento MB, Antunes J, Bentes RN, Coles GC. Anthelmintic resistant nematodes in Brazilian horses. Vet Rec. 2008;162:384–5.
29. Mercier P, Chick B, Alves-Branco F, White CR. Comparative efficacy, persistent effect, and treatment intervals of anthelmintic pastes in naturally infected horses. Vet Parasitol. 2001;99:29–39.
30. Tarigo-Martinie JL, Wyatt AR, Kaplan RM. Prevalence and clinical implications of anthelmintic resistance in cyathostomes of horses. J Am Vet Med Assoc. 2001;218:1957–60.
31. Boersema JH, Borgsteede FHM, Eysker M, Saedt I. The reappearance of strongyle eggs in feces of horses treated with pyrantel embonate. Vet Quart. 1995;17:18–20.
32. Borgsteede FHM, Boersma JH, Gaasenbeek CPH, Vanderburg WPJ. The reappearance of eggs in feces of horses after treatment with ivermectin. Vet Quart. 1993;15:24–6.
33. von Samson-Himmelstjerna G, Fritzen B, Demeler J, Schuermann S, Rohn K, Schnieder T, Epe C. Cases of reduced cyathostomin egg-reappearance period and failure of *Parascaris equorum* egg count reduction following ivermectin treatment as well as survey on pyrantel efficacy on German horse farms. Vet Parasitol. 2007;144:74–80.

9

Long term consistency and location specificity of equine gluteus medius muscle activity during locomotion on the treadmill

Rebeka R. Zsoldos[1*], Anna Voegele[2], Bjoern Krueger[3], Ulrike Schroeder[4], Andreas Weber[2] and Theresia F. Licka[4,5]

Abstract

Background: The equine m. gluteus medius (GM) is the largest muscle of the horse, its main movement function is the extension of the hip joint. The objective of the present study was to measure equine GM activity in three adjacent locations on GM during walk and trot on a treadmill, in order to document potential differences. Fourteen Haflinger mares were measured using surface electromyography and kinematic markers to identify the motion cycles on three occasions over 16 weeks. The electrodes were placed on left and right gluteus medius muscle over the middle of its widest part and 5 cm lateral and medial of it. For data processing, electrical activity was normalised to its maximum value and timing was normalised to the motion cycle. A Gaussian distribution approach was used to determine up to 10 modes of focussed activity, and results were analysed separately for stance and swing phase of the ipsilateral hindlimb.

Results: Fair reliability was found for mean mode values (Cronbach's alpha = 0.66) and good reliability was found for mean mode locations (Cronbach's alpha = 0.71) over the three data collection days. The magnitude of muscle activity identified as mean mode value was much larger at trot than at walk, and mean mode value was significantly different between stance phases of walk and trot for all electrode positions ($p < 0.01$). The pattern of muscle activity identified as mean mode location was significantly different for walk and trot at all electrode positions, both during stance and swing phases ($p < 0.001$). This indicates the different timing pattern between the gaits. Results of the three electrode positions on the same muscle during each gait were not significantly different when comparing the same measurement.

Conclusions: The middle of the equine GM does not show any indication of functional differentiation during walk and trot on a treadmill; this might be due to lack of segmentation as such, or due to lack of need for segmented use for these very basic main tasks of the muscle. The reliability of the sEMG measurements over several weeks was fair to good, an indication for the robustness of the methodology.

Keywords: Surface electromyography, Horse, Gluteus medius, Walk, Trot, Locomotion, Scar, Gaussian distribution

Background

The equine m. gluteus medius (GM) is the largest muscle of the horse, its main movement function is the extension of the hip joint. In the unloaded limb GM contraction swings the limb towards caudal for e.g. kicking out and in the weight bearing limb it pushes the body mass forward during locomotion. Besides that, this muscle also creates some abduction of the limb. As with all leg muscles, it has a function in stabilising the limb during the stance phase of locomotion, during dynamic movements such as the capriole (i.e. the horse leaping from the ground and kicking out with its hind legs) and during static postures such as the levade (i.e. the horse lifting its forelegs from the ground and balancing on its deeply bent hindlimbs).

The equine GM creates the visible contour of the equine croup, even though it is partly covered by the thin m. gluteus superficialis and by the gluteal fascia (the caudal extension of the thoracolumbar fascia) [1]. It is a well and frequently used candidate for surface electromyography (sEMG), and its electrical activity has been measured in a large number of studies. Furthermore, the

* Correspondence: rebeka.zsoldos@boku.ac.at
[1]Department of Sustainable Agricultural Systems, Section Livestock Sciences, University of Natural Resources and Life Sciences Vienna, Vienna, Austria
Full list of author information is available at the end of the article

importance of this muscle has been documented in several studies: Payne et al. [1] estimated the GM has the highest capacity for force production (11,900 N) in the horse due to its very large physiological cross sectional area and long fascicles. Also, lengths and contraction velocities of muscle fibres have a dramatic effect on such muscle force generation [2]. The use of the equine GM during increasing speed and treadmill gradient was documented [3–6], and both increases in speed and treadmill inclination were found to increase EMG activity due to the higher workload of the GM. In the human gluteus maximus, Bartlett et al. [7] also found that increased muscle activity is related to the speed and intensity of the movement, somewhat independent of the gait (walk or run). Recently, integrated EMG values for GM in Thoroughbred racehorses were documented to increase uphill, but not decrease downhill compared to flat work on the treadmill [6]. This may indicated the increased need for limb stabilisation downhill compared to flat or uphill work.

Compartmentalisation, or the division of muscles into anatomically and/or functionally distinct sub-volumes, is an important aspect of organization and control in many muscles [8]; this has been shown e.g. for the human deltoid muscle [9], where muscle fibres converge in a single insertion similar to the gluteus medius. Sub-regions are usually supplied by different "primary" branches of a motor nerve (and often also by different blood vessels). In principle this would allow detailed control of these regions by the central nervous system both during volitional and central pattern generated movements. Most commonly, the term compartmentalisation has been reserved to describe the situation in which different sets of motor units occupy different sub-volumes within a single muscle [10]. The presence of regional differences of muscle activities during locomotion has been shown in the longissimus dorsi muscle [11]. There each section is innervated by one spinal nerve and served by one artery (Ramus dorsalis of the Arteria costoabdominalis dorsalis) serves one section [12, 13]. The presence of such functional subunits has not been documented in the equine GM during locomotion. However, in the GM of two Thoroughbred horses regions with different fibre type compositions were identified, indicative of a degree of specialisation of these regions [14]. This fibre typing was continued in a variety of horse breeds, showing more Type I fibres (ideal for aerobic, slower, continuous work such as joint stabilisation) located deeply and more Type II fibres (better suited to anaerobic, fast bursts of work, such as wide range of motion movements) located superficially in the GM. As this fibre type composition changes with the muscle depth, this will also be the case for the electrical activity, which is influenced by the fibre type [15], however this influence is still under investigation. Besides the metabolic differentiation of fibre types, movement function obviously also depends on the direction of muscle fibres, which varies within the GM due to its different origins with a single insertion point. Different movement functions within a muscle are thought to aid spreading peak muscle load over a larger area as well as over more time to reduce the risk of injury and to improve muscle perfusion.

In very large muscles, such as the equine gluteus medius muscle (GM), regional variations potentially representing functional differences of muscle activities have to be expected. The aim of the present study was therefore to measure the surface muscle activity of the equine GM muscle during walk and trot on the treadmill in three adjacent locations and to document potential differences between these activities. The long term consistency of these findings was investigated by repeating measurements after 8 and 16 weeks.

The following hypotheses were examined: a) there are significant differences in the muscle activities between the three adjacent locations of surface EMG electrodes during locomotion b) the surface muscle activities do not change significantly over 16 weeks within any individual.

Methods

Horses used for the study were assessed for signs of back pain and/or lameness at walk and trot by two veterinarians (U.S. and T.F.L.) before and after each measurement and found to be without any such signs. Fourteen horses without clinical sign of back pain and lameness were used in this study (Haflinger mares, mean age 8 ± 3 years, confidence interval (CI) (6, 9), range 4–14 years; mean body mass was 463 ± 42 kg, CI (439, 487), range 396–526 kg; mean height at the withers was 131 ± 5 cm, CI (128, 134), range 125–145 cm). Haflinger mares heterozygous positive (polysaccharide storage myopathy type 1, $n = 7$) and negative (control, n = 7) for glycogen synthase (GYS1) mutation were included. As a part of the preceding study using the same data collections [16], the effect of diet was investigated; horses received 6 weeks of carbohydrate rich diet followed by hay only or vice versa, with a wash out period of two weeks between diets. The results of that study have been published [16], and no effect of diet or GYS1 was found; therefore, this factor is not further investigated in the present study. For that study surgical biopsies of GM (removing about 1 cm^3 of muscle) were taken unilaterally after the second and on the contralateral side after the third data collection, presenting the opportunity to investigate sEMG on one location before and after an intentional surgical muscle lesion. Horses were accustomed to the treadmill and the experimental set up at walk and trot, and prior to the measurements they were warmed up for two to three minutes. For the measurements, horses walked and trotted on the treadmill at their optimum speed [17]. Data collection was carried out on three different data collection days (DCD): first measurement (DCD 1), 8 weeks later the second measurement (DCD2) and 16 weeks after DCD1 the third measurement (DCD3).

Kinematic measurements

Seven reflective skin markers were positioned on each horse using adhesive tape; one on the forehead, one on the highest point of the withers, on the sacrum were placed and on the lateral side of each hoof to identify motion cycles. Three-dimensional kinematic data at walk (@W) and trot (@T) were collected using ten high-speed cameras recording at 120 Hz. Three-dimensional coordinates of each marker during the time course of each experiment were calculated from the data using kinematic software. These time series were then smoothed by use of a Butterworth low-pass filtered (cut-off frequency, 10 Hz). Based on the vertical movement of the left hind hoof, the motion cycles were cut, and left and right hind stance and swing phases were identified using a custom-made MATLAB script. The duration of each full motion cycle consisting of stance and swing phase was normalised to 100%. In the remainder of the paper, stance phase and swing phase always refer to the ipsilateral hind limb. For each horse and each trial, a minimum of six motion cycles at walk and a minimum of thirteen motion cycles at trot were considered. Per DCD, three consecutive trials were measured in each gait, resulting in a minimum of 18 motion cycles of walk and 39 motion cycles of trot per horse for further analysis.

Muscle activity measurements

Before electrode placement hair was removed with electric clippers, the skin was shaved, cleaned with isopropyl alcohol solution with slightly abrasive, roughly woven swabs to reduce the resistance. Surface electromyography activities were collected with wireless electrodes, with each sensor consisting of four parallel silver bars with an integrated amplifier, size 27 × 37 × 15 mm, mass 14.7 g. They were placed parallel to the fibre orientation over the left and right GM. The electrodes were positioned on the skin using the Delsys Adhesive Sensor Interface™ allowing for free skin contact of the silver bars without any solution or electrode gel, as specified by the manufacturer. The exact position of the central electrode was at the midpoint between origin and insertion in the middle of the widest section of the GM; this was identified by palpation of the tuber coxae, tuber sacrale and 3rd trochanter of the femur (Fig. 1).

Additional sensors were placed about 5 cm lateral and about 5 cm medial to this central electrode, resulting in three adjacent electrodes termed GM1 (medial on the muscle), GM2 (central on the muscle), and GM3 (lateral on the muscle) (Fig. 2). On DCD3, in all of the horses either the left or the right sided GM2 electrode was positioned over the biopsy scar as described above.

The resulting sEMG signal treated by removal of the direct current offset by subtraction of the mean of the

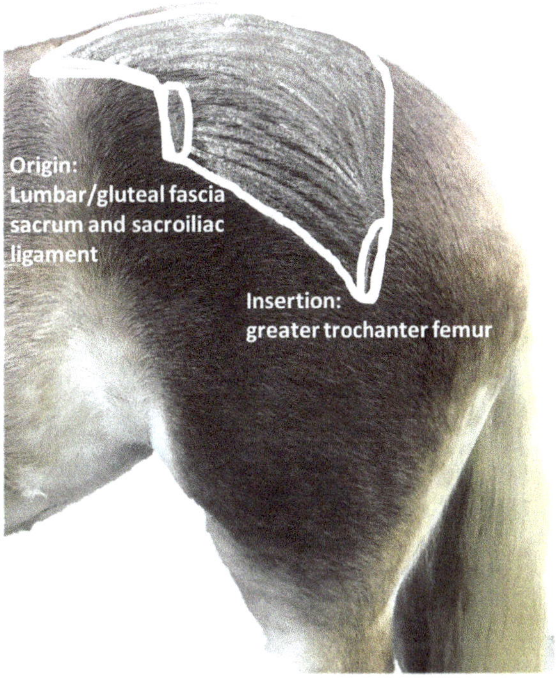

Fig. 1 Anatomical representation of the left equine gluteus medius muscle on a Haflinger horse. Origin and insertion of the left equine gluteus medius muscle are highlighted

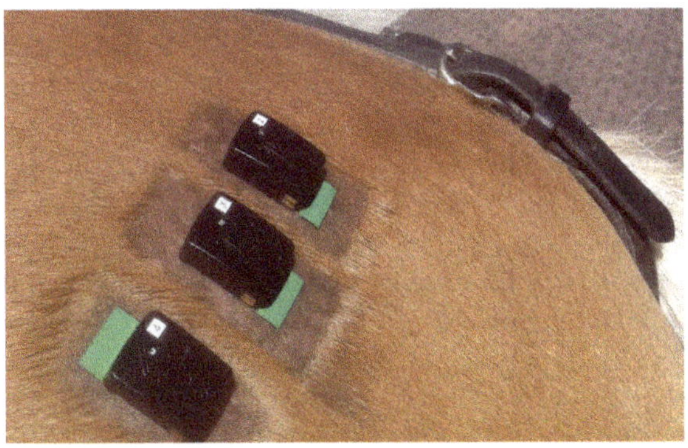

Fig. 2 Surface electrodes placed over the left gluteus medius muscle. The three electrodes placed over the left gluteus medius muscle of horse 13 are shown. The sensor number 2 is GM1 (medial to the middle of the muscle), sensor number 3 is GM2 (central on the muscle), and sensor number 4 GM3 (lateral to the middle of the muscle). For each electrode clipped and shaved the rectangles are visible; these were clipped once and remained visible for the rest of data collection days and were only re-shaved. At the last EMG measurement the middle electrode (GM2) was over the eight week old biopsy site

data. Then the sEMG was full-wave rectified and sampling rate was reduced to 120 Hz. A Butterworth low-pass filter was applied (fourth order; cut-off frequency 20 Hz), and the signal was normalised to its unique absolute maximum value at each electrode position for each DCD, resulting in a maximum value of 1 measured at least once per DCD.

The collections of all sEMG cycles of one horse together were used as the basis of Gaussian Mixture Models in the following way: the ten highest sEMG peaks in the set of all sEMG peaks were collected per motion cycle and grouped over the set of all motion cycles per horse, gait and electrode position. This set of peaks was then modelled by a Gaussian mixture distribution [18, 19].

Results were considered separately for stance phase (St) and swing phase (Sw) of the left (L) and right (R) sides. Mean mode locations (MML), representing the time points within the motion cycle where notable activity is located, were reported in percent of the overall motion cycle duration [18]. Mean mode values (MMV) of the gluteus during the motion cycle are also reported as an arbitrary unit representing the magnitude of the muscle activity (width and height of the trace); the maximum MMV is 1 [18]. Both MML and MMV are illustrated in Fig. 3.

Statistical analyses were done in SPSS. Normality of distributions was tested with Shapiro-Wilk tests. The analysis of parametric data consisted of a T-test for paired samples with Bonferroni corrections (comparisons of walk and trot; comparisons of left and right body side) and repeated measures ANOVA test with Bonferroni corrections (comparisons between the three measurement days). The results of parametric data are presented as mean ± standard deviation (SD). Non-parametric data was tested using a Wilcoxon Signed rank test (comparisons of walk and trot; comparisons of left and right body side) as well as a Friedman's test (comparisons between the three measurement days). The results of the non-parametric data are presented as mean and interquartile range. In all of the tests above, the level of significance was set at 0.05. Overall reliability of the measurements of MML and MVL on different days was calculated as Cronbach's alpha coefficients.

Results

A minimum of 8 modes were identified on all three locations for all horses in both gaits, with 8 horses showing 10 modes at all locations both at walk and at trot. At walk 4 measurements out of 126 measurements per side showed 9 modes and at trot 1 measurement showed 8 modes and 3 measurements showed 9 modes out of 126, leaving for both gaits 122 measurements with 10 modes. A qualitative difference in the shape of the activity curves at the three electrode positions was clearly visible in all horses (Fig. 4).

Mean mode value results - magnitude of mode activity

Similar GM activities were found for left and right sides in both gaits, and no significant differences between MMVs of the three electrode positions. Fair reliability was found for MMV over the three DCDs (Cronbach's alpha = 0.66). Ranges of MMV over all horses over all of the DCDs and all electrode positions were slightly larger at trot than at walk during the stance phase (7%–10% at walk, 8%–18% at trot) but slightly smaller at trot than at walk during the swing phase (11%–16% at walk; 7%–14% at trot).

The muscle activities measured during the stance phase at trot resulted in much larger MMVs compared to walk, and MMV was significantly different between the stance phases of walk and trot for all electrode positions ($p < 0.01$). This illustrates the larger effort of the GM during the stance phase at trot compared to walk. There was no significant difference between the MMVs at walk and trot during the respective swing phase (see Table 1), probably the effect of a faster but more passive forward swing of the hindlimb at trot. Significant differences were found when comparing results of DCD1 and DCD2 with DCD3: DCD1 GM1 MMVSt@T ($p = 0.001$), DCD2 GM1 MMVSt@T ($p = 0.002$), as well as DCD2 GM3 MMVSw@T ($p = 0.04$) (Table 1).

Mean mode location (MML) results – Pattern of mode activity

Good reliability was found for MML over the three DCDs (Cronbach's alpha = 0.71). Only four of the 36 DCD comparisons (3 locations, two sides with 6 values for each of the 3 DCDs) were significantly different. All significant differences were found for comparisons with DCD3 results: DCD1 right GM2 MMLSw@W ($p = 0.01$) DCD2 right GM2 MMLSw@W (p = 0.04), both involving results from electrodes over the biopsy site. Even though biopsies had been taken in half of the horses on the left side after DCD 2, no such differences were detected for the left GM2.

Ranges of MML over all horses over all of the DCDs and all electrode positions were slightly larger at walk than at trot and slightly larger during the stance phase (4%–27% at walk; 3%–20% at trot) than during the swing phase (3%–13% at walk; 4%–11% at trot).

Walk and trot resulted in significantly different MMLs at all electrode positions, both during stance and swing phases ($p < 0.001$), indicating the differences in timing pattern. No significant differences between the electrode positions were found.

The time difference between the left and right stance phase and swing phase MMLs was around 50% of the

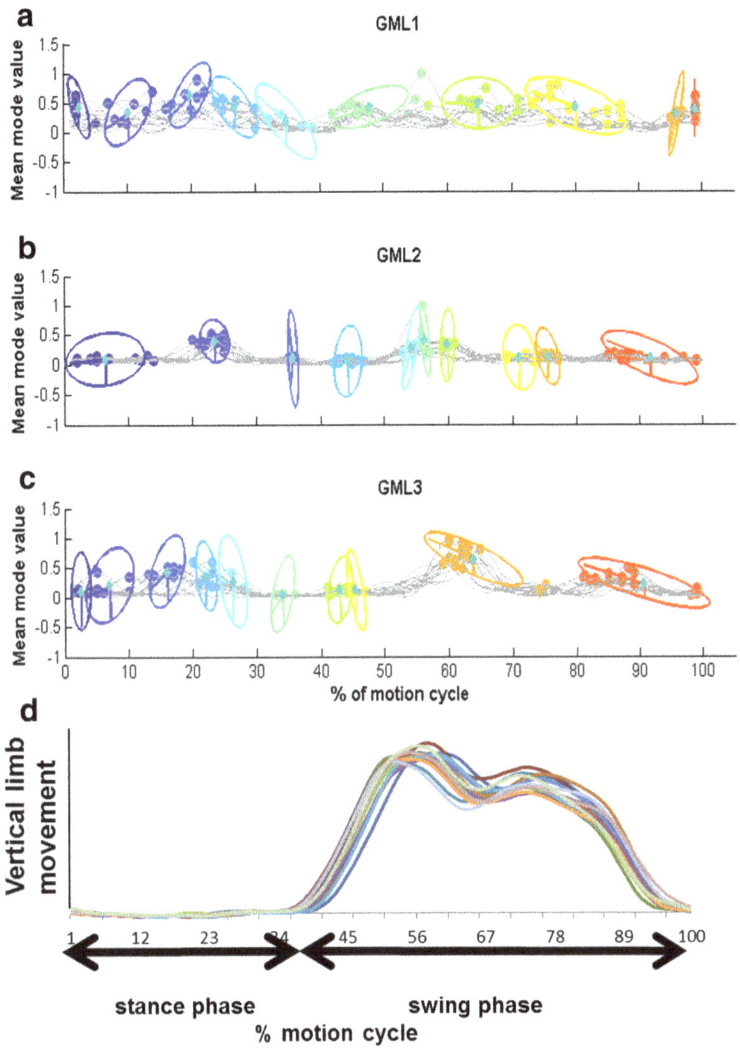

e

GML1	Mode 1	Mode 2	Mode 3	Mode 4	Mode 5	Mode 6	Mode 7	Mode 8	Mode 9	Mode 10
Mean Mode Location (% of motion cycle)	3	7	16	22	26	34	43	46	64	91
Mean Mode Value (all values normalized between 0 and 1)	0.12	0.22	0.45	0.37	0.28	0.06	0.13	0.13	0.65	0.24

Fig. 3 Activity of the left gluteus medius muscle of horse 11. Surface electromyography activity over 15 motion cycles at data collection day 2 trotting on a treadmill is shown; **a**) results of one electrode lateral to the centre of the muscle (GML1), **b**) one over the centre of the muscle (GML2), and **c**) one medial to the centre of the muscle (GML3), **d**) shows the mean vertical movement of the left hindlimb (HHL) during the same measurement and **e**) lists the numerical values for the ten modes evaluated

motion cycle at walk and trot (Table 2). This gait dependent shift between left and right hindlimb use at walk and trot was documented as statistically highly significant differences in MML ($p < 0.001$) in all electrode positions.

Discussion

In the present study no difference in GM muscle activities between the left and right sides were found; this indicates symmetrical movements. In an older study on hindlimb lame horses, left and right gluteal muscle activity differed significantly between the lame hindlimb and the sound hindlimb [20], further supporting that bilaterally similar GM activity is associated with absence of lameness.

The results of the present study obtained on a treadmill should not be applied to over ground locomotion without careful consideration. Horses were walking and trotting at their optimum speed on the treadmill [17], and this created a consistent movement pattern with the stance phase standardised to a higher degree than the swing phase. During the stance phase the push-off of the limb is not as

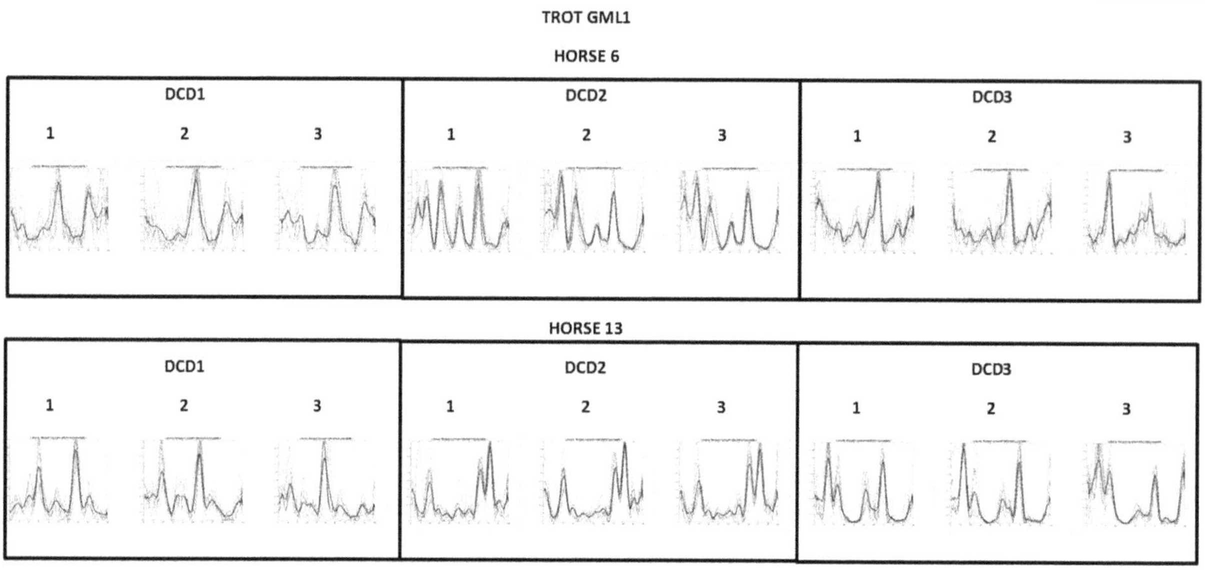

Fig. 4 Activity of the left gluteus medius muscle during trot measured three times over 16 weeks. The results obtained from the electrode lateral to the centre of the left gluteus medius muscle (GML1) of horse 6 (top row) and horse 13 (bottom row) are shown for data collection day 1 (DCD1, left three plots), at data collection day 2 (DCD2, middle three plots) and at data collection day 3 (DCD3, right three plots). For each DCD the trot measurements obtained (1, 2, 3) shown, with all motion cycles considered for data analysis. Mean curves of these trials are shown with red on these figures

strong as it would be during over ground locomotion [21, 22]. Similarly in rats, increased (passive) hip extension at initial limb contact with the treadmill was found when treadmill versus over ground walk was compared [23]. The use of treadmill set-ups for equine locomotion studies has been standard for many years as the laboratory settings with a treadmill allow for exactly repeatable conditions, and the use of measurement equipment is safe and easy in such a laboratory. However, over the last decade, the measurement of horses moving over ground has increased steadily, as the availability of new technologies (such as wireless sensors, and day light kinematic cameras) has expanded. The present study however used a data collection set up optimised for the submaximal exercise tests of the preceding study into the effects of the GYS1 mutation of half of the study horses [16] where the standardisation needed was only achievable on the treadmill. For a study into GM muscle use the selection of a treadmill set up or an over ground set up is relevant, as the function of GM during locomotion is mainly hip extension which is expected to be less forceful on the treadmill compared to over ground.

Also, the degree of GM abduction counteracting the gravitational adduction during the swing phase was expected to be electrode position specific, however this was not found in the present study. The activity level of the muscles was different during the swing and stance phase, and their relative contribution also changed with the gaits. Contrary to earlier assumptions, that the swing phase is mainly passive in any gait, considerable muscle activity during the swing phase was found especially when walking slowly; in humans this active swing phase control at walk was documented [24] and found to require slightly less than one-third of the net energy cost of walking [25]. In guinea fowl, Marsh et al. [26] reported that the swinging limb used an appreciable

Table 1 Mean mode value of the equine gluteus medius muscle activity

Mean mode value		WALK						TROT					
		Stance Phase			Swing Phase			Stance Phase			Swing Phase		
GM1	Mean ± SD	0.29	±	0.09	0.34	±	0.09	0.36	±	0.10	0.33	±	0.07
GM2	Mean ± SD	0.28	±	0.08	0.33	±	0.07	0.35	±	0.09	0.32	±	0.07
GM3	Mean ± SD	0.28	±	0.08	0.34	±	0.09	0.36	±	0.11	0.33	±	0.07

The magnitude of electrical activity (mean mode value) of the equine gluteus medius muscle during swing and stance phase is listed; the maximum value of this arbitrary unit is 1. As there was no statistically significant difference between the values of the left and right side, the average of both sides is given. Mean mode values are shown as (mean ± SD) for electrode lateral to the centre of the muscle (GML1), for electrode over the centre of the muscle (GML2), and for electrode medial to the centre of the muscle (GML3) at walk and trot during the three data collection days

Table 2 Mean mode location of left and right equine gluteus medius muscle activity per motion cycle

Mean mode location [%]		WALK				TROT			
		Left swing phase	Left stance phase	Right swing phase	Right stance phase	Left swing phase	Left stance phase	Right swing phase	Right stance phase
GM1	Median	76	25	25	76	68	16	16	67
	IQR	3	5	4	4	4	3	5	3
GM2	Median	77	25	26	76	67	16	17	68
	IQR	4	4	3	6	4	5	4	3
GM3	Median	76	25	25	76	68	16	17	67
	IQR	7	3	4	5	3	2	4	3

For walk and trot, the overall mean mode locations in percentage of the total duration of each motion cycle (mean ± SD) results (considering for the three data collection days over 16 weeks with three measurements on each data collection day) obtained from three electrodes (slightly lateral to the centre of the muscle: GM1, over the centre of the muscle: GM2, and slightly medial to the centre of the muscle GM3) are listed for each side

fraction of the energy in both walking and running. Pontzer [27] showed that the estimated leg-swing energy costs in quadrupeds were lower than in humans where they accounted for 15% (during walking) and 10% (during running) of the cost of locomotion. The results of the present study support this for the equine gluteus activity during walking and trotting on a treadmill: The swing phase MMV was roughly similar at walk and trot while the stance phase MMV was much larger at trot than at walk, creating a relatively smaller contribution of the swing phase muscle activity to the overall energy costs at trot. This is in contrast to a study into the function of the extrinsic hindlimb muscles in trotting dogs, where a small burst of activity of the gluteus superficialis muscle was found within the first 20% of the swing phase [28]. This timing of a relatively small burst of gluteal activity was also found in the gluteus medius activity in the walking and trotting horses in the present study. This may be due to the transfer of function from the gluteus medius muscle to the gluteus superficialis between these two species.

Most studies into sEMG of animals have used data obtained either on a single day of measurements or before and after interventions. Therefore, the reliability of sEMG measurements over 16 weeks (fair for MMV and good for MML) in the present study cannot be directly compared to other studies. However, these results do support the hypothesis of reliability of the EMG signal over many weeks. One study comparing the reliability of sEMG results of human trunk muscles in the same locations twice with a 1-week interval showed excellent reliability for the submaximal MVC while MVC as such had a lower reliability [29] and another study comparing sEMG reliability over two weeks showed high reliability of MVC of trunk muscle measurements [30]. Both these studies were undertaken over a shorter period of time than the present study, and they involved MVC measurements, which are inherently impossible in animals [31].

Compartmentalisation of large muscles has been described for muscles acting on a number of joints (such as the longissimus dorsi muscle of the horse [32]) as well as for muscles acting on a single joint with a single point of insertion (such as the human deltoid muscle). In the human deltoid muscle an even more detailed study showed that anatomical and functional segments exist in this muscle, with seven anatomical segments identified based on the intramuscular tendons and at least two electrode positions (posterior and middle deltoid) with different activities of sEMG have been described [33].

The equine gluteus medius muscle which was examined in the present study has previously been investigated for signs of compartmentalisation with differences in fibre types of muscle fibres depending on the depth of the muscle [14] However, functional compartmentalisation of the equine gluteus has not been investigated using EMG. In the equine longissimus muscle activities along the thoracolumbar spine were found to peak at different time points during the motion cycles [11, 32] associated with the need for spinal stabilisation at these locations. This muscle has a clearly segmented structure, and each segment is innervated by the respective spinal nerve. Ritruechai et al. [33] demonstrated based on computed tomography scans the complex architecture of the equine longissimus dorsi muscle. Its different regions thus perform different mechanical functions; and the recruitment of different compartments within this muscle depends on the mechanics of the movement [34]. Such a clear differentiation was not found for the equine gluteus medius between the three EMG locations used in the present study. However, the equine gluteus medius muscle showed subtle differences in timing and extent of muscle activity measured by sEMG at the three electrode positions in the present; and, furthermore, these differences were quite variable for the horses measured. Such small differences may result in slightly different force vectors within the same muscle. They may allow muscles to fulfil a variety of functions while reducing the risk of damage to the muscle, its tendon or its origin/insertion. Varying muscle fibre use is also expected to allow more efficient blood flow to and from the muscle.

In many muscles, such a functional differentiation is represented by different heads, bi- or multi-pennation, or an inner fascial layer separating muscle volumes; the function of the different heads of large muscles such as the equine triceps brachii and biceps brachii [35] has been documented. However, in the present study, no significant differences between patterns and between magnitudes of muscle activities of GM at the three electrode positions could be identified despite the large muscle volume investigated, and the first hypothesis is therefore rejected.

The three dimensional architecture and electrical activity of a thick muscle is not documented in its full complexity using sEMG. The main use of a muscle is reflected in its fibre type distribution, which is closely related to the lengths and velocities of muscle fibres, and the force that can be generated by it [2]. Preceding studies have shown that regional morphological variations of the longissimus dorsi muscle exist [31, 36] as the fascicles closer to the vertebral bodies experience smaller strains and strain rates than the more dorsal and lateral fascicles. In the GM, similar differences in fibre type distribution were found [14], and in the horses used for the present study, superficial muscle biopsies of GM were taken. These showed a relatively even distribution of about 30% each of Type 1, Type 2a and Type 2X fibres [16]; this result is similar in its Type1 versus Type 2 fibre distribution to a previous study [37], where the higher use of GM at the higher speeds was interpreted as a reflection on the relatively larger number of type IIB fibres (subsumed in the Type 2X fibres) in the periphery of those muscles. The connection of the muscle fibres of the deeper layers of thick muscles may be similarly related to intramuscular EMG readings of these layers, this is an option for future studies.

Over the last two decades GM was the equine limb muscle most commonly investigated using sEMG; however there is still no consensus over the most suitable electrode position on this muscle. In humans, the SENIAM project (Surface Electromyography for the Non-Invasive Assessment of Muscles) was established to provide such guidelines for sensor placement and signal processing [36] but in animals the location of the electrode positions is not yet standardized which is detrimental for the comparison of results [31].

The placement of electrodes over innervation zones is known to reduce the signal amplitude [38] this should therefore be avoided [39], however for the equine gluteus no such information is as yet available. Surface array methods have been used in humans to identify innervation zones, this may also be possible in animals, and should become available over the next decade.

Descriptions of the location for the placement of gluteus medius electrodes have not always been well specified; descriptions include the approximate middle of the muscular belly [40], over the assumed midpoint of the muscle belly [5] or the midway along the lengths of the muscle [20]. In the current study the electrodes were positioned in three closely adjacent locations, around the midpoint identified by palpation of the palpable bony prominences, with the aim of making at least one of the electrode positions comparable to the electrode positions used in the studies discussed above. This was achieved, as palpating the anatomical descriptions in the other studies [5, 20, 40] lead to a location very close to GM2 electrode.

The present study was subject to several methodological considerations.

The number of motion cycles available for analysis was different for walk and trot, as the data collection time for each trial was kept the same. On each data collection day, 18 motion cycles of walk and 39 motion cycles of trot were obtained. An alternative approach might have been to double the data collection time for the walk, resulting in roughly similar numbers. Such an approach might have detected more significant differences between walk and trot.

In most textbooks, the placement of sEMG electrodes over scars is discouraged [41]. However, in the present study one of the electrodes was positioned over the eight week old muscle biopsy scar on DCD3. Interestingly the results were not different from the results obtained from the other electrode at the same day, but there were differences detected in the pattern of activity (MML) when comparing with same position during preceding measurements. This was somewhat surprising, and may have been associated with the biopsy scar. It is possible, that the magnitude of the muscle activity is unaffected as the volume of muscle fibres sampled below the electrode is larger than the $1cm^3$ removed. On the other hand, the data processing chosen for the present study (normalisation to the maximum value) may have reduced the relevance of these fewer muscle fibres samples.

The present study was carried out with very little data processing, consisting of resampling, DC offset removal, rectification, and a low pass filter; this was followed by normalisation to the maximum value. Other studies reporting gluteal muscle activity in dogs during locomotion focussing on the description of the muscle function have applied different data processing steps, including rectification, generation of an EMG sample with 120 bins, and normalization of the EMG amplitude to the control trials [28].

Additionally, a high pass filter is used in many studies for sEMG data processing; however, this was not done in the present study based on the fact that the high pass filter employed may, in some instances reduce, or even remove pertinent information contained in the signal

[42]. Also, no clear indications for the need of a high pass filter (such as wavering baselines) were noted in the EMG signals.

Conclusions

The GM does not show any indication of segmentation when comparing three sEMG electrode positions during walk and trot on the treadmill; this might be due to a lack of segmentation per se (supported by anatomy) or due to a lack of need for segmented use for these very basic main tasks of the muscle. A scar of muscle biopsy (with about 1 cm^3 muscle tissue removed) did not create measurable differences in magnitude of muscle activity, but differences found in pattern of activity may be associated with the scar. The reliability of the sEMG measurements over several weeks was fair to good, an indication for the robustness of the methodology.

Abbreviations

@T: at trot; @W: at walk; CI: 95% confidence interval; DCD: Data collection day; GM: Gluteus medius muscle; GM1: Medial to the middle of the gluteus medius muscle; GM2: Central on the gluteus medius muscle; GM3: Lateral to the middle of the gluteus medius muscle; GYS1: Glycogen synthase mutation; IQR: Interquartile range; L: Left; MML: Mean mode locations; MMV: Mean mode value; R: Right; S.D.: Standard deviation; sEMG: Surface electromyography; St: Stance phase; Sw: Swing phase

Acknowledgements

Not applicable.

Funding

This work was supported by the Austrian Science Fund (FWF) and part of the GeMMQuad project (I1532-B23).

Authors' contributions

RRZS and TFL was involved in devising the aim and the experimental design of this study, in generating and analysing the data and the writing of this manuscript. AV contributed to the analysis of the data. US contributed to the gathering and analysis of the data. BK and AW assisted in the analysis of the data. All the authors assisted in the reviewing of this manuscript during the writing process and have read and accepted the final version of this manuscript. All authors read and approved the final manuscript.

Consent for publication

Not applicable.

Competing interests

The authors declare that they have no competing interests.

Author details

[1]Department of Sustainable Agricultural Systems, Section Livestock Sciences, University of Natural Resources and Life Sciences Vienna, Vienna, Austria. [2]Multimedia, Simulation and Virtual Reality Group, Institute of Computer Science II, University of Bonn, Bonn, Germany. [3]Gokhale Method Institute, Stanford, CA, USA. [4]Department for Companion Animals and Horses, University of Veterinary Medicine Vienna, Vienna, Austria. [5]Royal (Dick) School of Veterinary Studies, The University of Edinburgh, Edinburgh, Scotland, UK.

References

1. Payne RC, Hutchinson JR, Robilliard JJ, Smith NC, Wilson AM. Functional specialisation of pelvic limb anatomy in horses (Equus caballus). J Anat. 2005;206(6):557–74.
2. Arnold EM, Hamner SR, Seth AJ, Millard M, Delp SL. How muscle fiber lengths and velocities affect muscle force generation as humans walk and run at different speeds. J Exp Biol. 2013;216(11):2150–60.
3. Robert C, Valette JP, Denoix JM. The effects of treadmill inclination and speed on the activity of two hindlimb muscles in the trotting horse. Equine Vet J. 2000;32(4):312–7.
4. Robert C, Valette JP, Pourcelot P, Audigie F, Denoix JM. Effects of trotting speed on muscle activity and kinematics in saddlehorses. Equine Vet J. 2002;S34:295–301.
5. Crook TC, Wilson A, Hodson-Tole E. The effect of treadmill speed and gradient on equine hindlimb muscle activity. Equine Vet J. 2010;42(S38):412–6.
6. Takahashi T, Matsui A, Mukai K, Ohmura H, Hiraga A, Aida H. The effects of inclination (up and down) of the treadmill on the electromyogram activities of the forelimb and hind limb muscles at a walk and a trot in thoroughbred horses. J Equine Sci. 2014;25(4):73–7.
7. Bartlett JL, Sumner B, Ellis RG, Kram R. Activity and functions of the human gluteal muscles in walking, running, sprinting, and climbing. Am J Phys Anthropol. 2014;153(1):124–31.
8. Richmond FJ. Elements of style in neuromuscular architecture. Am Zool. 1998;38(4):729–42.
9. Sakoma Y, Sano H, Shinozaki N, Itoigawa Y, Yamamoto N, Ozaki T, Itoi E. Anatomical and functional segments of the deltoid muscle. J Anat. 2011; 218(2):185–90.
10. English AW, Weeks OI. Compartmentalization of single muscle units in cat lateral gastrocnemius. Exp Brain Res. 1984;56(2):361–8.
11. Licka TF, Peham C, Frey A. Electromyographic activity of the longissimus dorsi muscles in horses during trotting on a treadmill. Am J Vet Res. 2004; 65(2):155–8.
12. Budras, et al. Chapter 9: selected body systems in tabular form. 3. Peripheral nervous system. In: Budras KD, Sack WO, Röck S, editors. Anatomy of the horse, an illustrated text. Sixth ed. Hannover, Germany: Schluetersche Verlagsgesellschaft; 2011. p. 96.
13. Schaller O. Arthrologia. In: Schaller O, editor. Nomina Anatomica Veterinaria. Stuttgart, Germany: Enke Verlag; 1992. p. 302.
14. Bruce V, Turek RJ. Muscle fibre variation in the gluteus medius of the horse. Equine Vet J. 1985;17(4):317–21.
15. Schiaffino S, Reggiani C. Fiber types in mammalian skeletal muscles. Physiol Rev. 2011;91(4):1447–531.
16. Schröder U, Licka TF, Zsoldos R, Hahn CN, MacIntyre N, Schwendenwein I, Schwarz B, van den Hoven R. Effect of diet on Haflinger horses with GYS1 mutation (polysaccharide storage myopathy type 1). J Equine Vet Sci. 2015; 35(7):598–605.
17. Peham C, Licka T, Mayr A, Scheidl M, Girtler D. Speed dependency of motion pattern consistency. J Biomech. 1998;31(9):769–72.
18. Vögele AM, Zsoldos RR, Krüger B, Licka T. Novel methods for surface EMG analysis and exploration based on multi-modal Gaussian mixture models. PLoS One. 2016;11(6):e0157239.
19. Bishop CM. Chapter 9: mixture models and EM. In: Bishop CM, editor. Pattern recognition and machine learning (information science and statistics). New York, USA: Springer-Inc.; 2006. p. 423–50.
20. Zaneb H, Kaufmann V, Stanek C, Peham C, Licka TF. Quantitative differences in activities of back and pelvic limb muscles during walking and trotting between chronically lame and nonlame horses. Am J Vet Res. 2009;70(9):1129–34.

21. Weishaupt MA, Wiestner T, Hogg HP, Jordan P, Auer JA. Vertical ground reaction force–time histories of sound warmblood horses trotting on a treadmill. Vet J. 2004;168(3):304–11.
22. Buchner HHF, Savelberg HHCM, Schamhardt HC, Merkens HW, Barneveld A. Kinematics of treadmill versus overground locomotion in horses. Vet Quart. 1994;16(S2):87–90.
23. Pereira JE, Cabrita AM, Filipe VM, Bulas-Cruz J, Couto PA, Melo-Pinto P, Costa LM, Geuna S, Mauricio AC, Varejao AS. A comparison analysis of hindlimb kinematics during overground and treadmill locomotion in rats. Behav Brain Res. 2006;172(2):212–8.
24. Whittlesey SN, van Emmerik RE, Hamill J. The swing phase of human walking is not a passive movement. Mot Control. 2000;4(3):273–92.
25. Umberger BR. Stance and swing phase costs in human walking. J R Soc Interface. 2010;7(50):1329–40.
26. Marsh RL, Ellerby DJ, Carr JA, Henry HT, Buchanan CI. Partitioning the energetics of walking and running: swinging the limbs is expensive. Science. 2004;303(5654):80–3.
27. Pontzer H. Predicting the energy cost of terrestrial locomotion: a test of the LiMb model in humans and quadrupeds. J Exp Biol. 2007;210(3):484–94.
28. Schilling N, Fischbein T, Yang EP, Carrier DR. Function of the extrinsic hindlimb muscles in trotting dogs. J Exp Biol. 2009;212(7):1036–52.
29. Dankaerts W, O'Sullivan PB, Burnett AF, Straker LM, Danneels LA. Reliability of EMG measurements for trunk muscles during maximal and sub-maximal voluntary isometric contractions in healthy controls and CLBP patients. J Electromyogr Kines. 2004;14(3):333–42.
30. Marshall P, Murphy B. The validity and reliability of surface EMG to assess the neuromuscular response of the abdominal muscles to rapid limb movement. J Electromyogr Kines. 2003;13(5):477–89.
31. Valentin S, Zsoldos RR. Surface electromyography in animal biomechanics: a systematic review. J Electromyogr Kines. 2015;28:167–83.
32. Licka T, Frey A, Peham C. Electromyographic activity of the longissimus dorsi muscles in horses when walking on a treadmill. Vet J. 2009;180(1):71–6.
33. Ritruechai P, Weller R, Wakeling JM. Regionalisation of the muscle fascicle architecture in the equine longissimus dorsi muscle. Equine Vet J. 2008; 40(3):246–51.
34. Wakeling JM. The recruitment of different compartments within a muscle depends on the mechanics of the movement. Biol Lett. 2009;5(1):30–4.
35. Watson JC, Wilson AM. Muscle architecture of biceps brachii, triceps brachii and supraspinatus in the horse. J Anat. 2007;210(1):32–40.
36. Hermens HJ, Freriks B, Merletti R, Stegeman D, Blok J, Rau G, Disselhorst-Klus C, Hagg G. European recommendations for surface electromyography. Roessingh Res Develop. 1999;8:13–54.
37. Grotmol S, Totland GK, Kryvi H, Breistøl A, Essén-Gustavsson B, Lindholm A. Spatial distribution of fiber types within skeletal muscle fascicles from Standardbred horses. Anat Rec. 2002;268(2):131–6.
38. Beck TW, Housh TJ, Cramer JT, Mielke M, Hendrix R. The influence of electrode shift over the innervation zone and normalization on the electromyographic amplitude and mean power frequency versus isometric torque relationships for the vastus medialis muscle. J Neurosci Meth. 2008;169(1):100–8.
39. Mesin L, Merletti R, Rainoldi A. Surface EMG: the issue of electrode location. J Electromyogr Kinesiol. 2009;19(5):719–26.
40. Robert C, Valette JP, Degueurce C, Denoix JM. Correlation between surface electromyography and kinematics of the hindlimb of horses at trot on a treadmill. Cells Tissues Organs. 1999;165(2):113–22.
41. Hermens HJ, Freriks B, Disselhorst-Klug C, Rau G. Development of recommendations for SEMG sensors and sensor placement procedures. J Electromyogr Kines. 2000;10(5):361–74.
42. Nugent MM, Stapley PJ, Milner TE. Independent activation in adjacent lumbar extensor muscle compartments. J Electromyogr Kines. 2012;22(4):531–9.

10

Detection of *Strongylus vulgaris* in equine faecal samples by real-time PCR and larval culture – method comparison and occurrence assessment

A. Kaspar[1*], K. Pfister[1,2], M. K. Nielsen[3], C. Silaghi[1,4], H. Fink[5] and M. C. Scheuerle[1,6]

Abstract

Background: *Strongylus vulgaris* has become a rare parasite in Germany during the past 50 years due to the practice of frequent prophylactic anthelmintic therapy. To date, the emerging development of resistance in Cyathostominae and *Parascaris* spp. to numerous equine anthelmintics has changed deworming management and the frequency of anthelmintic usage. In this regard, reliable detection of parasitic infections, especially of the highly pathogenic *S. vulgaris* is essential. In the current study, two diagnostic methods for the detection of infections with *S. vulgaris* were compared and information on the occurrence of this parasite in German horses was gained. For this purpose, faecal samples of 501 horses were screened for *S. vulgaris* with real-time PCR and an additional larval culture was performed in samples of 278 horses. A subset of 26 horses underwent multiple follow-up examinations with both methods in order to evaluate both the persistence of *S. vulgaris* infections and the reproducibility of each diagnostic method.

Results: The real-time PCR revealed *S. vulgaris*-DNA in ten of 501 investigated equine samples (1.9%). The larval culture demonstrated larvae of *S. vulgaris* in three of the 278 samples (1.1%). A direct comparison of the two methods was possible in 321 samples including 43 follow-up examinations with the result of 11 *S. vulgaris*-positive samples by real-time PCR and 4 *S. vulgaris*-positive samples by larval culture. The McNemar's test (p-value = 0.016) revealed a significant difference and the kappa values (0.525) showed a moderate agreement between real-time PCR and larval culture.

Conclusions: The real-time PCR detected a significantly higher proportion of positives of *S. vulgaris* compared to larval culture and should thus be considered as a routine diagnostic method for the detection of *S. vulgaris* in equine samples.

Keywords: *Strongylus vulgaris*, Strongyle, Equine, Larval culture, Real-time PCR, Germany

Background

The increasing resistance of small strongyles (Cyathostominae), *Parascaris* spp. (Ascarididae) and *Oxyuris equi* (Oxyuridae) against anthelmintic drugs [1–5] requires major changes in equine parasite control. During the past 50 years, horses were mainly treated prophylactically multiple times per year [6–11]. This approach was based on recommendations, given mainly in order to avoid infections with *Strongylus vulgaris* [12]. Due to the emerging resistance of Cyathostominae to available anthelmintic drugs, an alternative approach based on selective anthelmintic therapy (SAT) has been implemented in some horse farms [13, 14]. The principle of SAT comprises a major change in the frequency of applying anthelmintics with an anti-strongyle spectrum based on the treatment of horses with a faecal egg count (FEC) of strongyle eggs above a certain threshold, determined with a quantitative faecal egg count method, such as the

* Correspondence: alexandra.kaspar@tropa.vetmed.uni-muenchen.de
[1]Comparative Tropical Medicine and Parasitology, Veterinary Faculty, Ludwig-Maximilians-University, Munich, Germany

McMaster [3, 15–18]. Nevertheless, for reliable detection of infections with other important equine parasite species like *Anoplocephala perfoliata*, *O. equi*, and *Strongylus* spp. (large strongyles, Strongylinae) further specific diagnostic tests should be applied [13].

The intensive anthelmintic treatment regime of the last decades might be responsible for the current low occurrence of *S. vulgaris* (0.2–1.3%) recently determined by larval culture in German horses [10, 19–21]. However, a reliable diagnostic method for the differentiation of *S. vulgaris* from Cyathostominae is essential because of the significant pathology caused by the migration of *S. vulgaris*-larvae in the mesenteric arteries. Associated pathogenic effects like intestinal infarction, peritonitis, verminous arteritis, thrombosis and embolism can provoke symptoms such as colic, hind-leg lameness, neurological abnormalities and lesions in the heart, liver or kidney which might be lethal in the worst case [22–24].

Currently, the standard method for the differentiation of Cyathostominae and Strongylinae is the determination of morphological differences of third-stage larvae (L3) after larval culture since their egg morphology is similar and thus impossible to distinguish [25, 26]. However, even though this method has the advantage of being cost-efficient, it is also very time-consuming due to the developmental period of the strongyle egg to the L3 and the subsequent morphological differentiation which should be performed by experienced qualified personnel [27–29].

To date, detection of *S. vulgaris* is possible with conventional and real-time PCRs in faecal samples as well as a serum Enzyme linked Immunosorbent Assay (ELISA) [30–32].

According to previously conducted studies, both the conventional and the real-time PCR provide high specificity and sensitivity [30, 33–37]. Moreover, Nielsen et al. [30] claimed the real-time PCR as a potential standard method for a reliable detection of *S. vulgaris* [30]. Therefore, the objectives of the present study were to evaluate the detection of *S. vulgaris* via real-time PCR in comparison to the standard diagnostic method (larval culture) and to add information about the occurrence of *S. vulgaris* in German horses detected by real-time PCR and larval culture.

Methods

Faecal samples

In total, 1455 equine faecal samples from 91 German farms were collected from March 2013 to May 2014. All samples were obtained from the Diagnostic Centre of the Chair of Comparative Tropical Medicine and Parasitology, LMU Munich, Germany. Every sample was analysed using a modified McMaster method with a detection limit of 20 eggs per gram faeces (EPG) and the combined-sedimentation-flotation method [17]. Samples with a minimum of 20 EPG were selected.

Assuming a German horse population of 1.1 million [38] with a prevalence for an infection with *S. vulgaris* of 1% [3, 10, 19–21], a minimum of 459 horses had to be investigated since in this case the probability to detect at least one infected horse was above 99% [39]. Thus, samples of 501 of 1455 horses (horse level, 34.4%) were investigated by real-time PCR for an infection with *S. vulgaris* (Fig. 1) in the present study.

For 278 of the 501 horses, a sufficient amount of faecal material was available for the preparation of a larval culture in order to differentiate the larvae.

In 26 of the 501 investigated horses, follow-up examinations were carried out 4–6 months after the initial examination. In this context, a total of 43 samples of the 26 horses from six different farms were analysed by real-time PCR and larval culture. Follow-up examinations were performed if either the real-time PCR-result was positive for *S. vulgaris* (Table 1) or non-determinable larvae were present in the first larval culture examination. In the case of two horses which were *S. vulgaris* positive in the first examination, the whole horse group was tested also for *S. vulgaris*. With these examinations, the dissemination of *S. vulgaris* should be verified at horses living on the same holding.

None of the horses that underwent a follow-up examination had been dewormed until the date of the retest. Thus, a total of 544 samples were analysed by real-time PCR and a total of 321 samples were analysed by larval culture (sample level, Fig. 1).

The total data pool was separated into two data sets for the different analyses either on sample or on horse level (Fig. 1). Furthermore, prevalence analyses were performed on horse and on farm level. As soon as one horse was tested *S. vulgaris*-positive, its farm of origin was classified as positive on the farm level.

As far as possible, information was collected on every investigated horse, namely age, breed, gender and the date of the last anthelmintic treatment.

Egg detection, concentration and isolation

Gastrointestinal strongyle eggs were isolated from faecal samples using a combined-sedimentation-flotation technique [17]. At first, 40 g of faeces were mixed with 500 ml of water, then filtered through a 300 μm sieve and sedimented at 10 °C for 12 h. The sediment was divided and transferred into two centrifuge tubes – one with saturated sugar solution (specific weight/solution density 1.27 +/- 0.01) and one with saturated zinc sulphate solution ($ZnSO_4$, specific weight/solution density 1.28 +/- 0.01; Carl Roth GmbH and Co. KG, Karlsruhe, Germany) and centrifuged for 5 min at 635 g.

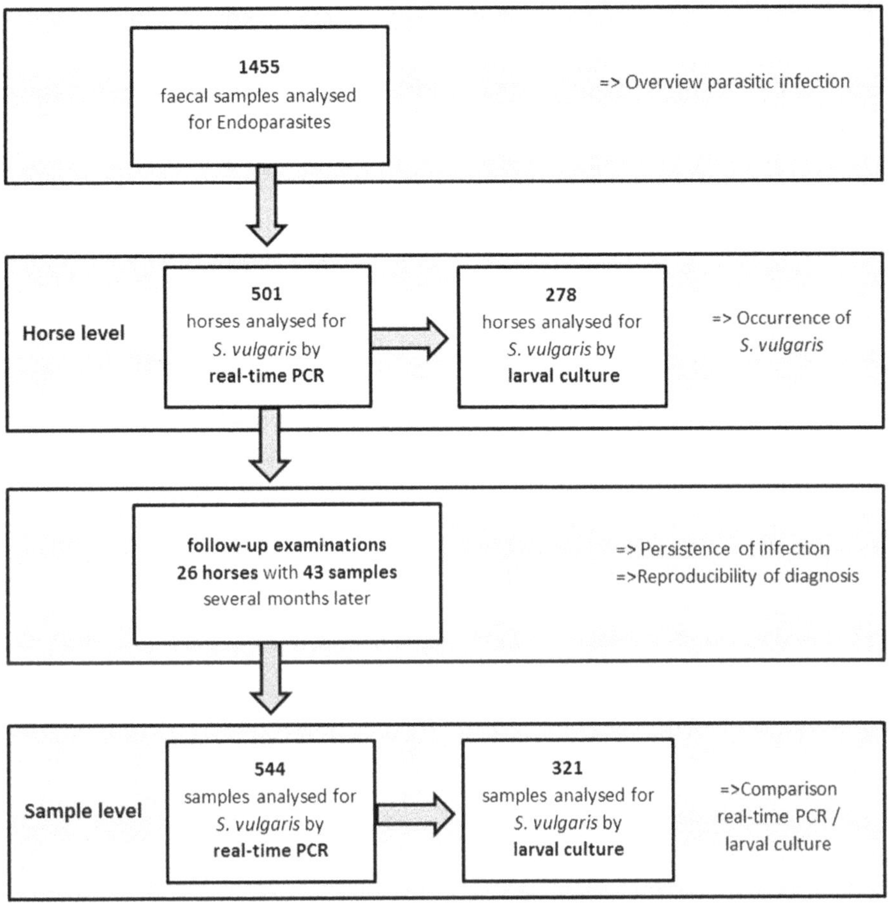

Fig. 1 Overview of the division of equine faecal samples and the corresponding motivation

In case of a positive microscopic diagnosis of strongyle eggs, the egg counts were classified as sporadic (0–300 eggs), numerous (301–3000) or plentiful (>3000).

The eggs were rinsed from the cover slip with distilled water, poured into a beaker and into a centrifuge tube

Table 1 Dates of initial and follow-up examinations of four real-time PCR-positive horses

Horse	1	2	3	4
Date of first examination	2013/05/27	2013/06/04	2013/08/06	2013/06/13
Date of follow-up examinations	2013/09/16	2013/10/30	2014/02/11	2014/02/07
		2013/10/31	2014/02/12	2014/02/08
		2013/11/01	2014/02/13	
		2013/11/02	2014/02/14	
		2013/11/03	2014/02/15	
		2013/11/04	2014/02/16	
		2013/11/05	2014/02/17	
			2014/02/18	
			2014/02/19	
			2014/02/20	
			2014/02/21	

(conical shape) which was then centrifuged at 635 g for 10 min. The remaining liquid was withdrawn by suction. The remaining sediment was transferred into a 1.5 ml Eppendorf tube and then centrifuged one more time at 10,621 g for 3 min. The remaining liquid was withdrawn by suction and the sediment was frozen in a dry state at −20 °C.

Larval identification

Larval cultures modified after Roberts and O'Sullivan [40] were carried out for the morphological determination of *S. vulgaris* larvae. Ten grams of faeces were incubated at room temperature for 14 days. During these 14 days, the samples were ventilated every day for 1 h and moistened to avoid desiccation of the faeces. The Baermann-Wetzel technique was used to isolate the L3 [17].

After the migration time of 24 h at room temperature, 10 ml of the liquid was drawn off from the Baermann funnel. For concentration of the larvae, the liquid was stored in a centrifuge tube at 10 °C for another 24 h. Subsequently, the supernatant was withdrawn by suction and 2 ml were left in the tube. All larvae contained in the sediment were counted (L/10 g = larvae per 10 g faeces) and morphologically determined via microscope. For immobilisation of the larvae, one drop of Lugol's iodine was added to one drop of the larval suspension and investigated immediately. The morphological criteria were determined according to Bürger et al. [41] and Boch et al. [42].

DNA extraction from isolated eggs

The extraction was carried out according to the manufacturer's instruction (QIAamp® DNA Mini and Blood Mini Handbook) according to the blood or body fluid spin protocol (Qiagen, Hilden, Germany). Briefly, before extraction the sediment containing the strongyle eggs was incubated in an ultrasonic bath with 150 µl PBS (phosphate buffered saline) for 4 min followed by a freezing step for 20 min at –20 °C and a repetition of the incubation step. The DNA was eluted with 200 µl elution buffer. DNA concentration (ng/µl) and quality of each sample were analysed with the Nano-Drop ND-1000 Spectrophotometer (PeqLab, Erlangen, Germany). The samples were stored at –20 °C until used for PCR analysis.

Molecular analyses

Real-time PCR

Samples were screened for *S. vulgaris* by real-time PCR using the AB-7500 Real Time PCR System (Applied Biosystems, Darmstadt, Germany). PCR targeted rDNA sequences of the second internal transcribed spacer (ITS-2: 169 bp) of Cyathostominae and *Strongylus* spp. nematodes. The forward primer Sv-f (5′-GTATACATTAAA-TAGTGTCCCCCATTCTAG-3′), the reverse primer Sv-r (5′-GCAAATATCATTAGATTTGATTCTTCCG-3′) and the modified probe Sv-p 5′-FAM-TGGATTTATTCT-CACTACTTAATTGTTTCGCGAC-BHQ1-3′ were used as previously described [30]. The 25 µl reaction volume consisted of 5 µl of template DNA, 2.25 µl (0.9 µM) of each primer (10 µM), 0.5 µl of the probe (0.2 µM) and 15 µl TaqMan Gene Expression Master Mix (Life technologies GmbH, Darmstadt, Germany). A negative and a positive control were added to each PCR reaction. The positive control (DNA extracted from adult worms which were morphologically identified as *S. vulgaris*) was provided by the M. H. Gluck Equine Research Center, University of Kentucky, USA. The following PCR protocol was used: an initial activation at 95 °C for 10 min followed by a set of 40 cycles, each consisting of 15 s at 95 °C and 60 s at 60 °C (AB Systems Standard protocol, Applied Biosystems, Darmstadt, Germany).

To verify the efficiency and the relative sensitivity of the real-time PCR, a standard curve was created according to the AB Systems manufacturer's instruction (Applied Biosystems, Darmstadt, Germany) with ten fold dilution steps of the positive control (threefold preparation). An analysis was performed up to a dilution step of 1:100,000.

Results were recorded as the mean PCR cycle number at which the fluorescence detection threshold had been exceeded (C_t). The threshold line was set at the optimal point in the linear phase of the amplification plot. The observation of an exponential rise up to a C_t of 37.5 was counted as positive according to the results of the standard curve.

Conventional PCR and Sequencing

In order to verify the results of the real-time PCR, an additional conventional PCR with subsequent sequencing of the amplification products was performed with real-time PCR-positive samples using the same primers (100 µM) (Eppendorf Mastercycler MWG Biotech, Ebersberg, Germany).

The reaction mix contained 5 µl of template DNA, 5 µl (1x) of buffer 10x, 1 µl (200 µM) of dNTPS, 0.5 µl (1 µM) of each primer (100 µM), 0.25 µl (1.25 U) of HotStar Taq DNA Polymerase Kit (Qiagen, Hilden, Germany) and water-ultra pure grade filled up to a total volume of 50 µl. The following protocol was used: 5 min at 95 °C for one cycle, 95 °C for 30 s, 49 °C for 30 s and 72 °C for 40 s for 45 cycles, followed by 72 °C for 5 min.

PCR products of the conventional PCR were analysed in 2% agarose gel stained with Gel Red™ nucleic acid stain, 10.000 x in water (both from Biotium, Hayward, USA). The DNA Gene ruler 100 bp Plus DNA ladder was used for sizing and quantification of the PCR products. The visualisation of gel images was performed with a gel documentation system (Peqlab, Erlangen, Germany).

The positive PCR products were purified using the QIAquick PCR Purification Kit (Qiagen, Hilden, Germany) following the manufacturer's protocol. Forward and reverse sequencing was performed by Eurofins MWG Operon (Ebersberg, Germany). Reverse sequences were reversed, complemented, and aligned to the forward sequences using online tools (Reverse Complement: http://www.bioinformatics.org/sms/rev_comp.html, Clustal Omega: https://www.ebi.ac.uk/Tools/msa/clustalo). Database searches and sequence comparisons were done with BLAST provided by the National Center for Biotechnology Information (BLAST: http://blast.ncbi.nlm.nih.gov/Blast.cgi).

Statistical Analyses

Statistical analyses were performed using IBM® SPSS® software version 22.0 (IBM Corporation, Armonk, USA) and Microsoft Excel 2010 (Microsoft Corporation, Seattle, USA). Since our data did not satisfy normality assumptions, classical non-parametric methods were used. The correlation between EPG and L/10 g faeces was calculated using a Spearman's rank correlation coefficient. In order to determine the statistical difference of the capability of the real-time PCR and the larval culture to detect a *S. vulgaris* positive sample, a McNemar test was performed for comparison of paired proportions and Kappa values were calculated for an evaluation of agreement between the two tests. All statistical analyses were interpreted as statistically significant up to the p-value < 0.05.

Results

Sample origin

Out of the 1455 samples, strongyle faecal egg counts (FECs) of 804 samples (55.3%) were positive ranging from <20 to 11,080 EPG. An EPG of >200 was recorded in 348 samples (23.9%).

Of the 804 faecal egg count positive samples, 544 samples (20–11,080 EPG; mean: 482 EPG; median: 200 EPG) from 501 horses were randomly selected for further investigation with real-time PCR. The number of investigated horses originating from 91 farms ranged from 1 to 112 per farm.

The FEC of the real-time PCR *S. vulgaris*-positive samples, ranged from 20 to 1720 EPG. The FEC of one sample could not be determined due to the low amount of faecal material available. However, it was included in the study due to the numerous strongyle eggs detected in the combined-sedimentation-flotation.

Of those 501 animals, the age was known for 374 and the gender was known for 425 horses.

The age of the horses ranged from 1 to 35 years with an average age of 12 years. Samples originating from 174 mares (40.9%), 246 geldings (57.9%) and five stallions (1.2%) were examined. On average for 398 horses, the last anthelmintic treatment had been performed 11 months (range 2–196 months) prior to enrolment into the present study.

Larval culture

Larval culture were performed and evaluated from 321 faecal samples originating from 278 horses (including follow-up examinations). A total of 451,832 Cyathostominae larvae (L3) were detected in 93.5% of the cultures (100.0% of the farm level). Between 1 and 31,251 larvae per 10 g faeces were counted. The EPG and the L/10 g faeces correlated with a Spearman's rank correlation of 0.826 ($p = 0.000$).

A total of 11 *S. vulgaris* larvae were detected in altogether four samples (sample level).

Regarding the horse level, *S. vulgaris* was found in 3 of 278 horses (1.1%). *S. vulgaris*-positive horses originated from 3 out of 62 investigated farms (4.8%).

The occurrence of different larval stages in a total of 321 larval cultures was documented in detail. Strongylinae larvae in first- (L1) and second-stage (L2) were found in 52 samples in a range of sporadic (max. 4 larvae/sample) and in four samples in a range of numerous (>15 larvae). Additionally, empty sheaths of Strongylinae larvae were found in five of the 321 samples (max. 3/sample). Larvae of *Triodontophorus* sp. were found in nine samples, larvae of *Trichostrongylus axei* in three samples and larvae of free-living nematodes were present in 15 samples.

Real-time PCR

The real-time PCR was positive in 13 of 544 investigated samples (2.4%; Table 2).

With regard to the horse level of 501 horses, the occurrence of *S. vulgaris* obtained by real-time PCR was 1.9% (10/501). *S. vulgaris*-positive horses originated from 10 of 91 investigated farms (10.9%). On each farm only one horse was tested positive for *S. vulgaris.*

The standard curve revealed an efficiency of 90.5% (slope -3.571; R^2 0.996). A positive analysis was performed with a dilution of up to 1:10,000 (y-Inter 37.428) of the positive control in order to estimate the relative sensitivity. Due to an unknown amount of DNA in the positive control, no reference value was available and it was not possible to analyse a quantitative potential of the real-time PCR. There was no statistically significant correlation between the number of *S. vulgaris* larvae within one sample and the corresponding C_t-value since the total amount of samples was too small. One PCR-

Table 2 Summarized results of real-time PCR-positive samples (FEC, combined-sedimentation-flotation, larval culture** and C_t-value)

Sample-No.	FEC	Combined-sedimentation-flotation*	Larval culture Cyathostominae**	Larval culture *S. vulgaris* **	C_t
21	20	+	16	3	27.58
144[a]	20	+	5	4	24.94
235	1040	+++	—	—	33.49
318	100	+	533	0	26.90
334	900	+++	1256	0	35.81
355	—	++	—	—	27.67
375	40	+	28	0	35.99
412	1720	+++	103	0	23.40
448[b]	20	+	34	0	36.95
451[c]	20	+	57	0	35.07
460	200	++	535	0	35.75
497	480	+++	1111	3	29.87
520	680	+++	3028	1	34.97

FEC = faecal egg count; Ct = cycle threshold
* + = sporadic eggs; ++ = numerous; +++ = plentiful
** Number of larvae of Cyathostominae or *S. vulgaris* /10 g faeces
— diagnostic method not performed
[a]follow-up examination of sample-No. 21, 2013/09/16
[b]sample-No. 375, 2014/02/12 and
[c]sample-No: 375, 2014/02/15

positive sample containing a single larva of *S. vulgaris* had a C_t-value of 34.97, which is higher compared to the C_t-values of 24.94 to 29.87 obtained by the other three PCR-positive samples containing three or four larvae in the culture.

Follow-up examinations

Four initially *S. vulgaris*-positive horses were retested in follow-up examinations. The follow-up samples of two horses tested positive on a daily basis. Horse 1 (Table 1) was tested positive in the single follow-up examination (Table 2) which was performed 4 months after the initial examination. The follow-up examination of horse 3 (Table 1) was carried out 6 months after the initial examination. In this context faecal samples were taken and analysed for *S. vulgaris* on 11 consecutive days. In addition to the initial investigation, *S. vulgaris* was also detected in the faecal samples collected on day 1 and 4 (Table 2). The other follow-up examinations of all horses that were retested were negative for *S. vulgaris.* However, the low number of follow-up examined horses in this study did not allow for a statistical analysis.

Comparison of real-time PCR and larval culture

A direct comparison between real-time PCR and larval culture could be drawn for 321 samples of the sample level. The results of the latter two methods differed significantly according to the McNemar's test (0.016) (Table 3). This is in line with Cohen's kappa which shows a value of 0.525 indicating only moderate agreement.

Conventional PCR and Sequencing

Interpretable sequencing results were obtained from all 13 isolates. Two of them were 100% homologous with a sequence of *S. vulgaris* from a horse from Australia (GenBank accession no. X77863.1) [33]. Eleven sequences showed a 99% similarity to the latter sequences containing a single nucleotide substitution (C instead of G at bp 61).

Sequences determined at the ITS-2 region in this work were deposited in GenBank under the following accession numbers: KT250609 – KT250621.

Table 3 Comparison of results of larval culture and real-time PCR

	PCR –	PCR +	Total
Larval culture –	310	7	317
Larval culture +	0	4	4
Total	310	11	321

κ-value: 0.525; 95% confidence interval: [0.219–0.831]; McNemar's test: *p*-value = 0.016

Discussion

S. vulgaris was detected in three of 278 investigated horses (1.1%) by larval culture and in 10 of 501 investigated horses (1.9%) by real-time PCR. This low occurrence of *S. vulgaris* is in line with previously conducted studies investigating faecal samples from German horses obtained by larval culture with a prevalence ranging from 0.2 to 1.3% [10, 19–21]. A comparable prevalence has also been reported in Switzerland [3].

In contrast, studies from other countries revealed a higher prevalence for *S. vulgaris* both by larval culture and by PCR. In a study from Denmark, Nielsen et al. [36] screened 663 horses from 42 farms via larval culture and real-time PCR with a result of 113 *S. vulgaris*-positive horses (17.5%). In an investigation of horses from Poland by necropsy, nematodes were isolated from the intestine and differentiated revealing a prevalence of 22.8% (165/725) for *S. vulgaris* [43]. An even higher prevalence of 41.3% (19/46) was detected in horses from Sardinia, Italy, via larval culture [44]. In comparison with data from Germany, Bracken et al. [45] found a higher prevalence of infection with *S. vulgaris* in Danish horses not only on the horse level with 13.6% (45/331) but also on the farm level with 72.2% (13/18). A comparable result on the farm level was reported by Nielsen et al. with a prevalence of 64.3% (27/42) determined by larval culture [18].

In the present study, *S. vulgaris* was detected in ten out of 91 investigated farms (10.9%). A low occurrence of *S. vulgaris* in farms from Germany was further demonstrated by the fact that both the real time PCR and the larval culture revealed a single *S. vulgaris*-positive horse per farm, solely. These findings are in agreement with another recent German prevalence survey, which reported a farm level of 1.04% (2/192) and detected only a single *S. vulgaris* positive horse per farm using larval culture [21].

The intensive anthelmintic treatment regime of the last decades as well as the long prepatent period of six to 7 months might be responsible for the current low occurrence of *S. vulgaris* in Germany [3, 46, 47]. Based on these arguments, a low occurrence for *S. vulgaris* under 5% was reported by Hertzberg et al. [3] for horses living in Switzerland. Despite the intensive anthelmintic treatment regime of the last decades, the occurrence of *S. vulgaris* still persists on a low level. A possible reason for this persistence could be a non-complete larvicidal efficacy of ivermectin, as suggested by Nielsen et al. due to his study results in 2014 [48].

Different surveys have pointed out that the number of Cyathostominae larvae in larval cultures is higher compared to the number of *S. vulgaris* larvae [21, 23, 29, 49]. For example, Ogbourne and Duncan [23] refer to a wide variety of Strongyle species in equine faecal samples comprising larvae of *S. vulgaris* in less than 10%. Bellaw and Nielsen [29] reported that approximately 1.0% of detected larvae were L3 of *S. vulgaris* (1486 *S. vulgaris* vs. 142,725 Cyathostominae larvae). In the present study, 0.16% of all larvae in 13 *S. vulgaris*-positive samples were L3 of *S. vulgaris* (11 *S. vulgaris* larvae vs. 6706 Cyathostominae larvae). Both the marginal number of counted *S. vulgaris*-larvae in the larval culture and the finding of only one *S. vulgaris*-positive horse per farm might be explained by a low infection rate and by a low shedding of *S. vulgaris* eggs of infected horses.

Accordingly, false negative results might occur in samples comprising a low number of *S.vulgaris* eggs. Thus, the dependence on the presence of eggs in the investigated faecal sample is the major disadvantage of the diagnosis of *S. vulgaris* with coprological methods like real-time PCR and larval culture [31, 32, 50]. Additionally, the relatively long prepatent period as well as the dependence of the development of the infectious L3 on environmental conditions is responsible for the seasonality of *S. vulgaris* which may lead to false negative results at certain times of the year [23, 51]. Various studies were able to proof a seasonal fluctuation of egg shedding of *S. vulgaris*, with a peak in summer and a depression in winter [52–55]. Yet, the study at hand was not able to confirm such an influence of seasonality on the detection of *S. vulgaris*-eggs due to a non-seasonal collection of samples and a low number of positive samples. However, the present study revealed a significantly improved detection rate of *S. vulgaris* by real-time PCR in comparison to the current standard larval culture method.

A molecular method for the detection of *Strongylus* spp. was first described by Campbell et al., in 1995. The specificity of the conventional PCR investigating the ITS-2 gene of *S. vulgaris* has already been analysed by detecting interspecific variations in the sequence of the ITS-2 gene between *S. vulgaris*, *S. edentatus* and *S. equinus* via conventional PCR and subsequent sequencing [33]. The specificity of the real-time PCR was confirmed by Nielsen et al. [30] via cross reaction testing between *S. vulgaris*, *S. edentatus*, *S. equinus* and a mixture of different Cyathostominae larvae. Therefore, a possible cross reaction with DNA of other equine strongylid nematodes was not expected. Besides the specificity, the sensitivity of the real-time PCR has also been analysed by Nielsen et al. resulting in a detection limit of a 0.5 strongyle egg-equivalent [30].

Due to the developmental period of up to 14 days for strongyle larvae, the larval culture is a time-consuming [29] though cost-efficient method for the detection of *S. vulgaris*. A sensitivity of 73% and a specificity of 84% determined by larval culture compared to necropsy data have been reported by Nielsen et al. [28]. However, this method is also dependent on the presence of *S. vulgaris* eggs in the investigated samples just like the real-time PCR. Furthermore, the larval culture has the additional disadvantage that false negative results might occur due to an inhibited development of L3 which might be caused by fluctuations of temperature, humidity, fungal growth and contamination with free-living nematodes. In the present study, the successful cultivation of Cyathostominae larvae was verified by the positive correlation between the "number of Cyathostominae larvae" and the "FEC" of the faecal sample. Cyathostominae larvae were successfully cultivated in 98% of the strongylid egg-positive samples. A possible explanation for an inhibited development of larvae might be an accidental partial freezing of samples during transportation which was reported for 12 FEC-positive samples which revealed only a limited number of larvae in the larval culture. The negative impact of low temperatures on the development of strongyle larvae has already been descripted by Ogbourne and Duncan [23], Hasslinger [56] and Enigk [51]. Furthermore, steps of the procedure following cultivation such as discharging of supernatant, purification, sedimentation and pipetting might also lead to a loss of larvae. The morphological differentiation of larvae via microscopic examination needs to be conducted by qualified and experienced personnel, since the detection of *S. vulgaris* larvae is difficult and time-consuming [28]. False negative results due to the latter aspects might easily occur especially since *S. vulgaris*-positive samples often comprise only a limited number of *S. vulgaris* larvae. The usage of an aliquot procedure with the design to investigate 100–200 larvae in order to save time might also contribute to false negative results.

With the intent to avoid this error, all larvae per 10 g/faeces were identified in the present study.

In order to maintain or even reduce the low occurrence of *S. vulgaris* in Germany and to prevent the introduction of *S. vulgaris* within and among farms, an appropriate optimisation of the diagnostic and management procedures is crucial [3].

Application of real-time PCR for detection of *S. vulgaris* as a routine method is possible in any laboratory with appropriate equipment for the real-time PCR-procedure [45]. Strongyle eggs recovered by the standard combined sedimentation/flotation method can directly be used for the DNA-extraction and subsequent real-time PCR. In the present study, a FEC of 20 EPG was sufficient for the detection of *S. vulgaris* by real-time PCR.

Since the follow-up examinations in this study revealed that not every single examination reproduce an initially *S. vulgaris*-positive result, the investigation of composite faecal samples per horse by real-time PCR for an adequate detection of *S. vulgaris* might be useful in the daily routine diagnostics. Moreover, all horses of a herd or at least "high-risk patients" like newcomers and horses with unknown deworming history should be examined individually [3]. Preliminary to an integration into a new farm, thorough examination for *S. vulgaris* is also essential for horses originating from countries with a high prevalence for *S. vulgaris* like Denmark, a part of Italy and Poland [36, 43, 44].

Conclusion

According to the results of the present study, the detection of patent infections with *S. vulgaris* via real-time PCR reveals a significantly improved detection rate improvement compared to the current standard method based on larval culture and subsequent morphological differentiation. Thus, the real-time PCR might be a reliable option for the detection of *S. vulgaris* in equine faecal samples in the routine diagnostic and in the integrated equine parasite control.

Abbreviations

C_t: Cycle threshold; DNA: Deoxyribonucleic acid; ELISA: Enzyme linked Immunosorbent assay; EPG: Egg per gram faeces; FEC: Faecal egg count; ITS: Internal transcribed spacer; L/10 g: Larvae per 10 g faeces; L1: Larvae-stage 1; L2: Larvae-stage 2; L3: Larvae-stage 3; PBS: Phosphate buffered saline; PCR: Polymerase chain reaction; SAT: Selective anthelmintic therapy; $ZnSO_4$: Zinc sulphate solution

Acknowledgements

The authors would like to thank all involved equine veterinarians and horse owners for supporting the study. Also, we highly appreciate the reliable assistance of Elisabeth Kiess, Kathrin Simon, Tim Tiedemann and Andrea Mihalkov in laboratory work.

Fundings

Funding for this work was provided by the diagnostic laboratory of the Institute of Comparative Tropical Medicine and Parasitology.

Authors' contributions

AK carried out the study, was responsible for the acquisition of samples, performed coproscopic and statistical analyses and drafted the manuscript. KP designed the study including conception and constitution. MN contributed significantly to the development of the manuscript and assisted with the interpretation of data. CS supported the realization of the PCR and the draft of the manuscript. HF performed the statistical analyses. MS conceived the study, participated in its design, coordination and analysis and contributed to drafting the manuscript. All authors have read and approved the final manuscript.

Competing interests

The authors declare that they have no competing interests.

Consent for publication

Not applicable.

Author details

[1]Comparative Tropical Medicine and Parasitology, Veterinary Faculty, Ludwig-Maximilians-University, Munich, Germany. [2]Present address: Parasite Consulting GmbH, Wendschatzstrasse 8, CH-3006 Berne, Switzerland. [3]Department of Veterinary Science, M.H. Gluck Equine Research Center, University of Kentucky, Lexington, KY, USA. [4]Present address: National Centre of Vector Entomology, Institute for Parasitology, Vetsuisse Faculty, University of Zurich, CH-8006 Zurich, Switzerland. [5]Department of Statistics, Ludwig-Maximilians-University, Munich, Germany. [6]Present address: ParaDocs Laboratory, Ismaning, Germany.

References

1. Kaplan RM. Drug resistance in nematodes of veterinary importance: a status report. Trends Parasitol. 2004;20(10):477–81.
2. Wolf D, Hermosilla C, Taubert A. *Oxyuris equi*: Lack of efficacy in treatment with macrocyclic lactones. Vet Parasitol. 2014;201(1–2):163–8.
3. Hertzberg H, Schwarzwald CC, Grimm F, Frey CF, Gottstein B, Gerber V. Helminthenmanagement beim adulten Pferd: Notwendigkeit einer Neuorientierung. Schweiz Arch Tierheilk. 2014;156(2):61–70.
4. Peregrine AS, Molento MB, Kaplan RM, Nielsen MK. Anthelmintic resistance in important parasites of horses: Does it really matter? Vet Parasitol. 2014; 201(1–2):1–8.
5. Nielsen MK, Reinemeyer CR, Donecker JM, Leathwick DM, Marchiondo AA, Kaplan RM. Anthelmintic resistance in equine parasites—Current evidence and knowledge gaps. Vet Parasitol. 2014;204(1–2):55–63.
6. Tolliver SC, Lyons ET, Drudge JH. Prevalence of internal parasites in horses in critical tests of activity of parasiticides over a 28-year period (1956-1983) in Kentucky. Vet Parasitol. 1987;23(3-4):273–84.
7. Lloyd S, Smith J, Connan RM, Hatcher MA, Hedges TR, Humphrey DJ, Jones AC. Parasite control methods used by horse owners: factors predisposing to the development of anthelmintic resistance in nematodes. Vet Rec. 2000; 146(17):487–92.
8. O'Meara B, Mulcahy G. A survey of helminth control practices in equine establishments in Ireland. Vet Parasitol. 2002;109(1–2):101–10.
9. Matthee S, Dreyer FH, Hoffmann WA, Niekerk FE. An introductory survey of helminth control practices in south africa and anthelmintic resistance on Thoroughbred stud farms in the Western Cape Province. J S Afr Vet Assoc. 2002;73(4):195–200.
10. Fritzen BM. Untersuchungen zum Vorkommen von Anthelmintika-Resistenz in nordrhein-westfälischen Pferdebeständen. PhD. Hannover: Tierärztliche Hochschule Hannover; 2005.
11. Hinney B, Wirtherle NC, Kyule M, Miethe N, Zessin K-H, Clausen P-H. A questionnaire survey on helminth control on horse farms in Brandenburg, Germany and the assessment of risks caused by different kinds of management. Parasitol Res. 2011;109(6):1625–35.
12. Drudge JH, Lyons ET. Control of internal parasites of the horse. J Am Vet Med Assoc. 1966;148(4):378–83.
13. Nielsen MK, Pfister K, von Samson-Himmelstjerna G. Selective therapy in equine parasite control—Application and limitations. Vet Parasitol. 2014; 202(3–4):95–103.
14. Menzel M, Becher AM, Greite L, Pfister K. Die Einführung der Methode der Selektiven Entwurmung in einem Pferdebestand. Tierärztliche Umschau. 2013;68:116–20.
15. Becher AM, Mahling M, Nielsen MK, Pfister K. Selective anthelmintic therapy of horses in the Federal states of Bavaria (Germany) and Salzburg (Austria): An investigation into strongyle egg shedding consistency. Vet Parasitol. 2010;171(1–2):116–22.
16. Gomez HH, Georgi JR. Equine helminth infections: Control by selective chemotherapy. Equine Vet J. 1991;23(3):198–200.
17. Pfister K, Beelitz P, Hamel D. Parasitologische Diagnostik. In: Moritz A, editor. Klinische Labordiagnostik in der Tiermedizin. 7th ed. Stuttgart: Schattauer; 2013. p. 628–99.
18. Nielsen MK, Vidyashankar AN, Olsen SN, Monrad J, Thamsborg SM. *Strongylus vulgaris* associated with usage of selective therapy on Danish horse farms—Is it reemerging? Vet Parasitol. 2012;189(2–4):260–6.
19. Hinney B, Wirtherle N-C, Kyule M, Miethe N, Zessin K-H, Clausen P-H. Prevalence of helminths in horses in the state of Brandenburg, Germany. Parasitol Res. 2011;108(5):1083–91.

20. Greite L. Untersuchungen zur Verbreitung von *Strongylus vulgaris* im Rahmen der selektiven Entwurmung bei Pferden in Süddeutschland. PhD. München: Ludwig-Maximilians-Universität; 2013.
21. Schneider S, Pfister K, Becher A, Scheuerle MC. Strongyle infections and parasitic control strategies in German horses - a risk assessment. BMC Vet Res. 2014;10(1):262–70.
22. McCraw BM, Slocombe JOD. *Strongylus vulgaris* in the horse: A review. Can Vet J. 1976;17(6):150–5.
23. Ogbourne CP, Duncan JL. *Strongylus vulgaris* in the horse: Its biology and veterinary importance, 2 edn. Farnham Royal: Commonwealth Institute of Parasitology; 1985.
24. Nielsen MK, von Samson-Himmelstjerna G, Pfister K, Reinemeyer CR, Molento MB, Peregrine AS, Hodgkinson JE, Jacobsen S, Kaplan RM, Matthews JB. The appropriate antiparasitic treatment: Coping with emerging threats from old adversaries. Equine Vet J. 2016;48(3):374–5.
25. Russell AF. The development of helminthiasis in thoroughbred foals. J Comp Pathol Ther. 1948;58(2):107–27.
26. Barth D. Parasitologische Diagnostik Teil I - Koprologische Untersuchungen. In: Edited by. Therapogen Praxisdienst. München: Therapogen-Werk, Zweigniederlassung der Sharp & Dohme GmbH; 1967. 11. Vol 2, edn.
27. van Wyk JA, van Wyk L. Freezing of sheep faeces invalidates *Haemonchus contortus* faecal egg counts by the McMaster technique. Onderstepoort J Vet Res. 2002;69:299–304.
28. Nielsen MK, Baptiste KE, Tolliver SC, Collins SS, Lyons ET. Analysis of multiyear studies in horses in Kentucky to ascertain whether counts of eggs and larvae per gram of feces are reliable indicators of numbers of strongyles and ascarids present. Vet Parasitol. 2010;174(1–2):77–84.
29. Bellaw JL, Nielsen MK. Evaluation of Baermann apparatus sedimentation time on recovery of *Strongylus vulgaris* and *S. edentatus* third stage larvae from equine coprocultures. Vet Parasitol. 2015;211(1–2):99–101.
30. Nielsen MK, Peterson DS, Monrad J, Thamsborg SM, Olsen SN, Kaplan RM. Detection and semi-quantification of *Strongylus vulgaris* DNA in equine faeces by real-time quantitative PCR. Int J Parasitol. 2008;38(3–4):443–53.
31. Andersen UV, Howe DK, Olsen SN, Nielsen MK. Recent advances in diagnosing pathogenic equine gastrointestinal helminths: The challenge of prepatent detection. Vet Parasitol. 2013;192(1–3):1–9.
32. Nielsen MK, Vidyashankar AN, Gravatte HS, Bellaw J, Lyons ET, Andersen UV. Development of *Strongylus vulgaris*-specific serum antibodies in naturally infected foals. Vet Parasitol. 2014;200(3–4):265–70.
33. Campbell AJD, Gasser RB, Chilton NB. Differences in a ribosomal DNA sequence of *Strongylus* species allows identification of single eggs. Int J Parasitol. 1995;25(3):359–65.
34. Hung G-C, Jacobs DE, Krecek RC, Gasser RB, Chilton NB. *Strongylus asini* (Nematoda, Strongyloidea): Genetic relationships with other strongylus species determined by ribosomal DNA. Int J Parasitol. 1996;26(12):1407–11.
35. Hung G-C, Chilton NB, Beveridge I, Gasser RB. A molecular systematic framework for equine strongyles based on ribosomal DNA sequence data1. Int J Parasitol. 2000;30(1):95–103.
36. Nielsen MK, Olsen SN, Lyons ET, Monrad J, Thamsborg SM. Real-time PCR evaluation of *Strongylus vulgaris* in horses on farms in Denmark and Central Kentucky. Vet Parasitol. 2012;190(3–4):461–6.
37. Andersen UV, Haakansson IT, Roust T, Rhod M, Baptiste KE, Nielsen MK. Developmental stage of strongyle eggs affects the outcome variations of real-time PCR analysis. Vet Parasitol. 2013;191(1–2):191–6.
38. FN DRV. FN - Zahlen, Daten, Fakten. Warendorf, Germany: Deutsche Reiterliche Vereinigung Bundesverband für Pferdesport und Pferdezucht; 2015.
39. Conraths FJ, Fröhlich A, Ziller M. Epidemiologische Untersuchungen in Tierpopulationen - Ein Leitfaden zur Bestimmung von Stichprobenumfängen. Wusterhausen und Greifswald-Insel Riems: Friedrich-Loeffler-Institut; 2011.
40. Roberts FHS, O'Sullivan PJ. Methods for egg counts and larval cultures for strongyles infesting the gastro-intestinal tract of cattle. Austr J Agric Res. 1950;1(1):99–102.
41. Bürger HJ, Stoye M. Parasitologische Diagnostik Teil II - Eizählung und Larvendifferenzierung. In: Edited by. Therapogen Praxisdienst. München: Therapogen-Werk, Zweigniederlassung der Sharp & Dohme GmbH; 1968. 24.Vol 3, edn.
42. Boch J, Supperer R. Veterinärmedizinische Parasitologie. vol. 6. Farnham Royal: Thomas Schnieder; 2006.
43. Studzińska MB, Tomczuk K, Demkowska-Kutrzepa M, Szczepaniak K. The Strongylidae belonging to *Strongylus* genus in horses from southeastern Poland. Parasitol Res. 2012;111(4):1417–21.
44. Pilo C, Altea A, Pirino S, Nicolussi P, Varcasia A, Genchi M, Scala A. *Strongylus vulgaris* (Looss, 1900) in horses in Italy: Is it still a problem? Vet Parasitol. 2012;184(2-4):161–7.
45. Bracken MK, Wøhlk CBM, Petersen SL, Nielsen MK. Evaluation of conventional PCR for detection of *Strongylus vulgaris* on horse farms. Vet Parasitol. 2012;184(2–4):387–91.
46. Herd RP. The changing world of worms: The rise of the Cyathostomes and the decline of *Strongylus vulgaris*. Comp Cont Educ Pract. 1990;12(5):732–6.
47. Kaplan RM. Anthelmintic resistance in nematodes of horses. Vet Res. 2002; 33:491–507.
48. Nielsen MK, Scare J, Gravatte HS, Bellaw JL, Prado JC, Reinemeyer CR. Changes in serum *Strongylus vulgaris*-specific antibody concentrations in response to anthelmintic treatment of experimentally-infected foals. Front Vet Sci. 2015;2.
49. Kuzmina TA. Contamination of the environment by strongylid (Nematoda: Strongylidae) infective larvae at horse farms of various types in Ukraine. Parasitol Res. 2012;110(5):1165–74.
50. Andersen UV, Howe DK, Dangoudoubiyam S, Toft N, Reinemeyer CR, Lyons ET, Olsen SN, Monrad J, Nejsum P, Nielsen MK. SvSXP: A *Strongylus vulgaris* antigen with potential for prepatent diagnosis. Parasit Vectors. 2013;6(1):84.
51. Enigk K. Die Widerstandsfähigkeit der Entwicklungsstadien der Strongyliden außerhalb des Wirtstieres. Arch f Tierheilkunde. 1934;67:363–76.
52. Poynter D. Seasonal fluctuations in the number of strongyle eggs passed by horses. Vet Rec. 1954;66:74–8.
53. Duncan JL. Field studies on the epidemiology of mixed strongyle infection in the horse. Vet Rec. 1974;94:337–45.
54. Ogbourne CP. Studies on the epidemiology of *Strongylus vulgaris* infection of the horse. Int J Parasitol. 1975;5(4):423–6.
55. Ogbourne CP. Variations in the fecundity of strongylid worms of the horse. Parasitol. 1971;63:289–98.
56. Hasslinger MA. Untersuchungen über den Einfluß verschiedener Temperaturen auf Eier und Larven von Pferdestrongyliden unter Laboratoriumsbedingungen sowie das Verhalten dieser exogenen Stadien auf der Weide. Berl Munch Tierarztl Wochenschr. 1981;94(1):1–5.

Comparison of image quality and in vivo appearance of the normal equine nasal cavities and paranasal sinuses in computed tomography and high field (3.0 T) magnetic resonance imaging

Joachim Kaminsky[1*], Astrid Bienert-Zeit[1], Maren Hellige[1], Karl Rohn[2] and Bernhard Ohnesorge[1]

Abstract

Background: Computed tomography (CT) is a well-established imaging technique in the diagnostics of equine sinunasal disease. High-field magnetic resonance imaging (MRI) is becoming more readily available in equine veterinary medicine. MRI is appreciated for its superior ability to depict soft tissues with high contrast. To compare the established technique of CT in the depiction of the equine nasal cavities, paranasal sinuses and adjoining anatomical structures to 3 Tesla MRI the nasal cavities and paranasal sinuses of 13 horses were examined using CT and 3 Tesla MRI.

Results: Comparison of CT and MRI images of the paranasal sinuses, nasal cavities and adjoining anatomical structures of 13 healthy horses showed that the inter-rater agreement for the CT examinations was higher than the inter-rater agreement for the MRI examinations. CT images proved to be significantly higher rated for the depiction of cortical bone, while MR images were higher rated for the appearance of soft tissues. For the distinction between different tissues or anatomical structures the MR images were significantly higher rated and especially T2-weighted sequences allowed for a good distinction between delicate structures. None of the MRI sequences produced an exact depiction of the lumen of the nasomaxillary aperture while the CT with a bone window allowed for a satisfying visualization.

Conclusion: The CT is an imaging modality that produces high quality images within a short time when examining equine nasal cavities and paranasal sinuses. The strength of CT lies in the high quality depiction of large and delicate structures with high radiodensity. High field MRI with a field strength of 3 Tesla produces images of high quality that allow for the distinction of delicate soft tissue structures but requires long examination times. The high field strength of 3 Tesla magnetic imaging introduces new possibilities in tomographic soft tissue imaging of the equine head but cannot match up with the CT in terms of visualization of bone and total examination duration. Therefore, clinicians should consider the exact imaging needs in clinical cases to decide whether a single examination or a combination of both imaging techniques may promise the greatest benefit for the patient.

Keywords: 3 Tesla, High field MRI, CT, Equine paranasal sinuses

* Correspondence: joachim-kaminsky@gmx.de
[1]Clinic for Horses, University of Veterinary Medicine Hannover, Foundation, Bünteweg 9, Hannover D-30559, Germany
Full list of author information is available at the end of the article

Background

Clinical diagnostics of the equine paranasal sinuses during pathology often require the use of an imaging technique to determine the cause and extent of the disease [1]. X-ray has proven to be a helpful tool in the diagnostics of equine paranasal disorders [1]. Nevertheless, superimposition of structures on x-ray images makes interpretation difficult and often inaccurate [2–6]. Tomographic imaging techniques abolish the disadvantage of superimposition and particularly CT has become a well-established imaging technique in the diagnostics of equine sinunasal disease due to its good bone to soft tissue contrast [7–13].

MRI is appreciated for its superior ability to depict soft tissues with high contrast [7]. High-field MRI of 1.5 Tesla is becoming more readily available in equine veterinary medicine [14–17] and even 3 Tesla MRI is starting to be used in equine veterinary diagnostics [18–20]. The higher field strength of 3 Tesla has only recently been introduced in equine veterinary medicine and opens up new possibilities but also bears disadvantages and challenges when compared to MRI of lower field strengths [18–20].

The aim of this study was to compare the established technique of CT in the depiction of the equine nasal cavities, paranasal sinuses and adjoining anatomical structures to 3 Tesla MRI in healthy horses. This work provides suggestions for which anatomical structures the MRI or rather the CT should be the technique of choice, and elucidates whether or not the high field strength of 3 Tesla MRI has benefits when compared to CT.

Methods

Study population

Thirteen healthy horses of different breeds (7 Warmbloods, 2 Arabian thoroughbreds, 3 Standardbreds, 1 Andalusian) were used to acquire the CT and 3 Tesla MR images. The group was composed of 8 mares, 4 geldings and 1 stallion. The age ranged from 4 to 20 years (mean ± SD, 11.6 ± 6.5). All horses showed no signs of disease related to the nasal cavities or the paranasal sinuses. Image acquisition was performed with horses under general anaesthesia. This study was approved by the Lower Saxony State Office for Consumer Protection and Food Safety's ethic committee. File reference: 33.9-42502-04-11/0592.

Computed tomography

For the CT examination the horses were positioned in right lateral recumbency on a stationary examination table. CT was performed using a Brilliance™ CT – Big Bore Oncology Scanner (Philips Medical Systems, Best, The Netherlands).

The entire head of each horse was scanned with an axial scan-mode with 140 kV, 500 mAs, a slice thickness of 1.5 mm and an image matrix of 1024. Using multiplanar reconstruction, two transverse image sequences, one with a bone window (CTbw; WL: 300, WW: 2800), and one with a soft tissue window (CTstw; WL: 50, WW: 500) were generated. Both sequences were oriented perpendicular to the hard palate.

Magnetic resonance imaging

Subsequently to the CT examination the horses were positioned in dorsal recumbency on a non-stationary table and cushioned by an inflatable mattress for the duration of the examination. Surface coils (Philips SENSE™ Flex M® and Philips SENSE™ Flex L®; Philips Medical Systems, Best, The Netherlands) were positioned around the sinunasal region and the head was fixed in position with a vacuum cushion (Fig. 1).

Magnetic resonance imaging was performed using a Philips Achieva™ 3.0 T X-Series® (Philips Medical Systems, Best, The Netherlands). Sequences were obtained in transverse planes orientated perpendicular to the hard palate and included T2 weighted turbo spin echo (T2w) (TR: 4500 ms, TE: 80 ms, NSA: 2, ST: 4.0 mm; ISS: 1.0 mm, matrix: 1024 x 1024) and proton density weighted turbo spin echo (PDw) (TR: 8300 ms, TE: 30 ms,

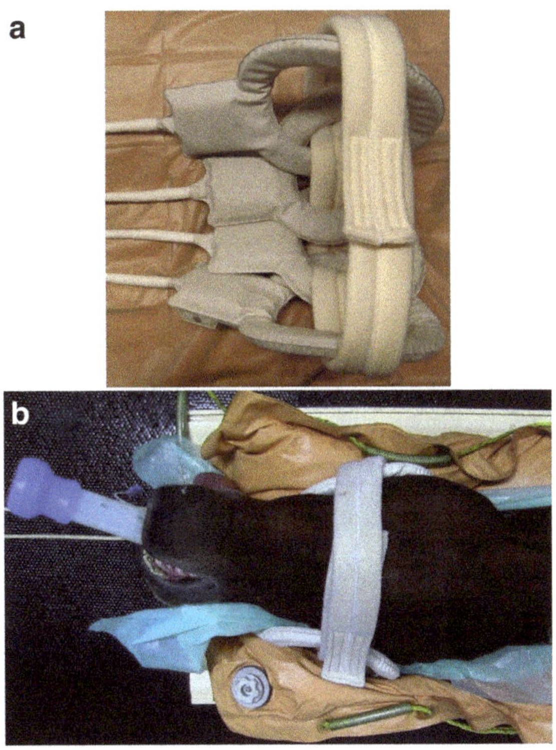

Fig. 1 Head positioning for MRI examination. **a** Surface coils arranged for positioning (Philips SENSE™ Flex M® and Philips SENSE™ Flex L®; Philips Medical Systems, Best, The Netherlands); **b** Head positioned on the examination table with surface coils placed around the ROI

NSA: 2, ST: 4.0 mm; ISS: 0.4 mm, matrix: 960 x 960) sequences. T1 weighted gradient echo (T1w) images (TR: 8 ms, TE: 4 ms, NSA: 2, ST: 0.9 mm; ISS: 0 mm, matrix: 720 x 720) were obtained as a 3-dimensional (3D) -dataset with isotropic voxels so that transverse planes could be reproduced using multiplanar reconstruction. MRI settings were chosen to gain maximum image quality within a reasonable time frame. In all horses and sequences it was attempted to picture the horse's head from a transverse line through the eyes to a plane right rostral to the first molar tooth.

The durations of the CT and MRI examinations, including the positioning of the horse on the examination table, as well as the time required for each MRI sequence were recorded. Additionally, the entire time of general anaesthesia was recorded.

Image interpretation

After acquisition of CT and MR images slices from different planes of the head were chosen. Five slices from each imaging technique and each horse were selected. To ensure comparability of the selected slices, predefined anatomical landmarks were used. The anatomical landmarks are shown in Fig. 2 (a-e).

The chosen images were evaluated by three independent veterinarians with experience in CT and MRI image evaluation. The evaluation was performed under the same conditions for all three observers. For image interpretation three DICOM certified TFT displays (EIZO FlexScan MX190S; EIZO Europe GmbH, Mönchengladbach, Germany and the imaging software eFilm Workstation® 1.5.3 (eFilm Medical, Toronto, Canada) were used. Observers were allowed to alter the window level and window width in all MRI sequences, but were not allowed to change level presets in the CT images so that image characteristics remained unaltered. Images were evaluated for general image quality characteristics (Table 1).

For each plane a set of anatomical structures, tissues or tissue borders was defined. In each of the chosen anatomical structures, tissues or tissue borders scores were given for contour distinction, tissue distinction or detail

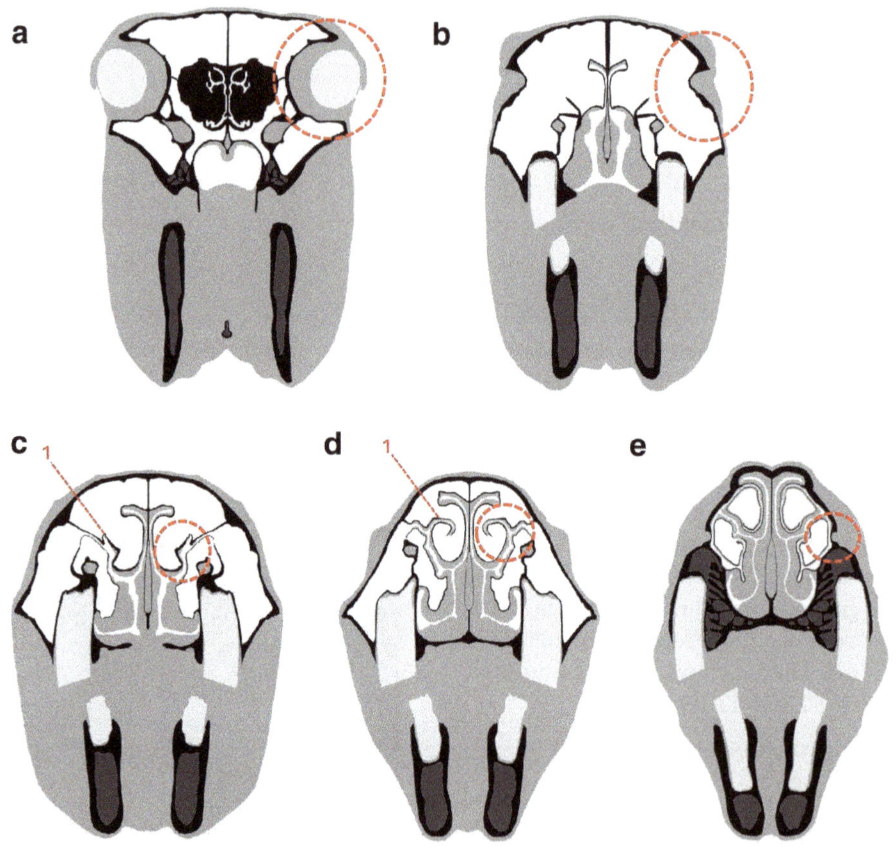

Fig. 2 Schematic images of chosen planes. **a** to **e** Dashed circle indicates the point of orientation. **a** Plane 1 positioned on a plane through the centre of the ocular bulbs; **b** Plane 2 positioned immediately rostral to the eyes; **c** Plane 3 positioned on the level of the nasomaxillary aperture, a 'hook' (red 1) [38] protruding dorsally from the spiral lamella of the dorsal concha in all horses was used as an anatomical landmark; **d** Plane 4 positioned on the level of the nasomaxillary aperture, where the 'hook' (red 1) [38] starts to extend mediodorsally; **e** Plane 5 positioned on the level of the infraorbital foramen

Table 1 Scoring system for the image quality parameters

Score	Image noise	Image sharpness	Image contrast
0-poor quality	very high image noise level	low image sharpness	low image contrast
1-moderate quality	high image noise level	moderate image sharpness	moderate image contrast
2-satisfying quality	moderate image noise level	satisfying image sharpness	satisfying image contrast
3-good quality	no image noise	high image sharpness	high image contrast

resolution (Table 2). All five imaging techniques were compared and evaluated in a graded score system between 0 and 3 with 0 representing poor and 3 representing excellent performance. Observers were provided with a scoring guideline stating the anatomical structure and the property that was to be rated for this particular anatomical structure.

For simplification the different evaluated anatomical parameters were pooled for similar characteristics and sorted into four groups as shown in Table 3.

Statistical analysis

Data collection was performed in spreadsheets using the software Office Excel 2003® (Microsoft Corporation, Redmond, WA, USA). For statistical analysis the software SAS® 9.3 (SAS Institute, Cary, NC, USA) was used.

Descriptive statistics were performed. Differences in inter-rater agreement were analysed using the McNemar-Bowker test of internal symmetry and calculating the Cohen's kappa coefficient.

To determine the variations between the five different imaging techniques a permutation test with 1000 simulations was performed for each evaluated parameter and between each technique. A P value <0.05 was considered statistically significant.

Results

A total of 13 horses were examined. In all horses the combined examination of CT and MRI was performed without complications. In the CT examination the entire head of each horse was scanned. The field of view in the MRI examination from caudal to rostral for all horses ranged from 217 to 300 mm (mean ± SD; 245.1 mm ± 27.0 mm). CT examinations took 10 to 18 min (mean 15 min) including positioning of the horse. MRI examinations took 61 to 71 min (mean 67 min) including positioning of the horse (Table 4).

Overall 325 images were evaluated and 702 parameters were graded by each evaluator.

Soft tissues

T2w was significantly superior to all other imaging techniques in displaying the venous plexus of the nasal mucosa (Fig. 3), whereas the palatine artery and the soft tissue inside the infraorbital canal were displayed with no significant difference in quality by T2w (Fig. 3) and PDw. T1w was the lowest rated MRI sequence in all three compared locations. When comparing CT to MR imaging, CTbw and CTstw were significantly lower rated than all MRI sequences, as no significant difference between the bone and soft tissue window was present (Table 5).

Contour distinction of cortical bone

When comparing the two CT window presets no significant difference in contour distinction was present (Fig. 4). Nevertheless, both CT window presets were significantly higher rated for the contour distinction of the vomer and the infraorbital canal in comparison to T2w and PDw, whereas no significant difference was present in contour distinction of the nasal and frontal bone plate. T2w and PDw showed no significant differences in all three evaluated locations. T1w was rated significantly lower than all imaging techniques in all evaluated locations (Table 6).

Resolution of anatomical detail

The detail resolution for cancellous bone was lowest rated in CTstw and T1w. No significant difference was detected between these two techniques in terms of detail resolution. When compared amongst each other, CTbw, T2w and PDw showed no significant difference in rating of detail resolution (Table 7).

Detail resolution for the ethmoid bone was rated with no significant difference in CTbw, CTstw, T2w (Fig. 5)

Table 2 Scoring system for the anatomical parameters

Score	Contour distinction	Tissue distinction	Detail resolution
0	not distinguishable	not distinguishable from each other	no details
1	distinguishable but often blurred	difficult to distinguish from each other	low details
2	well distinguishable and seldom blurred	well distinguishable from each other	high details
3	very well distinguishable and sharply defined	very well distinguishable from each other	excellent details

Table 3 Groups of evaluated anatomical parameters

Group	Characteristics	Anatomical parameters
1	soft tissue	venous plexus of the nasal mucosa
		palatine artery
		soft tissue inside the infraorbital canal
2	contour distinction of cortical bone	frontal and nasal bone plate
		vomer
		infraorbital canal
3	anatomical detail	cancellous bone
		ethmoid bone
		lumen of the nasomaxillary aperture
4	tissue distinction	distinction between air and mucosal lining of the rostral maxillary and conchofrontal sinus
		distinction between mucosal lining and cortical bone of the rostral maxillary and conchofrontal sinus
		distinction between venous plexus lining the nasal septum and cortical bone or cartilage
		distinction between bone and cartilage of the vomer and nasal septum

and PDw. Tw1 was rated significantly lower than all other imaging techniques (Table 7).

For the lumen of the nasomaxillary aperture the CTbw was the highest rated imaging modality (Fig. 6). All other imaging techniques showed no significant difference in detail resolution (Table 7).

Tissue distinction

For the distinction between air and mucosal lining, between mucosal lining and cortical bone and between the venous plexus and cortical bone or cartilage (Fig. 7), T2w and PDw showed no significant difference. However, both techniques were significantly superior to the other imaging techniques. T1w was the lowest-rated MRI sequence for all three features. CTbw and CTstw were rated significantly lower when compared to all MRI sequences for the distinction between air and mucosal lining as well as for the distinction between mucosal lining and cortical bone. Nevertheless, no significant difference between the bone and soft tissue window was detected in all three features. Additionally, CTbw, CTstw and T1w showed no significant difference in visualizing the distinction between cortical bone or cartilage and venous plexus.

The distinction between bone and cartilage was evaluated based on the distinction between the vomer and the cartilage of the nasal septum. Here, the median for all imaging techniques lay between 1 and 3 with no statistically significant differences between the five imaging techniques (Table 8).

Table 4 Range of duration for the examination parts

Part of examination	Range of duration (min)	Mean duration (min)
total time of general anaesthesia	76–89	84
CT examination (incl. positioning)	10–18	15
T2w	20–26	24
T1w	12–14	13
PDw	19–23	21
MRI examination (incl. positioning)	61–71	67

Image quality parameters

CTbw and CTstw showed significantly less image noise than the MRI sequences. However, no significant difference between the two CT presets was present. The image quality of the T2w sequence was significantly better than that of the other two MRI sequences, which showed no significant difference when compared amongst each other (Table 9).

The CTbw showed the significantly highest image sharpness of all imaging techniques followed by the CTstw and T2w, which showed no significant difference when compared amongst each other. PDw showed significantly lower image sharpness than the previous three techniques but was nevertheless significantly better than the T1w sequence (Table 9).

The T2w MRI sequence showed the significantly highest image contrast compared to all other imaging techniques. CTbw and CTstw showed no significant difference when compared amongst each other but produced a higher image contrast than the PDw. T1w showed the significantly lowest contrast of all imaging techniques (Table 9).

Rater agreement

Calculation of inter-rater agreement and internal symmetry revealed that CT examinations showed a higher

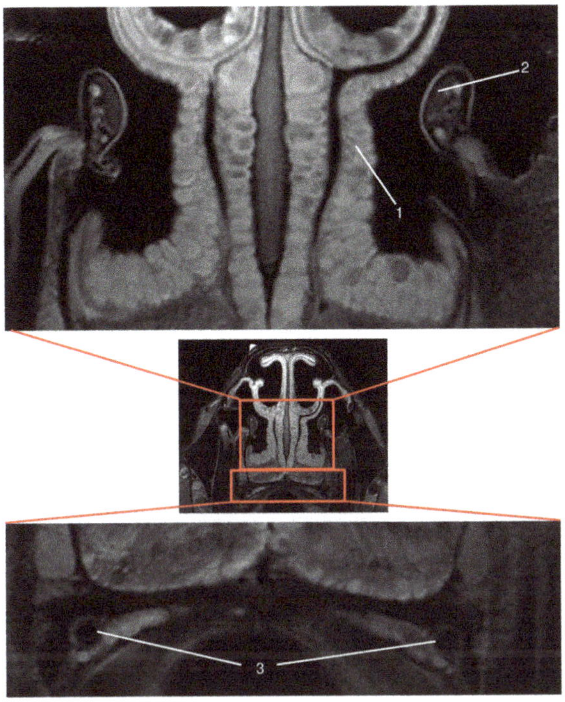

Fig. 3 T2w transversal image in plane 3. (1) venous plexus of the nasal mucosa, (2) infraorbital nerve, (3) palatine arteries

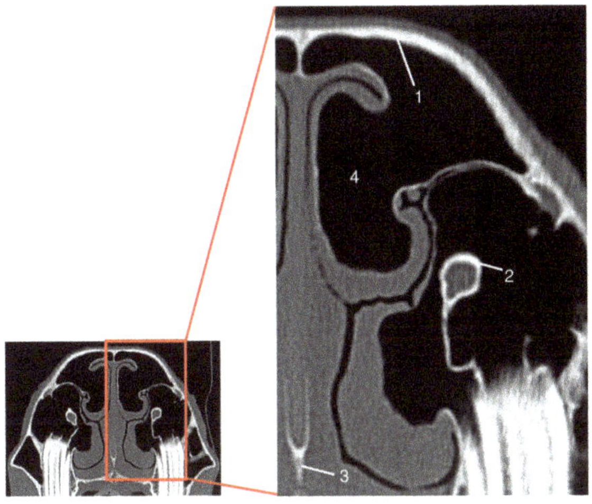

Fig. 4 CTbw transversal image in plane 3. (1) frontal bone plate, (2) infraorbital canal, (3) vomer, (4) conchofrontal sinus

inter-rater agreement than MRI examinations (Table 10) with a Kappa of 0.7019 for CTbw and 0.6353 for CTstw while T2w, T1w and PDw only showed a Kappa of 0.3339, 0.3704 and 0.333. In all comparisons of the three individual evaluators a significant internal symmetry was present. Nevertheless, these differences and tendencies were more pronounced in the MRI sequences. Rater 3 showed a tendency to evaluate the three MRI sequences with higher scores than the other two raters. Further assessment of those tendencies and differences showed that they did not affect the significance of the inter modality comparison.

Discussion

Various studies describe the appearance of the nasal cavities and the paranasal sinuses of the horse either using MRI [14, 16, 21] or CT [22–27]. However, none of these studies implement a direct comparison of CT and MRI under clinical conditions. Furthermore, all of the previous mentioned MRI studies were performed with MRI machines with magnetic field strengths lower than 3.0 T. In the present study a comparison of CT and 3.0 T MRI and their ability to depict different anatomical structures in the anatomical region of the equine nasal cavities and paranasal sinuses was conducted.

Table 5 Median of scoring results of all 3 evaluators for soft tissue appearance

Imaging technique	CTbw		CTstw		T2w		T1w		PDw		score
anatomical parameter	m	%	m	%	m	%	m	%	m	%	
venous plexus of the nasal mucosa	0	100	0	100	3	0	0	60.3	2	1.9	0
		0		0		7.7		39.1		39.7	1
		0		0		30.1		0.6		48.7	2
		0		0		62.2		0		9.6	3
palatine artery	0	100	0	100	2	0	1	3.8	2	0	0
		0		0		7.7		82.1		7.7	1
		0		0		44.9		14.1		56.4	2
		0		0		47.4		0		35.9	3
soft tissue inside the infraorbital canal	0	100	0	91.5	2	0	1	20.5	2	2.6	0
		0		8.5		11.1		65.8		18.8	1
		0		0		53.0		13.7		49.6	2
		0		0		35.9		0		29.1	3

m = median; % = distribution of awarded scores (0, 1, 2, 3) in percent

Table 6 Median of scoring results of all 3 evaluators for cortical bone appearance

Imaging technique	CTbw		CTstw		T2w		T1w		PDw		score
anatomical parameter	m	%	m	%	m	%	m	%	m	%	
frontal and nasal bone plate	3	0	3	0	2	1.9	1	14.1	2	9.0	0
		0		5.1		26.9		70.5		24.4	1
		41.0		37.2		43.6		15.4		42.3	2
		59.0		57.7		27.6		0		24.4	3
vomer	3	0	2	0	2	3.8	1	33.3	2	14.1	0
		2.6		6.4		33.3		60.3		28.2	1
		34.6		44.9		32.1		6.4		35.9	2
		62.8		48.7		30.8		0		21.8	3
infraorbital canal	3	0	3	0	2	2.6	0	74.4	2	4.3	0
		0		0.9		40.2		24.8		38.5	1
		0.9		47.0		37.6		0.9		40.2	2
		99.1		52.1		19.7		0		17.1	3

m = median; % = distribution of awarded scores (0, 1, 2, 3) in percent

The examinations for this study were performed under the same general anaesthesia to reduce the inflicted stress for the horses and lower the risks of general anaesthesia, notably the risk of a second recovery phase [28]. Furthermore, the technique of performing CT of the equine head in standing sedated horses is described [29, 30].

The time needed to scan the region of interest differed strongly between the two techniques. The CT was up to 48 min faster than the MRI. Those differences are due to the two different underlying physical principles of image generation. For the examination of clinical cases veterinarians should consider whether a combined examination of CT and MRI is necessary or if one of the two imaging modalities is more adequate and sufficient for the suspected diagnosis.

In the CT it was possible to scan the entire head in the above-mentioned time frame. The field of view for the MRI examination ranged from 21.7 to 30 cm. The receiver coils used in this study are flexible surface coils that were placed around the structures of interest. However, they only allow a field of view of about 30 cm in length. Slight mispositioning can lead to a loss of signal at the edges of this field and thereby to a loss of quality. For the examination of clinical cases it is therefore important to determine the region of interest beforehand and position the receiver coils accordingly. In this study the caudal edge of the coil was positioned at the level of the zygomatic process.

Transverse images of the head, as acquired in the present study, allow for a good overview of anatomical structures but should not be used as the single means of

Table 7 Median of scoring results of all 3 evaluators for detail resolution

Imaging technique	CTbw		CTstw		T2w		T1w		PDw		score
anatomical parameter	m	%	m	%	m	%	m	%	m	%	
cancellous bone	3	0	1	0	3	7.7	2	0	3	0	0
		0		56.4		2.6		41.0		17.9	1
		46.2		43.6		30.8		51.3		30.8	2
		53.8		0		59.0		7.7		51.3	3
ethmoid bone	3	0	1	2.6	2	0	1	5.1	2	15.4	0
		2.6		48.7		7.7		92.3		25.6	1
		43.6		35.9		53.8		2.6		53.8	2
		53.8		12.8		38.5		0		5.1	3
lumen of the nasomaxillary aperture	2	3.8	1	50.0	1	33.3	0	80.8	1	35.9	0
		15.4		34.6		57.7		19.2		53.8	1
		46.2		14.1		9.0		0		6.4	2
		34.6		1.3		0		0		3.8	3

m = median; % = distribution of awarded scores (0, 1, 2, 3) in percent

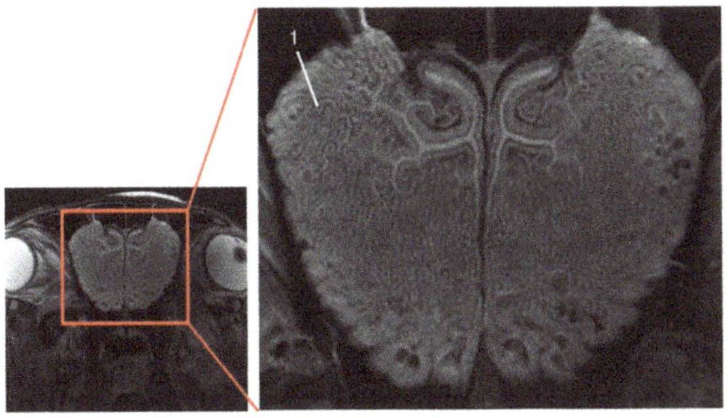

Fig. 5 T2w transversal image in plane 1; (1) ethmoid bone

orientation if pathologies are suspected. A dorsal and sagittal view of the region of interest should be acquired for a better understanding of the three-dimensional spatial relations.

The higher score results for all MRI sequences compared to CT presets in soft tissue imaging are expected results due to the physical basis of image acquisition [31]. The better soft tissue imaging properties of MRI may be of benefit in the diagnostics of tumours and tumour like lesions [17, 31–33]. GERLACH and GERHARDS (2008) described melanoma, adenocarcinoma, neuroendocrine carcinoma, sarcoma, ethmoid hematoma and paranasal cysts with the use of MRI. For detailed diagnostics of the tumour or neoplasia-like lesion, more sequences than those described in this study should be performed to evaluate the extent and quality of the tumour or tumour-like lesion [7, 34]. However, a reliable diagnosis of the quality of the mass can only be achieved by histopathological examination of the tissue.

The detail of cancellous bone was scored with no significant difference for CTbw, T2w, and PDw. In this tissue the strengths of MRI to depict soft tissues and CT to depict bone overlap since the MRI displays the fatty and fluid part of the cancellous bone and the CT displays the thin osseous trabecula.

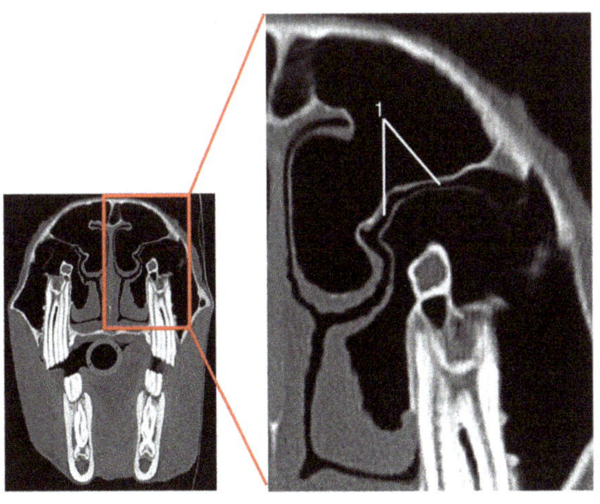

Fig. 6 CTbw transversal image in plane 3; (1) nasomaxillary aperture

CT is used in medical imaging diagnostics for its ability to produce a good delineation between soft tissue and bone [23, 31]. In this study, CT presets were scored significantly lower than T2w and PDw sequences for the delineation of mucosal lining and the frontal or nasal bone plate. The mucosal lining of the paranasal sinuses in its healthy state is thin and with low radiodensity so that conventional CT techniques are unable to depict it. MRI is capable of visualizing this soft tissue structure despite its small size and therefore is able to delineate bone and mucosa. In case of a pathologic swelling or hypertrophy of the paranasal mucosa, however, this also becomes visible in conventional CT [9].

T2w and PDw sequences also received higher score values for the depiction of the differences between the venous plexus of the nasal mucosa and the bone or cartilage of the nasal septum. These results can also be attributed to the MRI's superior ability to depict soft tissue.

In this study a resolution with a matrix of 1024 x 1024 resulted in highly detailed images in T2w MRI. In MRI, the high resolution correlates to the amount of time spent for one sequence [35]. This explains the higher acquisition times for the T2w, which was given the highest score values for image sharpness. The T1w sequence was given the lowest scores for image sharpness, which is consistent with the shortest acquisition times and the lower resolution with a matrix of 720 x 720. Furthermore, not only the resolution but also the image contrast is time-related in MRI. Any amount of time saved in a preset MRI sequence usually results in a loss of quality [35]. These properties may be of importance in clinical cases if delicate structures need to be depicted at high detail.

In this study, T2w, T1w and PDw sequences were chosen for comparison with CT images. These sequences are described to give a good anatomical overview [7, 14, 16]. For T2w and PDw sequences the higher magnetic field strength of the 3 Tesla magnet is advantageous since it allows for a higher signal to noise ratio and higher contrast [35–37]. Only through the properties of the 3.0 T magnet was high resolution imaging made possible in a reasonable timeframe for this study. The T1w sequence was performed as a gradient echo sequence a spin echo (SE) sequence to save time. Nevertheless, one downside of the 3.0 T MRI compared to magnets with lower field strengths is that in T1w SE sequences the time for acquisition is prolonged since the T1-relaxation time is extended [35–37]. T1w was the lowest-rated MRI sequence in the present study. This verifies the disadvantage of the 3.0 T MRI and

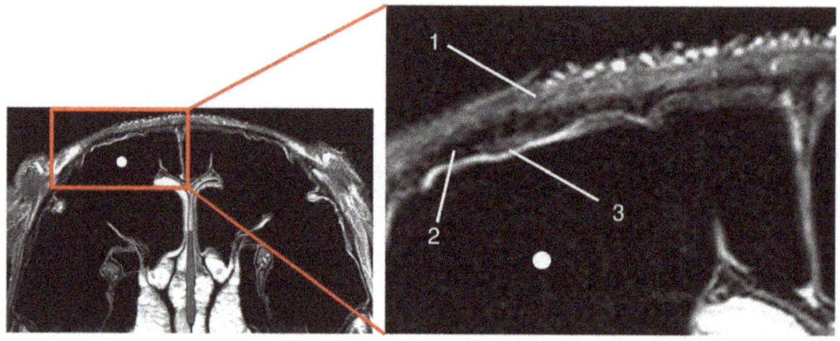

Fig. 7 T2w transversal image in plane 2; conchofrontal sinus (white spot); skin (1), signal free frontal bone plate (2), thin bright mucosa (3)

suggests that clinicians should rather acquire T2w images for orientation and first impressions of a pathological process.

The long examination times in MRI were chosen to create very high quality images to explore the possibilities of 3.0 T magnetic resonance imaging. In clinical use those high resolutions might be reduced to save time, since it is still possible to acquire images with a reasonable quality for diagnostics with shorter examination periods. Should high resolution images be required to evaluate particularly delicate structures, a reduction of the field of view might be another option to save time.

Although the T2w sequence produced images with the same resolution as the CT, the detail resolution of the nasomaxillary aperture was significantly better rated in CTbw than in the T2w. This is explicable through the different special resolutions of the two techniques. The CTbw has a slice thickness of only 1.5 mm compared to 4.0 mm of the T2w. The higher slice thickness leads to the partial volume effect which is an artefact that occurs when the properties of different tissues or structures are averaged in one voxel [35]. This artefact leads to a blurred appearance of delicate structures. If a congestion of the nasomaxillary aperture is suspected, the CTbw is the technique of choice as shown by BRINKSCHULTE (2012).

The graded scoring system applied in this study was designed to objectify the weak-and strongpoints of the resulting images from each imaging technique without focusing solely on technical data and numbers. For this reason, the scoring system relies on the assessment of the images through human validation. Consequently, the deviation in inter-rater agreement resulted from the

Table 8 Median of scoring results of all 3 evaluators for tissue distinction

Imaging technique	CTbw		CTstw		T2w		T1w		PDw		score
distinction between …	m	%	m	%	m	%	m	%	m	%	
… air and mucosal lining of the rostral maxillary and conchofrontal sinus	0	98.1	0	100	3	0	1	14.7	3	9.6	0
		1.3		0		9.6		61.5		7.1	1
		0.6		0		28.2		23.7		25.6	2
		0		0		62.2		0		57.7	3
… mucosal lining and cortical bone of the rostral maxillary and conchofrontal sinus	0	98.1	0	100	3	0.6	1	15.4	2	8.3	0
		1.9		0		10.3		70.5		10.9	1
		0		0		39.1		14.1		35.9	2
		0		0		50.0		0		44.9	3
… the venous plexus of the nasal septum and cortical bone or cartilage	2	7.1	2	1.9	3	0	2	1.3	3	0	0
		36.5		37.8		1.3		32.1		3.8	1
		45.5		48.1		9		66.7		17.9	2
		10.9		12.2		89.7		0		78.2	3
… bone and cartilage of vomer and nasal septum	2	0	2	0	3	0	1	12.8	3	2.6	0
		25.6		12.8		10.3		51.3		17.9	1
		43.6		82.1		25.6		35.9		15.4	2
		30.8		5.1		64.1		0		64.1	3

m = median; % = distribution of awarded scores (0, 1, 2, 3) in percent

Table 9 Median of scoring results of all 3 evaluators for image quality parameters

imaging technique	CTbw		CTstw		T2w		T1w		PDw		score
quality parameter	m	%	m	%	m	%	m	%	m	%	
image noise	3	0	3	0	3	0	2	0.5	2	2.6	0
		0		0		0		36.4		12.8	1
		2.6		0.5		28.7		53.8		51.3	2
		97.4		99.5		71.3		9.2		33.3	3
image sharpness	3	0	3	0	3	0.5	1	1.0	2	2.1	0
		0		4.6		5.6		81.5		10.3	1
		0		42.6		21.5		17.4		48.2	2
		100		52.8		72.3		0		39.5	3
image contrast	3	0	2	0	3	0	2	0.5	2	1.5	0
		0		3.6		2.6		318		8.7	1
		21		48.2		19		62.6		42.6	2
		79		48.2		78.5		5.1		47.2	3

m = median; % = distribution of awarded scores (0, 1, 2, 3) in percent

subjective impression that each evaluator expresses in the awarded scores. Despite these differing impressions, advantages and disadvantages of one imaging technique over the other were measurable. In contrast to the present study, a comparative study of MRI and CT of the equine fetlock joint showed a higher inter-rater agreement [20]. This study also used 3.0 T MRI. The different outcome may result from the more complex anatomy of the nasal cavities and paranasal sinuses, which render the interpretation in MRI more challenging.

Conclusion

The results of this study suggest that CT is still the imaging technique of choice if osseous structures are involved and short examination times are of importance. CT may allow for a differentiation of bone from soft tissues, but only if soft tissue structures possess a high radiodensity and size.

On the contrary, MRI depicts soft tissue structures in high detail and is able to differentiate between different soft tissue types. A benefit of the high resolution MR images is the ability to depict delicate structures such as the infraorbital canal and its contents so that previously undetected lesions may be visualized. 3.0 T MRI does not abolish the inherent problem of MRI that examinations take up a long time if more sequences and orientations are necessary to assess pathologic processes.

In conclusion, we show that both imaging techniques complement each other and that it is possible to examine horses' heads in CT and MRI under the same general anaesthesia without complications. A surgical intervention under the same general anaesthesia, however, will seldom be possible due to the prolonged time required for preparation and surgery.

Table 10 Inter-rater agreement with weighted Kappa (wk) and internal symmetry (p) ($p < 0{,}05$) for each modality

Modality		Compared raters			
		rater 1/2	rater 1/3	rater 2/3	all raters
CTbw	p	<0.0001	<0.0001	<0.0166	<0.0148
	wk	0.7999	0.8254	0.9495	0.7019
CTstw	p	<0.0001	<0.0001	<0.0001	<0.0137
	wk	0.7454	0.7072	0.9122	0.6353
T2w	p	0.0016	<0.0001	<0.0001	<0.0158
	wk	0.4815	0.3654	0.5802	0.3339
T1w	p	<0.0001	<0.0001	<0.0001	<0.0153
	wk	0.4855	0.4251	0.5405	0.3704
PDw	p	<0.0001	<0.0001	<0.0001	<0.0141
	wk	0.4629	0.4103	0.6276	0.333

Abbreviations

3D: Three dimensional; CT: Computed tomography; CTbw: CT image with bone window; CTstw: CT image with soft tissue window; ISS: Inter-slice spacing; MRI: Magnetic resonance imaging; NSA: Number of signal averages; PDw: Proton density weighted; SD: Standard deviation; SE: Spin echo sequence; ST: Slice thickness; T1w: T1 weighted; T2w: T2 weighted; TE: Echo time; TR: Repetition time; WL: Window level; WW: Window width.

Competing interests

None of the authors has any personal or financial relationships which could inappropriately influence or bias the content of this paper.

Authors' contributions

JK designed the study, optimized MR settings, acquired CT-and MR-data, analysed data, and wrote the manuscript. AB, MH and BO contributed to the study design, data analysis and interpretation and helped edit the manuscript. MH supported and supervised CT-and MR-data acquisition. KR performed the statistical analysis. All authors read and approved the final manuscript.

Authors' information
Joachim Kaminsky and Dr. Astrid Bienert-Zeit are joint first authors.

Author details
[1]Clinic for Horses, University of Veterinary Medicine Hannover, Foundation, Bünteweg 9, Hannover D-30559, Germany. [2]Institute for Biometry, Epidemiology and Information Processing, University of Veterinary Medicine Hannover, Foundation, Bünteweg 2, Hannover D-30559, Germany.

References

1. Bertone JJ, Biller DS, Ruggles A. Diagnostic techniques for evaluation of the paranasal sinuses. Vet Clin North Am Equine Pract. 1993;9:75–91.
2. Gibbs C, Lane JG. Radiographic examination of the facial, nasal and paranasal sinus regions of the horse. II. Radiological findings. Equine Vet J. 1987;19:474–82.
3. Tremaine WH, Dixon PM. A long-term study of 277 cases of equine sinonasal disease. Part 1: details of horses, historical, clinical and ancillary diagnostic findings. Equine Vet J. 2001;33:274–82.
4. Weller R, Livesey L, Maierl J, Nuss K, Bowen IM, Cauvin ER, et al. Comparison of radiography and scintigraphy in the diagnosis of dental disorders in the horse. Equine Vet J. 2001;33:49–58.
5. Park RD. Radiographic examination of the equine head. Vet Clin North Am Equine Pract. 1993;9:49–74.
6. Parente EJ, Franklin SH, Derksen FJ, Weishaupt MA, Chalmers HJ, Tessier C. Diagnostic techniques in equine upper respiratory tract disease. In: Auer JA, Stick JA, editors. Equine surgery. 4th ed. St. Louis: Elsevier Saunders; 2012. p. 536–57.
7. Kraft SL, Gavin P. Physical principles and technical considerations for equine computed tomography and magnetic resonance imaging. Vet Clin North Am Equine Pract. 2001;17:115–30.
8. Baptiste KE, Pleasant RS, Jones JC, Sponenberg DP, Sysel A, McLamb DJ. Paranasal sinus osteoma in an American miniature horse: computed tomographic evaluation and surgical management. Equine Practice. 1996;18:14–9.
9. Bienert A. Digitalradiographische, computertomographische und mikrobiologische Untersuchungen bei Backenzahnerkrankungen des Pferdes, Dissertation. Tierärztliche Hochschule Hannover: Hannover; 2002.
10. Cissell DD, Wisner ER, Textor J, Mohr FC, Scrivani PV, Theon AP. Computed tomographic appearance of equine sinonasal neoplasia. Vet Radiol Ultrasound. 2012;53:245–51.
11. Henninger W, Frame EM, Willmann M, Simhofer H, Malleczek D, Kneissl SM, et al. CT features of alveolitis and sinusitis in horses. Vet Radiol Ultrasound. 2003;44:269–76.
12. Veraa S, Dijkman R, Klein WR, Van den Belt AJ. Computed tomography in the diagnosis of malignant sinonasal tumours in three horses. Equine Vet Educ. 2009;21:284–8.
13. Manso-Díanz G, García-López JM, Maranda L, Taeymans O. The role of head computed tomography in equine practice. Equine Vet Educ. 2015;27:136–45.
14. Arencibia A, Vazquez JM, Jaber R, Gil F, Ramirez JA, Rivero M, et al. Magnetic resonance imaging and cross sectional anatomy of the normal equine sinuses and nasal passages. Vet Radiol Ultrasound. 2000;41:313–9.
15. Garrett KS, Woodie JB, Cook JL, Williams NM. Imaging diagnosis-nasal septal and laryngeal cyst-like malformationsin a Thoroughbred weanling colt diagnosed using ultrasonography and magnetic resonance imaging. Vet Radiol Ultrasound. 2010;51:504–7.
16. Gerlach K, Flatz K, Brehm W, Seeger J. Klinische Anatomie des Gesichtsbereiches des Pferdes in der Magnetresonanztomographie. Pferdeheilkunde. 2009;25:45–52.
17. Gerlach K, Gerhards H. Magnetresonanztomographische Merkmale von Zubildungen im Brereich der Nase, Nasennebenhöhlen und der angrenzenden Knochen: retrospektive Analyse von 33 Pferden. Pferdeheilkunde. 2008;24:565–76.
18. Cavalleri JM, Metzger J, Hellige M, Lampe V, Stuckenschneider K, Tipold A, et al. Morphometric magnetic resonance imaging and genetic testing in cerebellar abiotrophy in Arabian horses. BMC Vet Res. 2013;9:105.
19. Gutierrez-Crespo B, Kircher PR, Carrera I. 3 Tesla magnetic resonance imaging of the occipitoatlantoaxial region in the normal horse. Vet Radiol Ultrasound. 2014;55:278–85.
20. Hontoir F, Nisolle JF, Meurisse H, Simon V, Tallier M, Vanderstricht R, et al. A comparison of 3-T magnetic resonance imaging and computed tomography arthrography to identify structural cartilage defects of the fetlock joint in the horse. Vet J. 2013;199:115–22.
21. Junker C. Studien über die Abbildung von Kopf und Hals des Pferdes mittels Magnetresonanztomographie (MRT), Inaugural Dissertation. München: Ludwig-Maximilians-Universität München; 2002.
22. Morrow KL, Park RD, Spurgeon TL, Stashak TS, Arceneaux B. Computed tomographic imaging of the equine head. Vet Radiol Ultrasound. 2000;41:491–7.
23. Arencibia A, Vazquez JM, Rivero M, Latorre R, Sandoval JA, Vilar JM, et al. Computed tomography of normal cranioencephalic structures in two horses. Anat Histol Embryol. 2000;29:295–9.
24. Barbee DD, Allen JR, Gavin PR. Computed tomography in horses. Vet Radiol. 1987;28:144–51.
25. Brinkschulte M, Bienert-Zeit A, Lupke M, Hellige M, Staszyk C, Ohnesorge B. Using semi-automated segmentation of computed tomography datasets for three-dimensional visualization and volume measurements of equine paranasal sinuses. Vet Radiol Ultrasound. 2013;54:582–90.
26. Probst A, Henninger W, Willmann M. Communications of normal nasal and paranasal cavities in computed tomography of horses. Vet Radiol Ultrasound. 2005;46:44–8.
27. Smallwood JE, Wood BC, Taylor WE, Tate Jr LP. Anatomic reference for computed tomography of the head of the foal. Vet Radiol Ultrasound. 2002;43:99–117.
28. Johnston GM, Eastment JK, Wood JLN, Taylor PM. The confidential enquiry into perioperative equine fatalities (CEPEF): mortality results of Phases 1 and 2. Vet Anaesth Analg. 2002;29:159–70.
29. Saunders J, Nelson A, Vanderperren K. Particularities of equine CT. In: Schwarz T, Saunders J, editors. Veterinary computed tomography. 1st ed. Chichester, West Sussex: John Wiley & Sons Ltd; 2011. p. 421–6.
30. Porter EG, Werpy NM. New concepts in standing advanced diagnostic equine imaging. Vet Clin North Am Equine Pract. 2014;30:239–68.
31. Fatterpekar GM, Delman BN, Som PM. Imaging the paranasal sinuses: where we are and where we are going. Anat Rec (Hoboken). 2008;291:1564–72.
32. Gerlach K, Cronau M, McMullen R, Gerhards H. Magnetresonanztomographische Untersuchungen von Melanomen im Kopfbereich bei drei Pferden. Pferdeheilkunde. 2007;23:259–62.
33. Hoppe CT, Horstmann W, Gerhards H. Kernspintomographische Untersuchung von Erkrankungen der Kopfregion des Pferdes-drei Fallberichte. Pferdeheilkunde. 2003;19:143–50.
34. Tucker RL, Sampson SN. Magnetic resonance imaging protocols for the horse. Clinical Techniques in Equine Practice. 2007;6:2–15.
35. Weishaupt D, Köchli VD, Marincek B. Wie funktioniert MRI? 6th ed. Heidelberg: Springer Medizin Verlag; 2009.
36. Chang KJ, Kamel IR, Macura KJ, Bluemke DA. 3.0-T MR imaging of the abdomen: comparison with 1.5 T. Radiographics. 2008;28:1983–98.
37. Merkle EM, Dale BM. Abdominal MRI at 3.0 T: the basics revisited. Am J Roentgenol. 2006;186:1524–32.
38. Brinkschulte M. Morphologische Untersuchung der Apertura nasomaxillaris des Pferdes sowie deren Verzweigung in die Nasennebenhöhlen unter der Anwendung dreidimensionaler Rekonstruktion computertomographischer Schnittbildserien, Dissertation. Tierärztliche Hochschule Hannover: Hannover; 2012.

12

Intraocular lens power calculation for the equine eye

Ulrike Meister[1*], Christiane Görig[2], Christopher J. Murphy[3], Hubertus Haan[4], Bernhard Ohnesorge[1†] and Michael H. Boevé[1†]

Abstract

Background: Phacoemulsification and intraocular lens (IOL) implantation during cataract surgery in horses occur with increasing frequency. To reduce the postoperative refractive error it is necessary to determine the proper IOL power. In the present study retinoscopy, keratometry and ultrasonographic biometry were performed on 98 healthy equine eyes from 49 horses. The refractive state, corneal curvature (keratometry) and the axial location of all optical interfaces (biometry) were measured. The influences of breed, height at the withers, gender and age on values obtained and the comparison between the left and right eye were evaluated statistically. Corresponding IOL power were calculated by use of Binkhorst and Retzlaff theoretical formulas.

Results: Mean ± SD refractive state of the horses was + 0.32 ± 0.66 D. Averaged corneal curvature for Haflinger, Friesian, Pony, Shetland pony and Warmblood were 21.30 ± 0.56 D, 20.02 ± 0.60 D, 22.61 ± 1.76 D, 23.77 ± 0.94 D and 20.76 ± 0.88 D, respectively. The estimated postoperative anterior chamber depth (C) was calculated by the formula C = anterior chamber depth (ACD)/0.73. This formula was determined by a different research group. C and axial length of the globe averaged for Haflinger 9.30 ± 0.54 mm and 39.43 ± 1.26 mm, for Friesian 10.12 ± 0.33 mm and 42.23 ± 1.00 mm, for Pony 8.68 ± 0.78 mm and 38.85 ± 3.13 mm, for Shetland pony 8.71 ± 0.81 mm and 37.21 ± 1.50 mm and for Warmblood 9.39 ± 0.51 mm and 40.65 ± 1.30 mm. IOL power was calculated with the Binkhorst and Retzlaff theoretical formulas. Calculated IOL power for the several breeds ranged from 18.03 D to 19.55 D. The mean value across all horses was 18.73 D determined with Binkhorst formula and 18.54 D determined with Retzlaff formula.

Conclusions: Mean result of this study is: an 18.5 D IOL seemed to be the most appropriate to achieve emmetropia after IOL implantation in horses. Cataract surgery without IOL implantation results in hyperopic and visual compromised horses. Retinoscopy, keratometry and ultrasonographic biometry should be performed on every affected horse and postoperative visual outcome should be determined.

Keywords: Retinoscopy, Keratometry, Ultrasonographic biometry, Intraocular lens, Equine

Background

Cataract has been reported to occur in horses with an estimated incidence of 5% to 7% of all horses with otherwise clinically normal eyes [1]. Cataract can impair visual performance rendering the horse unfit for specific functions, results in devaluation of the individual and can predispose the animal to self-injury [2–4]. Visual function can be restored through surgical extraction of the cataractous lens. Reported methods include aspiration, intracapsular extraction, extracapsular extraction and phacoemulsification with aspiration. Without the implantation of an intraocular lens following cataract extraction horses were left aphakic and were markedly hyperopic [5–7]. Early reports suggested aphakic horses to perform quite well and visual performance to be functionally acceptable [4, 5, 8]. However in other studies significant visual impairment in aphakic horses were detected, which manifested as reduced night vision and reduced contrast sensitivity [3, 9]. In highly hyperopic eyes induced by aphakia, no object in space is accurately imaged on the retina with the degree of defocus

* Correspondence: Ulrike.Meister@t-online.de
†Equal contributors
[1]Stiftung Tierärztliche Hochschule Hannover, Klinik für Pferde, Bünteweg 9, 30559 Hannover, Germany
Full list of author information is available at the end of the article

worsening as objects are moved closer to the horse's eye. The mean refractive error in aphakic horses has been reported as approximately + 9.5 D [2–4]. In humans this degree of refractive error corresponds to a Snellen value of 20/ 1200 with 20/200 acuity in the best performing eye of an individual being designated as legally blind [3, 8]. It is reasonable to assume that induction of marked hyperopia associated with lens removal in the horse would dramatically affect visual performance. The degree of defocus scales directly with a decrease in measured visual acuity in humans and dogs [10].

While correction of the refractive error by implantation of an IOL during cataract surgery in humans and dogs is standard therapy, the number of horses, which undergoes phacoemulsification and IOL implantation is much smaller [11–16]. To the author's knowledge there are currently three different foldable IOLs for horses available. To assist the surgeon in achieving emmetropia in the patient after cataract surgery in humans, presurgical determination of required IOL power is performed with a wide variety of IOLs of differing dioptric power being commercially available. In veterinary patients biometry, keratometry and theoretical formulas have been employed to determine the ideal predicted power of IOL for several species [11, 12, 17–19]. It is important to note that many formulas used to predict required IOL power in human patients are inappropriate for veterinary patients as they incorporate mathematical constants (eg the A constant in the SRK formula) that are not direct applicable to veterinary species. A few studies were done to assess natural variation in the mentioned values in horses with healthy eyes [12, 17, 20, 21].

The purpose of this study was to determine the degree of refraction with retinoscopy, intraocular distances with A-mode (amplitude modulation) ultrasound with horse adjusted ultrasound velocities and corneal curvatures using a video-keratometer in healthy equine eyes. Furthermore the effect of breed, height at the withers, gender, age and differences between the left and right eye was determined. The values were used to calculate the expected IOL power with the Binkhorst and Retzlaff formulas and were compared with results of IOL implantations in reported studies.

Methods

Animals and ophthalmologic examination

The study was approved by the ethics commissions of the University of Veterinary Medicine Hannover, Foundation.

98 healthy equine eyes from 49 horses were examined. The horses were private-owned and examined in a darkened stable. Informed consent was obtained from all owners. All horses underwent a dilated (tropicamide 0.5%) ophthalmic examination, including pupillary light reflex, dazzle reflex, menace response, slit-lamp biomicroscopy[1] and indirect ophthalmoscopy.[2] Horses with ocular abnormalities were excluded from the study. Only manual restraint was used for all procedures. Following clinical evaluation retinoscopy, keratometry and A-mode ultrasonography were conducted on both eyes of each horse. Prior to the examinations the horses received 1 drop of tropicamide 0.5% ophthalmic solution[3] and directly in front of the ultrasonographic biometry 1 drop of lidocain ophthalmic solution[4] was instilled.

Retinoscopy

Streak retinoscopy was performed with a streak retinoscope[5] and 2 skiascopy bars.[6] Streak retinoscopy was performed about 30 min after the administration of tropicamide ophthalmic solution[3]. The horizontal and vertical meridians were refracted at a working distance of 67 cm using the technique described in a previous study [22]. Results of the horizontal and vertical measurements were averaged to calculate a mean refractive error of each eye.

Keratometry

Keratometry was conducted by using a custom-made video-keratometer.

The video-keratometer consisted of light-emitting diodes (LEDs) placed around a video camera, which recorded the reflected LED image from spherical surfaces. The spherical surfaces were either steel balls, for calibration purpose, or equine eyes producing Purkinje I images. The LEDs emitted in the near infrared, so invisible to the human and equine eye [3]. Switching on and off the LEDs failed to elicit a measurable response of the horse. The video camera was sensible in the near infrared. The light of each LED was projected by a small collimator to the spherical surface, so that the origin of the light appeared to be at infinity. As also mentioned in an earlier study [23] this is mandatory for the virtual image to be located optically at the focus point of the aerial mirror. In this study two horizontal collimators and two vertical collimators were used, so the horizontal and vertical radius of curvature of the equine eye could be measured separately. Each pair of collimators was separated 72 mm and arranged on a circle of 140 mm diameter around a video camera. As the surface acts as an aerial mirror, the distance of two reflected light points in the recording, called image height by opticians, was proportional by similar triangles to the radius of curvature of the sphere. The distance ring of the video camera was locked during all measurements and the apparatus was moved towards the horse eye (or steel ball) until the image was sharp. That was at approximately 300 mm distance from the collimator to the eye (or steel ball). As this method guaranteed always the same distance in repeated measurements, no other distance measurement was necessary, so that the measurement time was minimized for the horse.

By calibration with steel balls all constant factors like the aforementioned distance, the magnification factor of the video camera lens and any magnification of the video software are dumped in a single calibration value and need not be known in detail.

First of all the video-keratometer was calibrated by recording images of light reflected from steel balls of known radii of curvature. In the recordings the distance of the two horizontal and the two vertical reflections were measured in pixels and two (one for the horizontal case and one for the vertical case) calibration curves were generated showing pixel distance over steel ball radius. The calibration curves were linear.

Video-keratometry was performed on both eyes of every horse. The video-keratometer was moved to the horse's eye until the images were displayed sharp on the monitor of the video camera. The video camera recorded the two horizontal and two vertical light points for a few minutes. After these measurements the recordings were transferred to a computer program for analysis.

The distances of the two horizontal and the two vertical projected light points were measured in the video recordings and served, together with the calibration, to determine the horizontal and vertical radius of curvature and corneal astigmatism of the horse eye.

Three measurements of the radii from each plane (horizontal and vertical) were conducted and averaged. Horizontal corneal curvature (K1) and vertical corneal curvature (K2) was calculated by the eq. K [D] = (N − 1/ radius [m]). The effective corneal refractive index (N) was set at $N = 1.336$ [24]. Averaged corneal curvature (K) was calculated by $\frac{K1+K2}{2}$ in D.

IOL calculation

The detected data were used to calculate the predicted IOL power [D] by using the Binkhorst theoretical formula and the Retzlaff theoretical formula:

Binkhorst theoretical formula:

$$Pe = \frac{1336\,(4r-L)}{(L-C)(4r-C)}$$

Pe is the predicted IOL power in D, r is the averaged corneal radius in mm, L is the axial length (AxL) in mm and C is the expected postoperative anterior chamber depth in mm and was calculated by the formula C = ACD/0.73. A previous study [7] reported the mean preoperative-to-postoperative ACD ratio in equine eyes to be 0.73.

Retzlaff theoretical formula:

$$Pe = \frac{N}{L-C} - \frac{NK}{N-KC}$$

Pe is the predicted IOL power in D, N is the refractive index of aqueous and vitreous (1.336), L is the axial length (AxL) in m, C is the postoperative anterior chamber depth in m and K is the averaged corneal curvature in D.

A-mode ultrasonography

Measurements of anterior chamber depth (ACD), crystalline lens thickness (LT), vitreous body length (VBL) and axial length (AxL) of the globe was obtained with an ultrasonographic device[7] with a 10 MHz A-scan probe. Ultrasound velocities were set at 1532 m/s for the anterior chamber and the vitreous body and 1641 m/s for the lens. To achieve more accuracy, values for lens thickness were corrected by the equine-specific conversion factor of 1.008 [25]. Five A-scan recordings of both eyes of each horse were obtained vertically along the optical axis and the values for each eye were averaged. To obtain optimal scans the manual freeze mode was used and the images were frozen when all of the required echoes are present and of sufficient height.

Statistical analysis

As no significant difference was detected for any measurements between left and right eyes (tested with Wilcoxon-Mann-Whitney-U test), the data from both eyes were averaged for further statistical analyses. Values are reported as mean ± SD. Statistical analyses were made with computerized statistical software.[8] Normal distribution of data was tested by using the Kolmogorov-Smirnov test. Correlations between age, height at the withers, refractive state, r, ACD, LT, VBL and AxL were evaluated by calculation of Spearman correlation coefficients. The influence of gender and type of horse on refractive state, r and AxL were determined by using the Kruskal-Wallis test. Wilcoxon-Mann-Whitney-U tests were used for comparisons between the horizontal and vertical refractive state and the horizontal and vertical corneal radius in the same eye. Values of $P \leq 0.05$ were considered significant.

Results

The examined 49 horses contained 36 mares, 9 geldings and 4 stallions. Ages ranged from 2 to 25 years (mean age = 8.28 ± 5.36 years). Breeds included Haflinger ($n = 10$), Shetland pony ($n = 7$), Warmblood ($n = 13$), Friesian ($n = 8$) and Pony (breed unspecified; $n = 11$). Heights at the withers ranged from 85 to 178 cm (mean height = 145.06 ± 27.26 cm).

Retinoscopy

Retinoscopy was performed on both eyes of the 49 horses. Results of the horizontal and vertical measurements were averaged. The mean refractive state ±SD of the horses was + 0.32 ± 0.66 D. Emmetropia (− 1 D – 1D) was present in 91 (92.8%) of 98 eyes. Ametropia occurred in 7 (7.2%) eyes, with 4 (4.1%) being myopic with a maximum value of – 5.0 D and 3 (3.1%) being hyperopic with a maximum value of + 1.5 D. Anisometropia (refractive state of the left and right eye differing by more than 0.5 D) occurred in 6 horses (12.2%). In 4 of the anisometropic horses hyperopia was present in one eye, in one horse myopia was present in one eye and in one horse myopia was present in both eyes but of differing magnitudes. There were no significant differences between the horizontal and vertical meridians ($P = 0.068$). No significant influence was found for the breed of horse ($P = 0.074$), gender ($P = 0.761$), height at the withers ($P = 0.924$), age ($P = 0.917$), corneal radius ($P = 0.201$) or globe axial length ($P = 0.503$).

Keratometry

Horizontal and vertical corneal radii of curvature in the central corneal region were obtained for both eyes in all 49 horses. The appropriate horizontal (K1), vertical (K2) and averaged (K) corneal curvature were calculated and the corneal astigmatism was determined. The mean values ± SD for the different types of horses are shown in Table 1.

The horizontal corneal radius of curvature was significantly greater ($P < 0.001$) than the vertical corneal radius indicating the cornea to have an astigmatic curvature, being flatter horizontally than vertically. The differences in horizontal and vertical meridian ranged between 0.34 D and 0.95 D. There was no significant difference between the corneal curvature and the age ($P = 0.071$) of the horses. Corneal curvature did correlate positively, however, with the height of the withers ($P < 0.001$) and the axial length of the globe ($P < 0.001$). A significant influence also existed for the breed of horse ($P < 0.001$) and the gender ($P = 0.027$) on the corneal curvature.

Ultrasonographic biometry

Biometric values were obtained in both eyes of all horses by use of A-mode ultrasonography. The expected postoperative anterior chamber depth (C) was calculated using the formula $C = \frac{ACD}{0.73}$ [7]. Mean values ±SD are provided in Table 2.

Significant influences were found between AxL and ACD ($P < 0.001$), LT ($P < 0.001$), VBL ($P < 0.001$), corneal curvature ($P < 0.001$), height at the withers ($P < 0.001$), breed of horse ($P < 0.001$) and age ($P = 0.008$). The larger the AxL the greater the ACD, LT and VBL and the flatter the corneal curvature became and the taller and older the horses were the larger the AxL increased. The gender ($P = 0.026$) also influenced the AxL.

IOL calculations

For IOL calculation the Binkhorst theoretical formula and the Retzlaff theoretical formula were used. The expected mean IOL dioptric power was calculated using the values for AxL, K and C. The calculated mean IOL dioptric powers for the several breeds of horses are provided in Table 3.

Discussion

Values for corneal curvature, axial length of the globe and postoperative anterior chamber depth of each horse are necessary to calculate the appropriate IOL power with the Binkhorst and Retzlaff theoretical formulas. While streak retinoscopy and ultrasound biometry are relatively simple to perform and routinely used in equine cataract patients, it is rather difficult to determine keratometric data. In order to verify the visual outcome after IOL implantation a post operative evaluation via streak retinoscopy is necessary. Furthermore to establish the exact position of the IOL in the eye and to determine the postoperative anterior champer depth ultrasound measurements after IOL implantation should be performed [2, 26]. Moreover, these values could be used to improve the theoretical formulas with respect to equine cataract patients. In this study retinoscopy, videokeratometry and A-mode ultrasonography with horse adjusted ultrasound velocities were obtained on healthy equine eyes and the corresponding IOL powers were calculated using the Binkhorst and Retzlaff theoretical formulas.

Investigators of 1 study [27] found only changes in the refractive state during the 20 min following the application of topical tropicamid 1%. Hence in the present study streak retinoscopy was performed about 30 min after the application of tropicamide 0.5% ophthalmic solution. As in another study [28] emmetropia was specified as ±1.0 diopter. The mean refractive state ±SD of all horses was + 0.32 ± 0.66 D. In the examined population 92.8% of values for refractive state were within one diopter of emmetropia, 3.1% were hyperopic and 4.1%

Table 1 Corneal radius, corneal curvature and corneal astigmatism are depicted depending on breed

Breed	horizontal corneal radius (mm)	vertical corneal radius (mm)	averaged corneal radius (mm)	K1 (D)	K2 (D)	Mean K (D)	corneal astigmatism (D)
Haflinger	16.06 ± 0.47	15.53 ± 0.40	15.79 ± 0.42	20.94 ± 0.62	21.66 ± 0.55	21.30 ± 0.56	0.72 ± 0.36
Friesian	16.94 ± 0.50	16.66 ± 0.49	16.80 ± 0.48	19.86 ± 0.61	20.19 ± 0.62	20.02 ± 0.60	0.34 ± 0.29
Pony	15.16 ± 1.12	14.76 ± 1.34	14.96 ± 1.22	22.29 ± 1.61	22.94 ± 1.95	22.61 ± 1.76	0.65 ± 0.61
Shetland pony	14.46 ± 0.62	13.89 ± 0.57	14.18 ± 0.57	23.30 ± 1.00	24.25 ± 0.97	23.77 ± 0.94	0.95 ± 0.58
Warmblood	16.45 ± 0.66	15.99 ± 0.74	16.22 ± 0.69	20.46 ± 0.84	21.06 ± 0.96	20.76 ± 0.88	0.60 ± 0.35

Table 2 Anterior chamber depth, lens thickness, vitreous body length, axial length and calculated post operative anterior chamber depth are depicted depending on breed

Breed	Anterior Chamber Depth (ACD) (mm)	Lens Thickness LT) (mm)	Vitreous Body Length (VBL) (mm)	Axial Length (AxL) (mm)	C (calculated post operative ACD) (mm)
Haflinger	6.79 ± 0.39	10.91 ± 0.27	21.82 ± 0.94	39.43 ± 1.26	9.30 ± 0.54
Friesian	7.39 ± 0.24	11.70 ± 0.30	23.23 ± 0.97	42.23 ± 1.00	10.12 ± 0.33
Pony	6.34 ± 0.58	11.43 ± 0.82	21.18 ± 2.07	38.85 ± 3.13	8.68 ± 0.78
Shetland pony	6.36 ± 0.58	11.50 ± 0.50	19.45 ± 1.22	37.21 ± 1.50	8.71 ± 0.81
Warmblood	6.85 ± 0.37	11.79 ± 0.50	22.10 ± 0.92	40.65 ± 1.30	9.39 ± 0.51

were myopic. The results are congruous with another recent report [29]. Bracun found that 83.63% were emmetropic, 8.86% were hyperopic and 7.51% were myopic. However in another study [21] emmetropia was present in only 48.7%, hyperopia in 24.1% and myopia in 27.2% of the horses examined, but emmetropia was strictly defined as a refraction of 0 D. In the present study anisometropia (refractive power of the two eyes differs more than 0.5 D) occurred in 6 horses (12.2%).

The values obtained for the averaged corneal curvature (see Table 1) correspond well with one previous report employing photokeratometry [21] but differ by 3.5-8.5 D from two previously reported studies using photokeratometry and B-mode ultrasonography [12, 17]. Differences in corneal curvature values may be influenced by the differing methods employed. Mouney and co-workers documented a significant difference in values obtained by B-mode ultrasonography vs. photokeratometry of approximately 1.2 D [12]. Other differences between the current and previously reported studies include the state of sedation /anesthesia and presence of an auriculopalpebral nerve block [12, 21]. These differences highlight the need to establish standardized instrumentation and methods for performing keratometry.

A significant difference was found between the horizontal and the vertical corneal curvature, which are expected and normal and do not result in differences in the refractive error in the respective meridians. Corneal curvature depended significantly on the breed of horse, the height of the withers and the axial length of the globe. The taller the horse and the larger the axial length of the globe are, the flatter the corneal curvature. The age of the horses had no significant influence on the corneal curvature. In contrast to current findings, Grinniger and co-workers found a positive relationship between age and corneal radius of curvature [21]. A possible reason for the difference between the present study and this previous report is the range of ages included in the studies. In the prior report the group of the youngest horses was between 0 and 2.5 years old while the youngest horse in the present study was 2 years. A similar correlation between age and radius of curvature has been reported for cats. A dramatic decrease in corneal curvature in cats occurs between 9 weeks and 1 year of age [30]. The age of the horse should be considered for IOL selection in very young horses. Townsed et al. [31] found a reduced dioptric strength of the juvenile cornea of horses and a shortened vitreous chamber depth. Hence, to achieve emmetropia in the juvenile eye an IOL with a dioptric strength greater than that required by an adult equine is necessary. However the grown up horses is ametropic. Therefore, it is recommended to use IOL in young horses that will correct vision once the eye reaches its adult parameters.

Table 3 IOL calculations using Binkhorst and Retzlaff theoretical formulas

Breed	IOL dioptric power derived using Binkhorst formula	IOL dioptric power derived using Retzlaff formula
Haflinger	19.55 D	19.36 D
Friesian	18.20 D	18.03 D
Pony	18.27 D	18.07 D
Shetland pony	19.03 D	18.80 D
Warmblood	18.66 D	18.48 D

The present study found a significant correlation between gender and corneal curvature. Such a correlation was not found in a prior study [21]. Results may have been influenced by a much greater number of female ($n = 36$) vs. male ($n = 13$) horses being examined in this study.

A-mode ultrasonography is deemed to be the most accurate method for ocular biometry [32] with B-mode considered inappropriate for ocular biometry [33]. In human cataract surgery, A-mode ultrasound is routinely used in preoperative diagnostic and in postoperative monitoring [34]. Important parameters that can influence the accuracy of A-mode biometry are the values for ultrasound velocity used for determining exact distances. The ultrasound velocity in lens tissue differs considerably in different species. In dogs an underestimation of the lens thickness of 4% results when calculations of lens thickness are made on the basis of ultrasound velocity in the human lens. Use of inaccurate values for velocity can result in incorrect biometric values, which will, in turn, result in incorrect prediction of optimal IOL power to achieve emmetropia [35]. In a previous study [25] species-specific ultrasound velocity for equine aqueous humor, lens and vitreous body were determined. To obtain accurate A-mode biometric values it is recommended to use appropriate values for ultrasound velocity [25]. In the current study, the A-mode instrument utilized ultrasound velocities established for human patients. Hence, the values for lens thickness were corrected by using a conversion factor of 1.008. Intraocular distances (ACD, LT, VBL, AxL) determined in the present study are congruous with prior reports [12, 17, 21, 36].

Cataract surgery with subsequent IOL implantation following phacoemulsification is being performed with increasing frequency in horses [7]. It is a fact that IOL implantation significantly reduces postoperative refractive error [12]. In human medicine several IOL power calculation formulas were developed. They could be classified into theoretical formulas and regression formulas. While the theoretical formulas only were influenced by the variables corneal curvature, axial length, postoperative anterior chamber depth and aqueous and vitreous refractive index, the regression formulas were derived by linear regression analysis of data sets and incorporate mathematical constants [37]. These constants depend on several factors and were not transferable to veterinary species. To the authors` knowledge there are no regression formulas for veterinary medicine available. Hence, the appropriate IOL power for horses was calculated with the Binkhorst and Retzlaff theoretical formulas, despite the fact, that regression formulas show more accuracy than theoretical formulas in human patients [37, 38].

Mean predicted IOL dioptric power for the various types of horses was found to lie between 18.03 D and 19. 55 D (see Table 3). Compared to two other studies [12, 17] which calculated an IOL power of 29.91 D (with Binkhorst formula) and 29 D (with Retzlaff formula) and 22 to 23 D by the use of the Binkhorst formula values presented in this study were considerably smaller. Essential factors, which influence the predicted IOL power, are corneal curvature, axial length of the globe and predicted postoperative anterior chamber depth. In the present study, mean corneal curvature ranged over 20.02 – 23.77 D. These values differed from previous reports [12, 17] using different methods for obtaining keratometry results (16.46 D, 16.5 D and 15.3D) by 3.5-8.5 D. Additionally the postoperative anterior chamber depth (C) influences the predicted IOL power. C was calculated by the formula C = ACD/0.73, a value derived from a previous report [7]. In another study Mouney [12] used the measured ACD as the predicted C (6.8 mm). Postoperative ultrasound measurements of the PACD and refractive data obtained via streak retinoscopy are essential to determine actual position of the implanted IOL and to confirm the predictive power of any theoretical calculations. Other investigators [7], who implanted 30- and 25- diopter IOLs, suggested a 18-D IOL is the appropriate power to achieve emmetropia in adult horses. This prediction correlates well with the predicted IOL power in this study. Two previous in vivo studies [15, 16] implanted a 14 D IOL achieve emmetropia within 0.5 D. Currently, there are three foldable IOLs in two powers (+ 14 D and + 18 D) commercially available for horses. Based on current findings and determinations from previous studies [7, 15, 16, 26] the proper IOL power for use in horses lies between + 14 D and + 18.5 D. Additional studies using standardized methods that include preoperative ultrasonographic biometry and keratometry as well as post-operative refraction are necessary to determine the appropriate IOL power and the long-term effects of IOL implantation [7, 15, 16].

Conclusions

Aphakic horses have been documented to be markedly hyperopic and visually compromised. They have reduced night vision and reduced contrast sensitivity [3, 9]. Soundness is very important for the pleasure and performance horse. A visually compromised horse may show abnormal behavior and the rider faces an unpredictable, frightened animal [39]. Of the commercially available the IOL with 18 D appears to be the most appropriate to achieve emmetropia after IOL implantation in adult horses [7].

Endnotes

[1]Kowa-SL 15, Kowa Optimed Deutschland GmbH, Bendemannstr. 9, 40,210 Düsseldorf, Germany

[2]Heine Omega 500, Heine Optotechnik, Herrsching, Germany

[3]Mydriaticum Stulln, Pharma Stulln GmbH, Werksstr. 3, 92,551 Stulln, Germany

[4]Lidocaine, AST Farma B.V., Wilgenweg 7, 3421 TV Oudewater, The Netherlands

[5]Heine Beta 200, Heine Optotechnik, Herrsching, Germany

[6]Oculus Optikgeräte, Wetzlar, Germany

[7]I[3]SYSTEM-ABD, Innovative Imaging Inc., Sacramento, Calif.

[8]R Core Team (2016). R: A language and environment for statistical computing. R Foundation for Statistical Computing, Vienna, Austria. URL https://www.R-project.org/

Abbreviations

ACD: Anterior chamber depth; A-mode: Amplitude modulation; AxL: Axial length; B-mode: Brightness modulation; C: Postoperative anterior chamber depth; cm: Centimeter; D: Dioptre; IOL: Intraocular lens; K: Averaged corneal curvature; K1: Horizontal corneal curvature; K2: Vertical corneal curvature; LED: Light-emitting diode; LT: Lens thickness; m: Meter; MHz: Megahertz; mm: Millimeter; n: Number; N: Refractive index (1.336); Pe: Predicted IOL power; r: Averaged corneal radius; SD: Standard deviation; VBL: Vitreous body length

Acknowledgements
The authors gratefully acknowledge the support of the Acri.Tec AG represented by Ingeborg Fromberg, which provided equipment to conduct the examinations essential for this article.

Funding
There was no funding.

Authors' contributions
UM helped to design the study and to perform the measurements, analyzed the data and drafted the manuscript. CG designed and performed the measurements and helped to draft the manuscript. MB, BO, CM and HH participated in the design of the study and helped to draft the manuscript. All authors read and approved the final manuscript.

Consent for publication
Not applicable.

Competing interests
The authors declare that they have no competing interests.

Author details
[1]Stiftung Tierärztliche Hochschule Hannover, Klinik für Pferde, Bünteweg 9, 30559 Hannover, Germany. [2]De Graafschap dierenartsen, Schimmeldijk 1, 7251 MX Vorden, The Netherlands. [3]UC Davis Health, 2315 Stockton Blvd, Sacramento, CA 95817, USA. [4]HENSOLDT Optronics GmbH, Carl-Zeiss-Str. 22, 73447 Oberkochen, Germany.

References

1. Matthews AG. Lens opacities in the horse: a clinical classification. Vet Ophthalmol. 2000;3(2-3):65–71.
2. Colitz CMH, RJ MM. Diseases and surgery of the lens. In: Gilger BC, editor. Equine ophthalmology. 2nd ed. Maryland Heights: Elsevier/Saunders; 2010. p. 282–314.
3. Miller PE, Murphy C. Equine vision. In: Gilger BC, editor. Equine ophthalmology. 2nd ed. Maryland Heights: Elsevier/Saunders; 2010. p. 397–433.
4. Millichamp NJ, Dziezyc J. Cataract phacofragmentation in horses. Vet Ophthalmol. 2000;3(2-3):157–64.
5. Dziezyc J, Millichamp NJ, Keller CB. Use of phacofragmentation for cataract removal in horses: 12 cases (1985-1989). J Am Vet Med Assoc. 1991;198(10): 1774–8.
6. Whitley RD, Meek LA, Millichamp NJ, McRae EE, Priehs DR. Cataract surgery in the horse: a review of six cases. Equine Vet J. 1990;22(10):85–90. Supplement
7. McMullen RJ, Davidson MG, Campbell NB, Salmon JH, Gilger BC. Evaluation of 30-and 25-diopter intraocular lens implants in equine eyes after surgical extraction of the lens. Am J Vet Res. 2010;71(7):809–16.
8. Fife TM, Gemensky-Metzler AJ, Wilkie DA, Colitz CM, Bras ID, Klages DC. Clinical features and outcomes of phacoemulsification in 39 horses: a retrospective study (1993-2003). Vet Ophthalmol. 2006;9(5):361–8.
9. Farrall H, Handscombe MC. Follow-up report of a case of surgical aphakia with an analysis of equine visual function. Equine Vet J Suppl. 1990;10:91–3.
10. Murphy CJ, Mutti DO, Zadnik K, Ver Hoeve J. Effect of optical defocus on visual acuity in dogs. Am J Vet Res. 1997;58(4):414–8.
11. Gaiddon J, Rosolen SG, Steru L, Cook CS, Peiffer R. Use of biometry and keratometry for determining optimal power for intraocular-lens implants in dogs. Am J Vet Res. 1991;52(5):781–3.
12. Mouney MC, Townsend WM, Moore GE. Association of height, body weight, age, and corneal diameter with calculated intraocular lens strength of adult horses. Am J Vet Res. 2012;73(12):1977–82.
13. Davidson MG, Murphy CJ, Nasisse MP, Hellkamp AS, Olivero DK, Brinkmann MC, Campbell LH. Refractive state of aphakic and pseudophakic eyes of dogs. Am J Vet Res. 1993;54(1):174–7.
14. Edelmann ML, McMullen R, Stoppini R, Clode A, Gilger BC. Retrospective evaluation of phacoemulsification and aspiration in 41 horses (46 eyes): visual outcomes vs. age, intraocular lens, and uveitis status. Vet Ophthalmol. 2014;17:160–7.
15. Harrington JT, McMullen RJ, Clode AB, Gilger BC. Phacoemulsification and +14 diopter intraocular lens placement in a Saddlebred foal. Vet Ophthalmol. 2013;16(2):140–8.
16. Townsend WM, Jacobi S, Bartoe JT. Phacoemulsification and implantation of foldable +14 diopter intraocular lenses in five mature horses. Equine Vet J. 2012;44(2):238–43.
17. McMullen RJ, Gilger BC. Keratometry, biometry and prediction of intraocular lens power in the equine eye. Vet Ophthalmol. 2006;9(5):357–60.
18. Sanchez RF, Becker R, Dawson C, Escanilla N, Lam R. Calculation of posterior chamber intraocular lens (IOL) size and dioptric power for use in pet rabbits undergoing phacoemulsification. Vet Ophthalmol. 2017;20(3):242–9.
19. Gilger BC, Davidson MG, Howard PB. Keratometry, ultrasonic biometry, and prediction of intraocular lens power in the feline eye. Am J Vet Res. 1998;59(2):131–4.
20. Plummer CE, Ramsey DT, Hauptman JG. Assessment of corneal thickness, intraocular pressure optical corneal diameter, and axial globe dimensions in miniature horses. Am J Vet Res. 2003;64(6):661–5.
21. Grinninger P, Skalicky M, Nell B. Evaluation of healthy equine eyes by use of retinoscopy, keratometry, and ultrasonographic biometry. Am J Vet Res. 2010;71(6):677–81.
22. Davidson MG. Clinical retinoscopy for the veterinary ophthalmologist. Vet Comp Ophthalmol. 1997;7(2):128–37.
23. Mutti DO, Zadnik K, Adams AJ. A video technique for phakometry of the human crystalline lens. Invest Ophthalmol Vis Sci. 1992;33(5):1771–82.
24. Ofri R. Optics and physiology of vision. In: Gelatt KN, editor. Vet Ophthalmol. 4th ed. Ames: Blackwell Pub; 2007. p. 183–219.
25. Meister U, Ohnesorge B, Korner D, Boeve MH. Evaluation of ultrasound velocity in enucleated equine aqueous humor, lens and vitreous body. BMC Vet Res. 2014;10:250.
26. McMullen RJ Jr, Utter ME. Current developments in equine cataract surgery. Equine Vet J. 2010;22(37):38–45. Supplement
27. McMullen RJ, Davidson MG, Gilger BC. The effect of 1% tropicamide-induced mydriasis and cycloplegia on spherical refraction of the adult horse. Vet Ophthalmol. 2014;17(2):120–5.
28. Murphy CJ, Howland HC. Owl eyes: accommodation, corneal curvature and refractive state. J Comp Physiol. 1983;151(3):277–84.
29. Bracun A, Ellis AD, Hall C. A retinoscopic survey of 333 horses and ponies in the UK. Vet Ophthalmol. 2014;17:90–6.
30. Moodie KL, Hashizume N, Houston DL, Hoopes PJ, Demidenko E, Trembly BS, Davidson MG. Postnatal development of corneal curvature and thickness in the cat. Vet Ophthalmol. 2001;4(4):267–72.
31. Townsend WM, Wasserman N, Jacobi S. A pilot study on the corneal curvatures and ocular dimensions of horses less than one year of age. Equine Vet J. 2013;45(2):256–8.
32. Mattoon JS, Nyland TG. Small animal diagnostic ultrasound. 2nd ed. Philadelphia: Saunders; 2002.
33. Hamidzada WA. Effect of preservation on the dimensions of ocular components of the one-humped camel (Camelus dromedarius). Med Sci Res. 1999;27(2):141–4.
34. Mettenleiter EM. Sonographic diagnosis (B-mode technique) for the eyes in horses. 1. Methods and normal findings. Tierarztl Prax. 1995;23(5):481–8.
35. Görig C, Varghese T, Stiles T, van den Broek J, Zagzebski JA, Murphy CJ. Evaluation of acoustic wave propagation velocities in the ocular lens and vitreous tissues of pigs, dogs, and rabbits. Am J Vet Res. 2006; 67(2):288–95.
36. Badial PR, Cisneros-Alvarez LE, Brandao CVS, Ranzani JJT, Tomaz M, Machado VM, Borges AS. Ocular dimensions, corneal thickness, and corneal curvature in quarter horses with hereditary equine regional dermal asthenia. Vet Ophthalmol. 2015;18(5):385–92.

37. Sanders D, Retzlaff J, Kraff M, Kratz R, Gills J, Levine R, Colvard M, Weisel J, Loyd T. Comparison of the accuracy of the Binkhorst, Colenbrander, and SRK implant power prediction formulas. J Am Intraocul Implant Soc. 1981;7(4):337–40.
38. Retzlaff J. A new intraocular lens calculation formula. J Am Intraocul Implant Soc. 1980;6(2):148–52.
39. Whitley RD, Meek LA. Cataract-surgery in horses. Compend Contin Educ Pract Vet. 1989;11(11):1396–401.

13

Monitoring training response in young Friesian dressage horses using two different standardised exercise tests (SETs)

Cornelis Marinus de Bruijn[1], Willem Houterman[1], Margreet Ploeg[3], Bart Ducro[4], Berit Boshuizen[1,2], Klaartje Goethals[2], Elisabeth-Lidwien Verdegaal[5] and Catherine Delesalle[2*]

Abstract

Background: Most Friesian horses reach their anaerobic threshold during a standardized exercise test (SET) which requires lower intensity exercise than daily routine training. Aim: to study strengths and weaknesses of an alternative SET-protocol. Two different SETs (SETA and SETB) were applied during a 2 month training period of 9 young Friesian dressage horses. SETB alternated short episodes of canter with trot and walk, lacking long episodes of cantering, as applied in SETA. Following parameters were monitored: blood lactic acid (BLA) after cantering, average heart rate (HR) in trot and maximum HR in canter. HR and BLA of SETA and SETB were analyzed using a paired two-sided *T*-test and Spearman Correlation-coefficient ($p^* < 0.05$).

Results: BLA after cantering was significantly higher in SETA compared to SETB and maximum HR in canter was significantly higher in SETA compared to SETB.
The majority of horses showed a significant training response based upon longitudinal follow-up of BLA. Horses with the lowest fitness at start, displayed the largest training response. BLA was significantly lower in week 8 compared to week 0, in both SETA and SETB. A significantly decreased BLA level after cantering was noticeable in week 6 in SETA, whereas in SETB only as of week 8. In SETA a very strong correlation for BLA and average HR at trot was found throughout the entire training period, not for canter.

Conclusions: Young Friesian horses do reach their anaerobic threshold during a SET which requires lower intensity than daily routine training. Therefore close monitoring throughout training is warranted. Longitudinal follow up of BLA and not of HR is suitable to assess training response. In the current study, horses that started with the lowest fitness level, showed the largest training response. During training monitoring HR in trot rather than in canter is advised. SETB is best suited as a template for daily training in the aerobic window.

Keywords: Friesian, Standardized exercise test, Lactic acid, Heart rate, Trot, Longitudinal

Background

The basic principle of training is to expose a horse to a gradually increasing intensity of work. This slow and gradual build-up of workload enables the horse to adapt both physically and mentally. The latest few years, there is an increasing interest of horse riders to objectively measure fitness and training response in horses being trained. For this purpose telemetric devices are used, equipped with Global positioning system (GPS) that enable horse riders to monitor heart rate, velocity and performed distance during training [1]. When, throughout the training period these horses are subjected to a SET one can follow-up training response objectively and adjust the training schedule accordingly [2, 3]. However, little is known scientifically at this point concerning appropriate training protocols and applicable SET test protocols, for each individual sport discipline, for example dressage versus show jumping. Moreover, it can be expected that a breed specific approach is needed within each sport discipline [4–7].

* Correspondence: Catherine.Delesalle@UGent.be
[2]Department of Comparative Physiology and Biometrics, Faculty of Veterinary Medicine, Ghent University, Salisburylaan 133, Merelbeke 9820, Belgium

Friesian horses are most commonly used for dressage and carriage driving. Especially dressage has become very popular and Friesian horses are nowadays active in high level international dressage competition such as the World Equestrian Games and the Olympics. Selective breeding on one hand and solid knowledge about proper training of these horses has become increasingly important. In a previously performed study, it was demonstrated that unlike warmblood horses, Friesian horses tend to reach their anaerobic threshold during a SET test (SETA) which requires lower intensity exercise than daily routine training. Moreover, a tendency towards familial clustering with respect to the physiological response to that SET test was reported and a genetic background for poor performance was suggested for certain Friesian breeding lines [8].

Anaerobic training on a daily basis without sufficient recovery time will not have the desired training effect. On the contrary, complications such as muscle soreness and overreaching are expected to occur in such conditions [8]. Munster et al., performed a longitudinal training follow-up of 6 weeks duration in 66 young Friesian horses selected by the Dutch Royal Friesian Studbook. These horses were subjected to a SET test (SETA) at start (week 2) and end of the training period (week 6). The first SET test encompassed solely heart rate monitoring without blood lactic acid analyses. Whereas in the SET test performed at the end of the study, both HR and blood lactic acid monitoring was performed. It was shown that many young Friesian horses reached their anaerobic lactate threshold (4 mmol/L) during SETA, that actually even represented a lower intensity exercise grade than standard daily routine training performed in the studied horses. Therefore it was suggested that Friesian horses might require a different and more careful training approach, compared to other horse breeds to prevent overreaching [8].

In the Munster study, horses were categorized as either low-, moderate-, or non-training responder based upon longitudinal follow-up. Despite the fact that all studied horses showed a positive training response, the study showed a striking heterogeneity in degree of training response between horses and the authors proposed a genetic background for their findings because of a trend of familial clustering in the type of training response seen in the study. Existence of poor athletic capability of certain Friesian breeding lines was suggested.

However, more research is needed to explore this hypothesis. The main objective of the current study was (1) to compare time profiles of HR (average and maximum) and blood lactic acid levels between the SET used in the Munster study (SET A) and an alternatively designed SET (SET B) and; (2) to evaluate training response in the studied horses and (3) to check whether HR values obtained throughout the SETs can be correlated with blood lactic acid levels. The ultimate goal of the current study was to gain more insight in order to be able to formulate advice based upon evidence based research concerning proper training of young Friesian dressage horses.

Methods

Study population and applied study protocol

Two different SETs (SETA and SETB) were applied throughout a 2 month training period in 9 young Friesian dressage horses (of 3 and 4 years old, and with similar training level at the start of the study). A detailed lay-out of both SETs is provided in Tables 1 and 2. SETB alternated short episodes of canter with trot and walk in both directions and thus lacked long episodes of continuous cantering as in SETA. Total cantering time was the same for both SETs.

All participating horses were clinically examined and checked for absence of lameness before the start of the study. Throughout the training period, the riders were asked to daily record appetite, and possible presence of adverse signs such as lameness, disease or reluctance to work in a logbook.

All horses were subjected to the same training protocol, comprising of 30–45 min of basic dressage training comparable to the workload in SETB, three times per week. All horses were subjected to SETA (at day 1 of week 0, 2, 4, 6, and 8) (See Table 1 for SET layout) and SETB (at day 5 of week 0, 2, 4, 6 and 8) (See Table 2 for SET layout), during which HR (GPS equipped Polar RC 3) and BLA (Lactate Pro hand held analyzer) were monitored throughout the SETs. For each SET these parameters were monitored at following time points: BLA after cantering, average HR in trot and maximum HR in canter. Blood was collected by briefly stopping the horses during the SET and collecting a blood sample from the jugular vein in a plain syringe. Blood was collected at the start of training $[Lactate]_{start}$, directly after the left canter $[Lactate]_{canter1}$ and right canter $[Lactate]_{canter2}$ and at the end $[Lactate]_{end}$ during SET A and B. Time points of blood collection are marked [Lactate] in Tables 1 and 2. Plasma lactate concentrations were measured with a

Table 1 Outline of SETA

Time (min)	Exercise	Description	BLA sampling
00:00–03:00	Walk	±2 m/s	$[Lactate]_{start}$
03:00–05:00	Left trot	left hand ±3.5 m/s	
05:00–07:00	Right trot	right hand ±3.5 m/s	
07:00–09:00	Left canter	left hand ±5 m/s	$[Lactate]_{canter1}$
09:00–11:00	Right canter	right hand ±5 m/s	$[Lactate]_{canter2}$
11:00–21:00	Walk	Recovery ±2 m/s	$[Lactate]_{end}$

Table 2 Outline of SETB

Time (min)	Exercise	Description	BLA sampling
00:00–03:00	Walk	±2 m/s	$[Lactate]_{start}$
03:00–05:00	Left trot	left hand ±3.5 m/s	
05:00–06:00	Left canter	left hand ±5 m/s	
06:00–07:00	Left trot	left hand ±3.5 m/s	
07:00–08:00	Left canter	left hand ±5 m/s	$[Lactate]_{canter1}$
08:00–09:00	Walk	left hand ± 2 m/s	
09:00–11:00	Right trot	right hand ±3.5 m/s	
11:00–12:00	Right canter	right hand ±5 m/s	
12:00–13:00	Right trot	right hand ±3.5 m/s	
13:00–14:00	Right canter	right hand ±5 m/s	$[Lactate]_{canter2}$
14:00–24:00	Walk	Recovery ±2 m/s	$[Lactate]_{end}$

hand held analyzer: Akray Lactate pro. Lactate concentrations <0.8 mmol/L were below the detection limit and were set at 0.8 mmol/L. Ambient temperature and humidity were recorded during the measurements [9, 10].

The horse rider was equipped with a polar watch to enable him/her to control speed during SET tests.

Data processing and statistical analysis

The following heart rate parameters were calculated from the Polar heart rate curves for both SETs (SETA and SETB): average heart rate during trot (SETA: $HRav_{trot}$, SETB: $HRav_{trot1}$, $HRav_{trot2}$, $HRav_{trot3}$ and $HRav_{trot4}$), maximum heart rate during left canter (HR_{maxLC}) and maximum heart rate during right canter (HR_{maxRC}). For SETA, also average heart rate during left canter (HR_{LC}) and right canter (HR_{RC}) were calculated. BLA at the beginning of the set was recorded as $[Lactate]_{start}$, after left canter as $[Lactate]_{canter1}$, after right canter as $[Lacate]_{canter2}$ and at the end of the SET as $[Lactate]_{end}$.

Comparisons between HR and BLA of SETA and SETB were analyzed using a paired two-sided *T*-test adjusted with a Tukey-Kramer correction. Correlations between HR and BLA were analyzed using a Spearman Correlation Coefficient. All analyses were performed in SAS version 9.4 (SAS Institute., USA). *P* values less than 0,05 were considered significant. A Spearman Correlation Coefficient higher than 0.6 was considered a strong correlation in this study.

All nine horses were categorized as clinically healthy, without any sign of lameness or illness and remained that way throughout the study. Two horses were lost to follow up in the last training week due to circumstances. Ambient temperature during the study ranged between 10 and 18 °C. Relative humidity ranged between 75 and 87%.

Results

Comparison BLA and HR in SETA versus SETB

BLA levels at start of the SETs were less than 0.8 mmol/L for every horse in every week during the study. Analysis of the BLA in SETA and SETB revealed that the blood lactic acid time profile significantly differed between the two different SETs (see Fig. 1). Mean BLA levels $[Lactate]_{canter1}$ and $[Lactate]_{canter2}$ were significantly higher in SETA, in which the horses canter nonstop during 4 min (see Table 1), when compared to SETB, in which the workload was similar but more gait variation was applied in week 2, week 4 and week 6 (see Fig. 1). Blood $[Lactate]_{end}$, after 10 min of walking at the end of each SET didn't differ significantly between SETs at any week.

The mean HR_{maxLC} in SET A was 172 ± 23 beats/min and in SETB was 165 ± 23 beats/min. Mean HR_{maxRC} in SETA was 177 ± 24beats/min and 165 ± 20 beats/min in SETB. Maximum HR during left canter and during right canter was significantly higher in SETA in week 0 and week 2 and for right canter in week 8 (see Fig. 2). In week 4 and 6 no significant difference in maximum HR during canter was observed.

The smaller intervals of cantering performed in SETB lead to lower heart rate peaks. Mean heart rate in trot was not significantly different between SETA and SETB (130 ± 14 beats/min).

Training response–blood lactic acid time profile

The training response is reflected by a decrease in BLA concentration during SETS over the 8 week training period. In week 0 mean BLA $[Lactate]_{canter1}$ was 3.9 ± 2.1 mmol/L and 3.5 ± 1.5 mmol/L in respectively SETA and SETB. In week 8 mean BLA after canter1 was significantly lower 2.4 ± 0.5 mmol/L in SET A and 1.7 ±

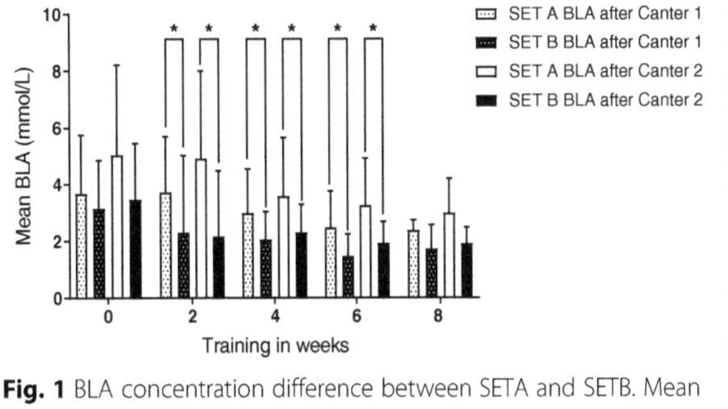

Fig. 1 BLA concentration difference between SETA and SETB. Mean BLA and SD are depicted and significant differences ($P < 0.05$) are marked by an *asterix*. Analysis of the BLA in SETA and SETB revealed that the blood lactic acid time profile significantly differed between the two different SETs. Mean BLA levels after canter1 and after canter 2 were significantly higher in week 2, week 4 and week 6 in SETA, in which the horses canter nonstop during 4 min, when compared to SETB, in which the workload was similar but more gait variation was applied

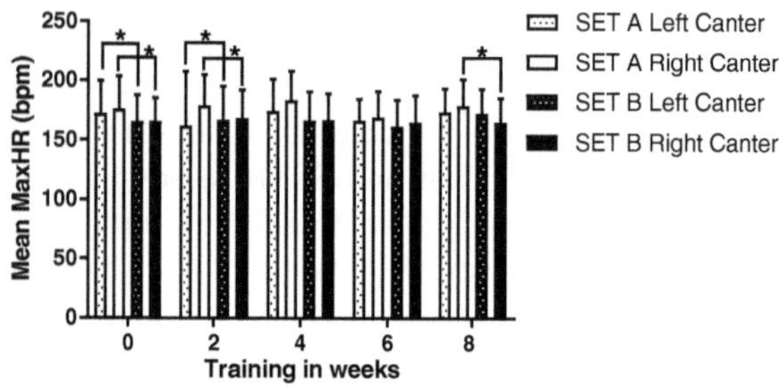

Fig. 2 Maximum Heartrate difference between SETA and SETB. Mean of the maximum HR and SD are depicted and significant differences ($P < 0.05$) are marked by an *asterix*. Maximum HR during left canter and during right canter was significantly higher in SETA in week 0 and week 2 and for right canter in week 8. In week 4 and 6 no significant difference in maximum HR during canter was observed

0.8 mmol/L in SET B (borderline significant). A similar response was expressed in BLA $[Lactate]_{canter2}$ with mean lactate concentrations of 5.0 ± 3.2 mmol/L and 3.8 ± 1.9 mmol/L in week 0 compared to the significantly lower 3.0 ± 1.2 mmol/L and 1.9 ± 0.6 mmol/L in week 8 (see Fig. 3). Mean BLA $[Lactate]_{end}$ also decreased in both SETs: 2.7 ± 2.0 mmol/L and 1.3 ± 0.7 mmol/L in week 0 compared to 0.9 ± 0.2 mmol/L and 0.9 ± 0.2 mmol/L in week 8.

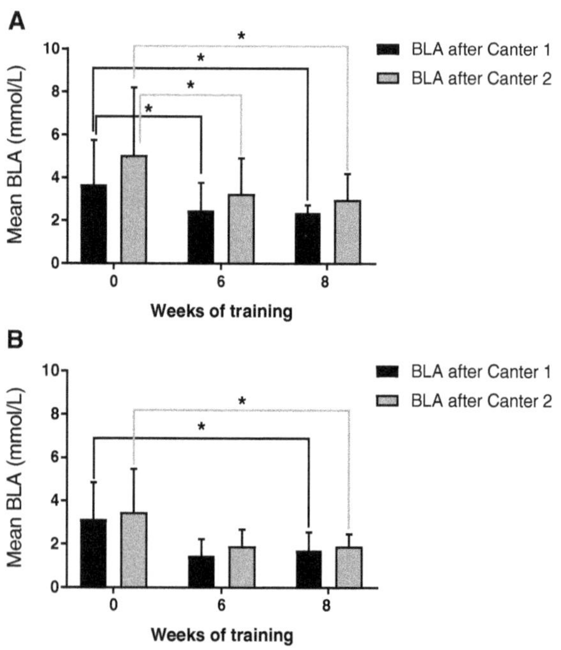

Fig. 3 Training response assessed by SETA and SETB. Training response was shown by comparing SET BLA results of week 8 with BLA in week 0. SET BLA levels after both canter 1 and after canter 2 are shown. Mean BLA and SD are depicted and significant differences ($P < 0.05$) are marked by an *asterix*. Both SETs were compared and it was shown that SETA could distinguish a positive training response sooner, already in week 6, than SETB, which could only show a positive training response in week 8

Seven out of 9 horses showed a significant positive training response and this response could only be assessed by means of longitudinal follow-up of $[Lactate]_{canter1}$ and $[Lactate]_{canter2}$ (See Table 3).

Interestingly, in the current study horses with the highest blood lactic acid concentration during SETs at the start of the study displayed the largest training response.

SETA was more suitable to early assess training response. A significantly decreased BLA level after cantering was already noticeable in week 6 in SETA, whereas in SETB only as of week 8.

Training response–heart rate time profile

Throughout the trial in none of the horses a significant decrease in average or maximum HR was found, not for SET A, nor for SET B. Only trends were seen. The heart rate time profile did show a moderate decrease, however not as consistent as the lactate parameters. In **SETA** results for HR_{maxLC} and HR_{maxRC} showed a slightly lower mean in week 8 compared to week 0. HR_{maxLC} was 177 ± 30 beats/min in week 0 and 174 ± 19 beats/min in week 8. HR_{maxRC} was 182 ± 28 beats/min in week 0 and 179 ± 22 beats/min in week 8.

Results were similar for HR_{LC} (169 ± 30 beats/min in week 0 and 165 ± 19 beats/min in week 8) and HR_{RC} (173 ± 29 beats/min in week 0 and 171 ± 21 beats/min in week 8), also not significant. Mean HR_{trot} was lower in week 8 than in week 0 (132 ± 14 beats/min and 136 ± 22 beats/min) however the difference was not significant.

In **SETB** HR_{maxLC} was slightly higher in week 8 (165 ± 24 beats/min) compared to week 0 (167 ± 19 beats/min). HR_{maxRC} did show a mild decrease from 166 ± 21 beats/min in week 0, to 164 ± 20 beats/min in week 8. Mean HR_{trot} decreased from 131 ± 14 beats/min to 130 ± 12 beats/min. In SETB none of the heart rate parameters showed a significant training response which was comparable the HR results presented for SETA.

Table 3 Overview of percentage of horses showing a positive training response. The training response is defined by a significant decrease in BLA during SETs after 8 weeks of training

SET	Blood sampling moment	% of horses showing positive training response	% of horses showing no positive training response
SETA	BLA canter 1	67	33
	BLA canter 2	78	22
SETB	BLA canter 1	89	11
	BLA canter 2	89	11

Correlations between heart rate and blood lactic acid for different SETs and gaits

Correlations between HR parameters and BLA $[Lactate]_{canter1}$, $[Lactate]_{canter2}$ at start and at the end of both SETA and SETB were studied in week 0, 2, 4, 6 and 8.

For SETA the correlations between HR in every gait and BLA$[Lactate]_{canter2}$ are very strong in week 0 (Fig. 4 panel A). Results of the analysis of BLA after $[Lactate]_{canter2}$ and the maximum and average HRs throughout SETA are shown in Fig. 4 panel a and in Additional file 1: Table S1. For SETA, only for trot significant correlations were found throughout the complete study. As depicted in Fig. 4 Panel a there was a strong correlation between HR in trot (HR_{trot}) with 1) BLA $[Lactate]_{canter2}$, with 2) BLA $[Lactate]_{canter1}$ (data not shown) and with 3) BLA $[Lactate]_{end}$ (data not shown).

Correlations were analyzed between HR parameters and 1) BLA $[Lactate]_{canter1}$ and 2) BLA $[Lactate]_{end}$ for SETA and SETB, and comparable results were found.

The correlation of BLA $[Lactate]_{canter2}$ with HR parameters becomes less strong in week 2 and even disappears in week 4 and 6. However, in week 8 this correlation is strong again.

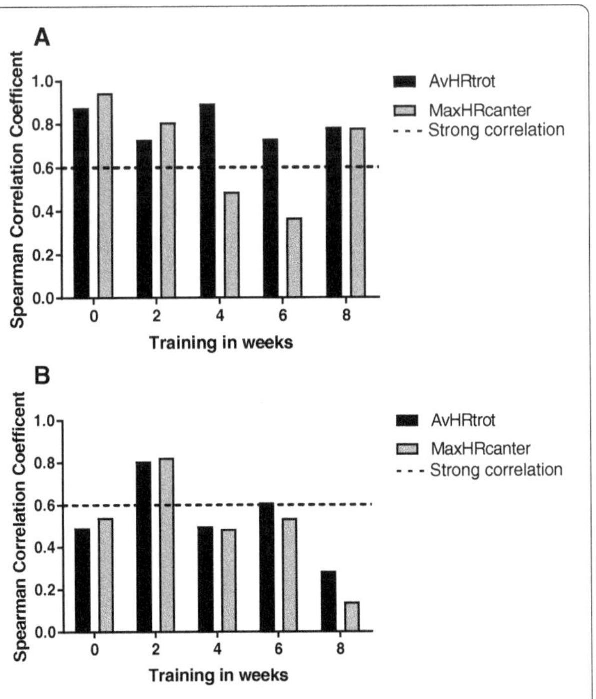

Fig. 4 Correlation between BLA and HR for SETA and SETB. Correlations between average HR during trotting or maximum HR while cantering and BLA after cantering are shown for weeks 0, 2, 4, 6 and 8. The correlation between BL and HR is shown for both SETA in panel (**a**) and SETB in panel (**b**). A Spearman Correlation Coefficient higher than 0.6 is considered a strong correlation in this study. It was shown that throughout the training period the HR in trot was strongly correlated to the BLA values after cantering. For SETA this effect was more pronounced than in SETB

In SETB there were only a few strong correlations found between BLA and HR parameters and existing correlations were all weaker than those found in SETA, as is shown in Fig. 4, panel b and Additional file 1: Table S2. For SETB only in week 2 a consistent strong correlation was found between all HR parameters and BLA $[Lactate]_{canter2}$. In the other training weeks there was no strong correlation present, except for week 6 ($HRav_{trot3}$ and BLA $[Lactate]_{canter2}$).

Importantly, BLA and average HR at trot have a very strong correlation throughout the entire training period (see Fig. 4, panel a), which was not the case for canter. This correlation was more pronounced and present throughout a longer period in SETA when compared to SETB (see Fig. 4, panels a and b).

Discussion

In the current study, again, most Friesian horses crossed their anaerobic threshold when performing SETA. And so, this finding is in accordance with previous reports, suggesting that Friesian horses reach their anaerobic threshold quicker than other horse breeds [8]. Other research encompassing anaerobic lactate threshold showed that most horses do not reach their anaerobic threshold at velocities reached in a standard canter like the one used in SETA [11–13]. Friesian horses are renowned for their impressive trot with high knee action, however, cantering is a much more difficult gait for them. In that respect, the relatively long four minutes of continuous canter applied in SETA might cause the young horses to reach their anaerobic threshold of 4 mmol/L much earlier. Another possible explanation as to why Friesian horses tend to cross their anaerobic threshold already at low intensity exercise, is their muscle fibre composition. It is possible that Friesian horses, just like Quarter horses predominantly possess fast twitch fibers in most important locomotor muscles [7]. Up until now, no study has been performed on muscle fibre typing in Friesian horses.

The horses in this study had higher blood lactate concentrations and higher peak heart rates in SETA when compared to SETB. As mentioned previously, SETB alternated short episodes of canter with trot and walk in both directions and thus lacked long episodes of continuous cantering as in SETA. Total cantering time was the same for both SETs. It is to be expected that shorter cantering intervals limit lactate accumulation and heart rate peaks. Indeed, the frequent changes in gait and directions incorporated into SETB, allow for intermittent recuperation. Our findings show that SETB provides a good template for training young Friesian dressage horses in the aerobic window. It resembles the approach of interval training, which essentially entails conditioning by applying frequent alternates between varying degrees of effort and incorporating periods of recovery in between efforts. This approach trains the equine body to efficiently switch between aerobic and anaerobic pathways [14, 15]. SETA and SETB were always performed with a 2 days interval, but always in the same order: SET B was always preceded by SETA, unfortunately no block design was applied.

Both SETA and SETB were suitable to assess training response in the studied horses, however SETA had a greater discriminative value than SETB with that respect, since a distinct training response could already be identified as of week 6 in SETA, whereas in SETB only as of week 8. Both BLA levels and HRs decreased progressively during both SETs over the course of the 8 weeks of training. However, this decrease was only significant for BLA. These results show that training response can only be properly assessed by longitudinal follow-up of BLA values. This is in accordance with previous research. [12 Several studies have shown that aerobic capacity in horses can improve without significant heart rate decrease as an effect of training [11, 16–19]. Moreover, HR is not only influenced by effort, but even so by emotional excitability, reason for which many studies on equine welfare have included the follow up of heart rate variability in their study design [20, 21]. In the study performed by Munsters et al., no BLA levels were determined at start of the training study, only a longitudinal follow up of HRs was performed.

However, BLA level determination requires blood collection, and it is not feasible to advocate this during each training day. Heart rate monitoring is a non-invasive and elegant technique and the results of our study show a high and consistent correlation between BLA levels after cantering and average HR at trot, which was not the case for average HR at canter. This finding could support the hypothesis that canter is quite a difficult and energy demanding gait for Friesian horses. The fact that canter is an exciting event for horses and not only an effort probably partly explains why correlations between BLA and HR at canter are less consistent. And the current study only encompasses 9 horses. However, the strong correlation between average HR at trot and BLA after canter in all 9 Friesian horses suggests that longitudinal follow-up of HR during trot is an interesting approach to monitor training response of young Friesian dressage horses. Munsters et al. found a significant correlation between mean HR in canter and BLA after canter [8]. Munoz et al. found significant correlations between HR, BLA and stride length in canter in Andalusian horses but not in Anglo-Arabian and Arabian horses [16]. The results of our study suggest that HR in trot is a useful parameter to monitor while training Friesian horses. However, more research is needed to further explore this approach.

Striking in our study is the fact that horses with the worst training parameters at the beginning of the training, also showed the largest training response. Horses that had the highest heart rates and blood lactic acid concentrations in week 0, also showed the largest response to training. The results of our study thus show that caution is warranted when categorizing a young Friesian horse as predisposed to poor performance and thus unfit for a professional sports career, based upon poor parameter results obtained during a SET performed at start of the training. Our study also shows that the duration of follow-up is important and that depending on the used SET, the necessary duration of follow-up differs. In our study this was 6 weeks when using SETA and 8 weeks when using SETB.

Our study population was too small to assess possible influence of pedigree on training response. And keeping in mind the very narrow genetic basis of the Friesian breed, such studies would entail the need for a very large study population of several of hundreds of horses.

Conclusion

Young Friesian horses do reach their anaerobic threshold during a SET test which requires lower intensity than daily routine training. Therefore close monitoring throughout training is warranted. Longitudinal follow up of BL and not of HR is suitable to assess training response and necessary duration of follow up depends on the applied SET test. SETA has the highest value to follow-up training response, whereas SETB provides a solid template to train Friesian horses in the aerobic window. Monitoring HR in trot and incorporating longer periods of trot during training of young Friesian horses is advised. Finally, caution is warranted to categorize a young Friesian dressage horse as unfit for a sportive career based upon poor SET parameter results at start of the training.

Additional file

Additional file 1: Table S1. Correlation between heart rate and $[\text{Lactate}]_{canter2}$ for SETA. **Table S2.** Correlation between HR and $[\text{Lactate}]_{canter2}$ for SETB. *Correlations heart rate and blood lactic acid.* Correlations between HR parameters and BLA $[\text{Lactate}]_{canter1}$, $[\text{Lactate}]_{canter2}$ at start and at the end of both SETA and SETB were studied in week 0, 2, 4, 6 and 8. Results of the analysis of BLA after $[\text{Lactate}]_{canter2}$ and the maximum and average HRs throughout SETA are shown in supplementary Table 1. Results of the analysis of BLA after $[\text{Lactate}]_{canter2}$ and the maximum and average HRs throughout SETB are shown in supplementary Table 2. Significant correlation coefficients are marked by an asterix (*). (PDF 22 kb)

Abbreviations

BLA: Blood lactic acid; GPS: Global positioning system; HR: Heart rate; HR_{avTrot}: Average heart rate during trot; HR_{LC}: Average heart rate during left canter; HR_{maxLC}: Maximum heart rate during left canter; HR_{maxRC}: Maximum heart rate during right canter; HR_{RC}: Average heart rate during right canter;

SET: Standardized exercise test; SETA: Standardized exercise test A; SETB: Standardized exercise test B

Acknowledgements
The authors would like to thank Peter Spahn, Susan Bouwman, Ids Hellinga and The Fenway Foundation for their collaboration.

Funding
This work has been financially supported by the Royal Dutch Friesian Studbook (KFPS) and The Fenway Foundation.

Authors' contributions
CD, CB and LV designed the study. Data collection was performed by WH and BB. Data analyses and interpretation was performed by CD, KG and BB. CB, WH, BB and CD were involved in creating the manuscript. MP, BD and LV critically revised the manuscript. All authors read and approved the final manuscript.

Competing interests
The authors declare that they have no competing interests.

Consent for publication
Not applicable for this study.

Author details
[1]Wolvega Equine Hospital, Stellingenweg 10, Oldeholtpade 8474 EA, The Netherlands. [2]Department of Comparative Physiology and Biometrics, Faculty of Veterinary Medicine, Ghent University, Salisburylaan 133, Merelbeke 9820, Belgium. [3]Department of Pathobiology, Faculty of Veterinary Medicine, Utrecht University, Utrecht, The Netherlands. [4]Department of Animal Breeding and Genetics, Wageningen University, Wageningen, The Netherlands. [5]Equine Health and Performance Centre, School of Animal and Veterinary Sciences, Roseworthy Campus, University of Adelaide, Adelaide, SA 5371, Australia.

References

1. Vermeulen A, Evans D. Measurements of fitness in Thoroughbred racehorses using field studies of heart rate and velocity with a global positioning system. Equine Vet J. 2006;36:113–7.
2. Leleu C, Cotrel C, Courroucé-Malblanc A. Relationships between physiological variables and race performance in French Standardbred trotters. Vet Rec. 2005;156:339–42.
3. Evans D. Physiology of equine performance and associated tests of function. Equine Vet J. 2007;39:373–83.
4. Davie A, Evans D. Blood Lactate Responses to Submaximal Field Exercise Tests in Thoroughbred Horses. Vet J. 2000;159:252–8.
5. Riber C, Castejon F, Rubio D, Tovar P, Vinuesa M. A Comparative Study of Aerobic Capacity and Fitness in Three Different Horse Breeds (Andalusian, Arabian and Anglo-Arabian). J Vet Med. 1994;41:645–52.
6. Muñoz A, Riber C, Santisteban R, Vivo R, Agüera S, Castejón F. Investigation of Standardized Exercise Tests According to Fitness Level for Three-Day Event Horses. J Equine Sci. 1998;9:1–7.
7. Rivero JLL, Hill EW. Skeletal muscle adaptations and muscle genomics of performance horses. Vet J. 2016;209:5–13.
8. Munsters CC, van den Broek J, van Weeren R, Sloet van Oldruitenborgh-Oosterbaan MM. Young Friesian horses show familial aggregation in fitness response to a 7-week performance test. Vet J. 2013;198:193–9.
9. Marlin DJ, Scott CM, Schroter RC, Harris RC, Harris PA, Roberts CA, Mills PC. Physiological responses of horses to a treadmill simulated speed and endurance test in high heat and humidity before and after humid heat acclimation. Equine Vet J. 1999;31:31–42.
10. White SL, Williamson LH, Kaykuth PL, Cole SP, Andrews FM, Geiser DR. Heart rate response and plasma lactate concentrations of horses competing in the cross-country phase of combined training events. Equine Vet J Suppl. 1995;20:47–51.
11. Sloet van Oldruitenborgh-Oosterbaan MM, Wensing T, Barneveld A, Breukink HJ. Value of standardised exercise tests and blood biochemistry in the selection and training of breeding stallions. Vet Rec. 1991;129:356–9.
12. Munoz A, Santisteban R, Rubio MD, Vivo R, Aguera EI, Escribano BM, Castejon FM. The use of functional indexes to evaluate fitness in Andalusian horses. J Vet Med Sci. 1997;59:747–52.
13. Munsters CC, van den Broek J, Welling E, van Weeren R, van Oldruitenborgh-Oosterbaan MM. A prospective study on a cohort of horses and ponies selected for participation in the European Eventing Championship: reasons for withdrawal and predictive value of fitness tests. BMC Vet Res. 2013. doi: 10.1186/1746-6148-9-182.
14. Lindner A, Signorini R, Brero L, Arn E, Mazzini R, Enriques A. Effect of conditioning horses once, twice, or thrice a week with high-intensity intermittent exercise on V4. J Equine Vet Sci. 2012;32:153–7.
15. Taylor L, Duberstein K, Turner K. The effects of interval versus continuous conditioning on physiological and minematic parameters of equine fitness. J Equine Vet Sci. 2015;35:388.
16. Munoz A, Santisteban R, Rubio MD, Riber C, Aguera EI, Castejon FM. Locomotor response to exercise in relation to plasma lactate accumulation and heart rate in Andalusian and Anglo-Arabian horses. Vet Res Commun. 1999;23:369–84.
17. Bruin G, Kuipers H, Keizer HA, Vander Vusse GJ. Adaptation and overtraining in horses subjected to increasing training loads. J Appl Physiol. 1994;76:1908–13.
18. Hickson RC, Hagberg JM, Ehsani AA, Holloszy JO. Time course of the adaptive responses of aerobic power and heart rate to training. Med Sci Sports Exerc. 1981;13:17–20.
19. Lindner A, Mosen H, Kissenbeck S, Fuhrmann H, Sallmann H. Effect of blood lactate-guided conditioning of horses with exercises of differing durations and intensities on heart rate and biochemical blood variables. J Anim Sci. 2009;87:3211–7.
20. Kowalik S, Janczarek I, Kedzierski W, Stachurska A, Wilk I. The effect of relaxing massage on heart rate variability in purebred Arabian racehorses. Anim Sci J. 2016. doi:10.1111/asj.12671.
21. Janczarek I, Wilk I, Zalewska E, Bocian K. Correlations between the behaviour of recreational horses, the physiological parameters and summer atmospheric conditions. Anim Sci J. 2015;86:721–8.

14

Characterization and differentiation of equine experimental local and early systemic inflammation by expression responses of inflammation-related genes in peripheral blood leukocytes

Anne Mette L. Vinther[1*], Peter M. H. Heegaard[2], Kerstin Skovgaard[2], Rikke Buhl[1], Stine M. Andreassen[1] and Pia H. Andersen[3]

Abstract

Background: Local inflammation may progress into systemic inflammation. To increase our understanding of the basic immunological processes during transition of equine local inflammation into a systemic state, investigation into the equine systemic immune response to local inflammation is warranted. Therefore, the aim of this study was to investigate the innate peripheral blood leukocyte (PBL) immune response to local inflammation in horses, and to compare this response with the PBL immune response during the early phase of acute systemic inflammation. Expression of 22 selected inflammation-related genes was measured in whole blood leukocytes from 6 horses in an experimental cross-over model of lipopolysaccharide- (LPS-) induced acute synovitis (3 μg LPS intraarticularly; locally inflamed [LI] horses) and endotoxemia (1 μg LPS/kg intravenously; systemically inflamed [SI] horses). Multiple clinical and hematological/biochemical examinations were performed, and serial blood samples were analyzed by reverse transcription quantitative real-time PCR. Post-induction expression profiles of all genes were compared between study groups using principal component analysis (PCA) and hierarchical clustering.

Results: Moderate synovitis and mild systemic inflammation of approximately 24 h duration was confirmed by clinical and paraclinical observations in LI and SI horses, respectively. In the LI group, samples obtained 3–16 h post-injection showed distinct clustering in the PCA compared with baseline levels, indicating a transcriptional response to local inflammation in PBLs in this time interval. There was no clinical or hematological indication of actual systemic inflammation. There was a clear separation of all LI samples from all SI samples in two distinct clusters, indicating that expression profiles in the two study groups were different, independent of time since LPS injection. Co-regulated genes formed four clusters across study groups which were distinctly differently regulated. Only few of individual genes displayed different expression between the study groups at all times after LPS injection.

Conclusions: Local inflammation in horses initiated an innate transcriptional response in PBLs, which differed from the transcriptional response during the early phase of systemic inflammation. This study may provide new insights into the immunobiology of PBLs during the transition of local inflammation into a systemic state.

Keywords: Cytokines, Endotoxemia, Gene expression, Horses, Inflammatory transition, Innate immunity, Leukocytes, Lipopolysaccharide, Local inflammation, Systemic inflammation

* Correspondence: aml.vinther@gmail.com
[1]Department of Large Animal Sciences, Faculty of Health and Medical Sciences, University of Copenhagen, Taastrup, Denmark
Full list of author information is available at the end of the article

Background

Acute inflammation denotes a series of generalized vascular and cellular events that is induced by the innate immune system in response to invading pathogens and tissue trauma [1, 2]. Systemic inflammation arises secondary to local inflammation if pro-inflammatory substances such as pathogen-associated molecular patterns (PAMPs), danger-associated molecular patterns (DAMPs), and perhaps pro-inflammatory cytokines are not successfully contained by local inflammatory processes, and if these pro-inflammatory substances overwhelm the initial systemic host defenses [1, 3, 4]. Activated peripheral blood leukocytes (PBLs), especially neutrophil granulocytes and monocytes, and endothelial cells are regarded key immunological cells in the initiation and progression of systemic inflammatory processes [5, 6].

In the equine clinic, systemic inflammation is a frequent sequela to diseases such as infectious colitis, metritis, peritonitis, pleuropneumonia, and intrauterine/neonatal infections [7]. There is however not always a clear association between type of disease and the development of systemic inflammation; some horses suffering from local inflammation develop systemic inflammation while others apparently do not. The importance of closely monitoring patients at risk of developing systemic inflammation is widely recognized, as rapidly initiated targeted therapy is crucial to the treatment of systemic inflammatory conditions [7, 8]. However, diagnosing early acute systemic inflammation can be a challenge, as signs of disease during the transition of local inflammation into a systemic state inevitably will contain a mix of host responses to both local and systemic inflammatory processes. Also, local inflammation may initiate clinical and paraclinical signs which usually are thought of as evidence of systemic involvement, even though systemic inflammation does not appear to be present. This has been demonstrated in experimental studies of local inflammation in horses in which alterations in body temperature, whole blood white blood cell (WBC) count, and in some cases serum acute phase reactants (APRs), were present in response to LPS-induced synovitis [9–12] and *E. coli*-induced endometritis [13]. A systemic response like this might be due to release of locally produced pro-inflammatory cytokines to circulation, as cytokines are well known initiators of both fever [14], neutrophilia [15, 16], and acute phase responses [17]. Another, more indirect, source of such cytokines might be PBLs and endothelial cells that are activated by pro-inflammatory substances released from locally inflamed tissues. From a clinical point of view, the most specific sign of the early acute stage of systemic inflammation regardless of cause seems to be the presence of pronounced neutropenia [18], indicating that activated neutrophils have marginated in the vasculature or have sequestrated in the perivascular tissue on a generalized level [19, 20].

In human medicine, research into systemic inflammation is focused on complementing clinical and standard paraclinical parameters with immunological markers to optimize future diagnosis and prognosis of systemic inflammatory conditions [21, 22]. In both human and equine research, the traditional way of investigating immune processes during the onset of systemic inflammation is by quantitating inflammatory mediators in serum or their corresponding gene expression in PBLs in response to experimentally induced systemic inflammation [23–27]. However, very little is known about the systemic immune response to local inflammation. To increase our understanding of basic immunological processes during transition of equine local inflammation into a systemic state, investigation of the PBL immune response to local inflammation is warranted. Such investigations might further indicate if specific PBL inflammatory mediators have a diagnostic potential in monitoring equine patients at risk of developing systemic inflammation.

Therefore, the aim of this study was to investigate the innate PBL immune response to local inflammation in horses, and to compare this response with the innate PBL immune response during the early phase of acute systemic inflammation. This was done by measuring the expression of 22 selected inflammation-related genes in whole blood leukocytes in an equine cross-over model of LPS-induced acute synovitis and endotoxemia. This study is part of a larger experiment investigating inflammatory processes during equine LPS-induced acute synovitis and endotoxemia. Clinical results and expression levels of single genes for the endotoxic horses have previously been presented in Vinther et al. [28].

Methods

Horses

Six healthy, adult (7.7 ± 5.4 years), Danish Warmblood or Danish Warmblood cross bred horses (2 geldings and 4 non-pregnant mares) with an initial body weight of 490 ± 39 kg were included in the study. The horses were owned by the University of Copenhagen. Prior to the experiment a full clinical examination and hematological and biochemical analyses of blood and radiocarpal synovial fluid were performed, and only horses without clinical or hematological/biochemical signs of acute or chronic inflammation [29] were included in the study. All horses were dewormed with ivermectin (Maximec®) and vaccinated against equine influenza and tetanus (ProteqFlu-Te®) prior to the study. Approximately 2 weeks before initiation of the study, horses were housed at the research station in individual 3 x 4 m box stalls, where they were fed an equine commercial grain mixture twice daily and had access to water and hay ad libitum.

Experimental design

A cross over study involving experimentally induced endotoxemia and lipopolysaccharide-(LPS) induced synovitis as models for systemic and localized inflammation, respectively, was performed. Horses were randomly assigned to start in one of the two study groups, and the experiments were performed simultaneously on one horse from each study group at a time. All horses had an acclimatization period in the climate controlled experimental facilities (maintained at a temperature of 12 ± 1 °C) of 5 days before initiation of the experiment. To eliminate possible effects of LPS tolerance there was a 4 week wash out period in between treatments [30]. LPS was derived from *Escherichia coli* strain 055:B5 (#L2880, Sigma-Aldrich Denmark). Using a polyurethane jugular catheter (Milacath®, 14G, 150 mm), endotoxemia was induced over one minute with 1 µg/kg LPS diluted in isotonic saline (Baxter A/S, Denmark) to a total volume of 15 mL (systemically inflamed [SI] horses). Synovitis was induced in the left carpal joint by sterile intraarticular injection (Eickinject®, 21G, 40 mm) of 3 µg LPS diluted in 1 mL Ringer's Acetate (Baxter A/S, Denmark) (locally inflamed [LI] horses). The LPS doses were established on the basis of the biological action of this particular batch of LPS in previous experiments, where the 3 ug per joint dose resulted in a clinically moderate but fully reversible synovitis that lasted for approximately 24 h, and the IV dose of 1 ug/kg resulted in a systemic inflammation with fully reversible clinical signs of inflammation that lasted for approximately 24 h. Due to intraarticular LPS injection failure in one LI horse, this group only consisted of 5 horses.

Clinical examinations

Examinations comprising general condition, rectal temperature (RT), heart rate (HR), respiratory rate (RR), mucosal membrane color, capillary refill time (CRT), and lameness score (LI horses only) were performed multiple times during the 24-h experimental period for both SI and LI horses (Additional file 1). Lameness was assessed independently by two observers using the AAEP lameness score [31]. Two hours after intraarticular LPS injection the left carpal joint of the LI horses was evaluated for swelling and heat, and synovial fluid total protein and WBC were measured. 2.2 mL of synovial fluid was obtained by aseptic arthrocentesis (EDTA tubes, BD Company), and synovial fluid total protein and WBC was measured by use of a refractometer[1] and a haemocytometer[2], respectively. Additional arthrocenteses at post-injection hour (PIH) 4, 8, 16, and 24 were performed as part of a study without relevance for this experiment.

Blood sampling

Dependent on specific requirements for analyses at each time point for sampling, 7.5–31.5 mL of blood was collected in syringes (Kruuse) through the indwelling jugular venous catheter and immediately transferred to the appropriate blood tubes. The first 5 mL of blood were drawn in a separate syringe and discarded. Baseline samples (PIH 0) were taken immediately before LPS injection. Blood for WBC and total neutrophil granulocyte and lymphocyte counts was collected in EDTA tubes (BD Company) at PIH 0, 1, 2, 3, 4, 6, 8, 12, 16, 20, and 24. Blood for APRs was collected in serum tubes (serum amyloid A (SAA), haptoglobin, iron) or citrate tubes (fibrinogen) (BD Company) at PIH 0, 2, 4, 8, 16, and 24. SAA was measured by immunoturbidometry[3]. Haptoglobin was measured by a biochemical peroxidase assay in duplicate[4]. Iron was measured by colorimetric spectrophotometry[5]. Fibrinogen was measured by the Clauss method in an automated coagulometric analyzer[6]. A standardized volume of 2.5 mL of blood for RNA extraction was collected in PAXgene Blood RNA Tubes (Qiagen/BD Company), which allows for later recovery of total RNA from all cells presents in the blood sample. According to the manufacturer's instruction PAXgene™ blood tubes were gently inverted 8–10 times after sampling and kept at room temperature for 2 to 24 h, before storage at -80 °C until RNA extraction. Samples from LI horses were collected at PIH 0, 1, 2, 3, 4, 6, 8, 10, 12, 16, 20, and 24. For SI horses, genes of interest were previously found to be maximally differentially expressed within the first 8 h after LPS injection [28]. Thus, only SI samples for PIH 1-8 were included in this study, serving as references for the expression profiles during acute systemic inflammation in these particular horses.

Target genes

Relative gene expression in PBLs was measured by quantification of specific mRNA for the following 22 genes: *IL1B, IL6, IL8, IL10, IL15, IL17, IL18, IL1RN, TNF, TLR4, CD14, ITGAM, ITGAX, SELL, MAPK14, CASP3, BCL2L1, MMP8, TIMP1, IL6ST, CCL5*, and *SOD2* (Additional file 2). As presented in Vinther et al. [28]; these 22 genes were significantly regulated in the SI horses at one or more time points during the first 8 h after intravenous LPS injection. Significant regulation was defined as a statistically significant change in gene expression of at least 2.5 fold compared with baseline levels.

RT-qPCR

Total RNA extraction and quality analysis, cDNA synthesis, specific primer design, pre-amplification and exonuclease treatment, and qPCR were performed as previously described [28]. Briefly, total cellular RNA was extracted from PAXgene™ blood samples using PAXgene Blood miRNA Kits (Qiagen/BD Company) and treated with RNase-Free DNase sets (Qiagen/BD Company). Concentration and purity of total extracted RNA was determined by spectrophotometric analyses[7]. Samples containing less than 20 ng RNA/μL were evaporated (37 °C) to increase RNA concentration using a SPD111V SpeedVac Concentrator[8]. RNA integrity was estimated using an Agilent 2100 Bioanalyzer[9] and RNA 6000 Nano Kits (Agilent Technologies). Each total RNA sample was assigned an RNA Integrity Number (RIN) from 1-10, with 10 being non-degraded RNA [32]. Mean RIN value ± SD of all samples was 8.9 ± 0.9.

After extraction RNA was DNase treated a second time and 300 ng of total RNA per sample converted into first-strand cDNA by reverse transcription using a Tprofessional TRIO 3x48[10] and QuantiTect Reverse Transcription Kits (Qiagen/BD Company) according to the manufacturer's instructions. For assay validation, two separate cDNA replicates were performed for each RNA sample and a non-reverse transcriptase control was included. The cDNA samples were diluted (1:7.7) in low TE-buffer (VWR-Bie & Berntsen) and stored at -20 °C until pre-amplification procedures.

Gene specific primer pairs were designed as previously described [28] and synthesized at TAG Copenhagen (Denmark) or Sigma-Aldrich (Denmark). Transcript IDs, primer sequences, amplicon lengths, reaction efficiencies, and correlation coefficients are shown in Additional file 2. Primer amplification efficiencies, correlations and dynamic ranges were acquired from standard curves based on 4 separate dilution series (3 representing the whole sample material and 1 representing samples with high levels of *IL6* to specifically meet the dynamic range of this particular gene) of pooled cDNA [28]. Efficiencies between 0.91 and 1.12 and correlations above 0.97 were accepted.

Stocks of 200 nM primer pair mix in low TE-buffer (VWR-Bie & Berntsen, Denmark) were prepared combining equal amounts of all primers used in the study. Primer pair mix, TaqMan PreAmp Master Mix (Applied Biosystems), and cDNA were mixed and incubated at 95 °C for 10 min followed by 16 cycles of 95 °C for 15 s and 60 °C for 4 min. After pre-amplification cDNA was exonuclease treated. An aliquot of the pre-amplified cDNA was saved for preparation of dilution series, and finally cDNA was diluted in low TE-buffer before quantitative real-time PCR (qPCR).

Quantitative PCR was performed in 48.48 Dynamic Array Integrated Fluidic Circuits (IFC) on a BioMark Fluidigm thermocycler[11] combining 48 samples with 48 primer sets in 2304 separate, simultaneous reactions [33]. Each chip included a non-template control (NTC), a non-reverse transcriptase control, and cDNA dilution pools. Expression data (Cq values in heat maps) and melting curves were acquired using Fluidigm Real-Time PCR Analysis software 3.0.2 (Fluidigm). NTCs and melting curves were used to monitor for non-specific amplification or sample contaminations, and non-reverse transcriptase controls were used to assess potential DNA contamination. For both of these control samples, a minimum of 5 Cq-values between potential signals and a given sample signal were required, and only genes with a single melting peak (within 1 °C) were accepted for further data analyses.

Pre-processing and quantification of data

Expression data were exported to GenEx5[12] for pre-processing and quantification. Dilution pools were used as inter plate calibrators. Data was corrected for primer efficiencies for each primer assay individually. Using NormFinder [34] and GeNorm [35] the most stably expressed reference genes (ACTB, TBP, DIMT1, SDHA, HPRT1, and B2M) out of 7 putative candidates were selected for normalization. cDNA technical replicates were averaged. Finally, relative expression levels were established. For each gene of interest the expression level was set to 1 for the sample with the lowest level of expression. Expression levels for the specific gene of interest in all other samples irrespective of horse and time for sampling were then calculated relative to this sample during data transformation from Cq (log2) to relative quantities (relative fold change, linear scale).

Statistical analyses

Clinical and paraclinical data

Clinical and paraclinical data are presented as mean values except for lameness score, which is presented as a range of scores of individual horses. Selected clinical and paraclinical data are depicted graphically using SigmaPlot v. 12.5. Statistical analyses of selected clinical and hematological responses (RT, HR, RR, WBC counts, APRs) were analyzed with a two-way ANOVA random effect (mixed effect) model (R v. 3.1.2) where time and treatment (and their interaction) were fixed effects that influence the mean. A serial Gaussian correlation structure was included to allow for time-dependent correlation within each treatment on each horse and to allow for variation among horses. Distribution of data was evaluated by residual plots and data was log2-transformed to improve the model fits. The t-test was used to compare each time point with the baseline at PIH 0 and to compare time points between treatments groups (APRs only). The Bonferroni correction was used to account for testing multiple disease parameters, and statistical significance was defined as $P < 0.0063$ (significance level 0.05/8 parameters). No correction was made for testing at multiple time points as number of time point varied between parameters.

Expression data

All analyses on expression data were performed on log2-transformed, autoscaled data using GenEx5. Data was autoscaled to have similar weights in the analyses regardless of magnitude and variation in expression levels of the 22 genes [36]. To disclose multivariate responses in the study groups, the method of principal component analysis (PCA) was employed, grouping samples or genes based on gene co-expression patterns [37]. In the present study, two PCAs were performed to visualize clusters of individual samples. The first PCA included expression data of all LI samples divided into the following subgroups: PIH 0, PIH 1-2, PIH 3-4, PIH 5-8, PIH 10-16, and PIH 20-24. The second PCA included expression data of SI (PIH 1-8) samples, LI (PIH 3-16) samples, and baseline samples for both study groups.

Based on expression level group means, differences in expression profiles of LI (PIH 3-16) samples in relation to SI (PIH 1-8) samples were further visualized in a heat map combining dendrograms of samples and genes. The dendrograms were performed by using agglomerative clustering, Euclidean distance as the distance measure and Ward's method as clustering algorithm.

Results

Clinical and paraclinical responses to endotoxin

During the acclimatization period all horses were bright, alert, and responsive, and considered healthy on clinical and hematological examinations (data not shown). A graphical presentation of baseline and post-injection results of selected clinical and paraclinical parameters including statistics for both study groups are presented in Figs. 1 and 2. SI horses started to show signs consistent with endotoxemia such as anorexia, mild to moderate colic, fever, tachycardia, and pronounced neutropenic leukopenia within the first two hours post-injection. WBC started to increase after PIH 2 due to an increasing number of circulating neutrophil granulocytes, and at PIH 4 rectal temperature peaked. Within the first 8 h after LPS injection all horses except one had hyperemic, dark red, or cyanotic mucosal membranes and a CRT of 3–4 s. At PIH 8, a mildly elevated rectal temperature and heart rate, an increasing WBC count, decreased appetite, depression, and mild colicky behavior characterized the clinical status. After 24 h horses had an elevated heart rate and WBC count but were all, except for one horse, bright, alert, and responsive. All horses recovered without treatment.

At PIH 2, LI horses were 4 to 5° lame (5° scale), and had mild tachycardia and tachypnea (Fig. 1). There was swelling of the left carpal joint, palpable heat, and elevated synovial fluid total protein (35.2 ± 9.7 g/L) and WBC count (61.2 ± 31.6 x 10^9 cells/L) (reference values of non-inflamed synovial fluid total protein and WBC approximately 0.2 g/L and 87-167 x 10^6 cells/L, respectively [38]). During the following 2–6 h horses expressed varying degrees of depression, decreased appetite, mild colic, increased WBC counts, tachycardia, and tachypnea. Rectal temperature was slightly elevated at PIH 5-12. Lameness score was 4 to 5° at PIH 8. At PIH 24, horses were 1 to 3° lame and were bright, alert, and responsive. During the entire disease period all horses had a normal mucosal membrane color and CRT ≤ 2 s. None of the horses received analgesic treatment at any time during the study.

As seen in Fig. 2, serum SAA concentration for SI horses increased from 8 h post-injection until the end of the study whereas for LI horses SAA levels increased at 16 and 24 h of the onset of the study. Iron concentration was decreased after 16 h for both groups. In the LI group, haptoglobin and fibrinogen concentration was increased at PIH 24, while no significant changes were seen for these two acute phase proteins in the SI group. SAA concentration was different between SI and LI horses at PIH 8. Otherwise acute phase proteins did not differ between study groups within the 24-h observational period.

Expression analyses

In the two study groups PCA was used to investigate the innate gene expression response in PBLs based on co-expression patterns of individual samples. Figure 3a and b were obtained by plotting all relevant samples by their two first principal components obtained from all 22 genes. Figure 3a shows distinct clustering of samples from LI (PIH 3-4), LI (PIH 5-8), and to a certain extend LI (PIH 10-16) horses compared with baseline samples. Figure 3b includes all samples from SI horses (PIH 1-8) together with the samples from LI horses in which co-expression patterns in Fig. 3a showed distinct clustering from baseline samples (PIH 3-16). There was a distinct clustering of SI samples from LI samples regardless of time since LPS injections. This clustering was not seen in baseline samples before LPS injection. All samples from SI horses clustered separately from pre-injection samples.

A heat map combining two dendrograms was constructed to visualize the expression matrix of samples and genes together to further characterize clustering within samples and genes for SI (PIH 1-8) horses and LI (PIH 3-16) horses based on average expression levels (Fig. 4). As seen in the upper dendrogram, samples grouped in three main clusters: SI (PIH 1-3), SI (PIH 4-8), and LI (PIH 3-16). The response levels of genes, going from green (low expression) to red

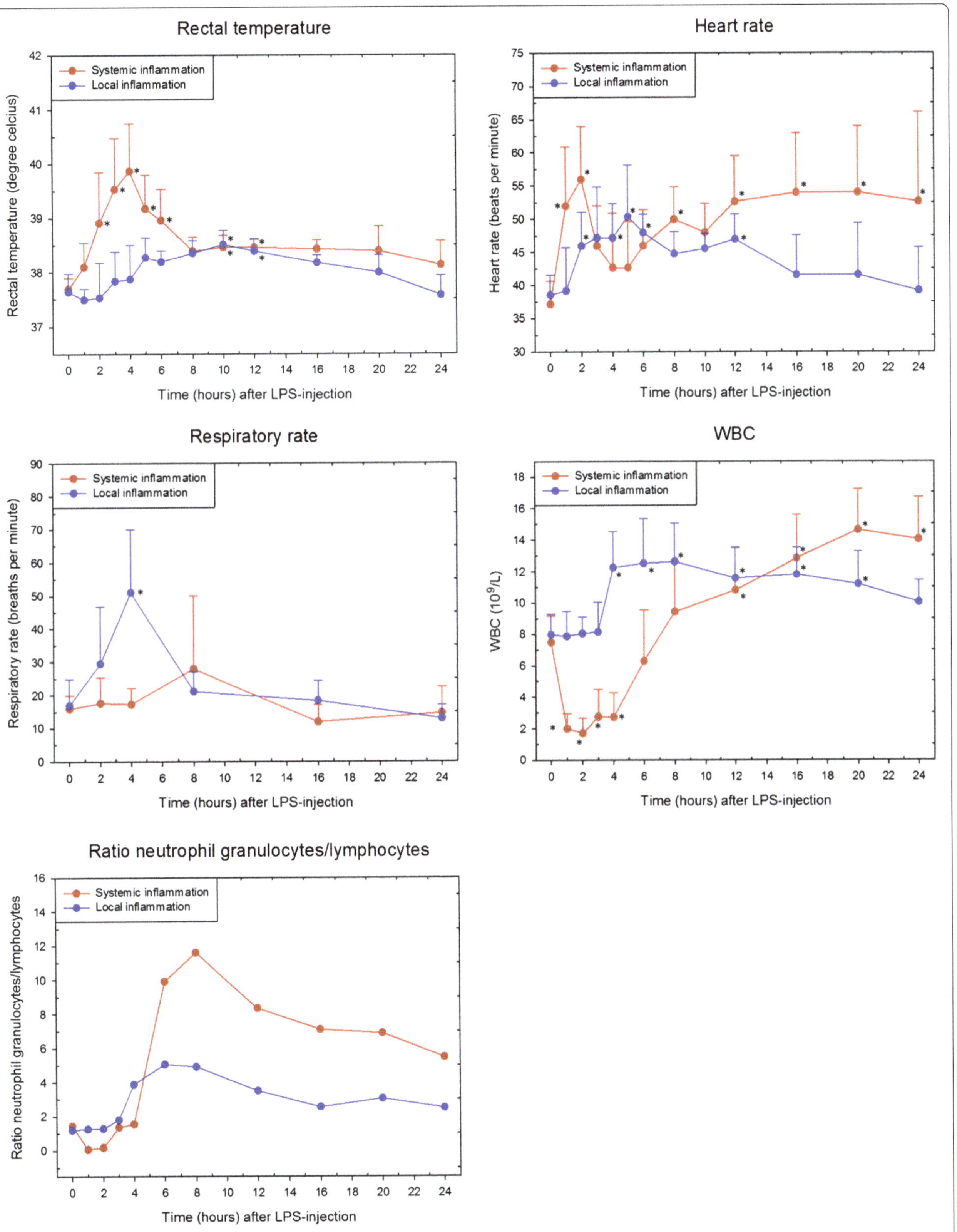

Fig. 1 (See legend on next page.)

(See figure on previous page.)
Fig. 1 Clinical and hematological responses to LPS. Rectal temperature, heart rate, respiratory rate, white blood cell (WBC) counts, and neutrophil granulocyte/lymphocyte ratio in horses with induced systemic (red) and local (blue) inflammation, respectively. All values except neutrophil granulocyte/lymphocyte ratio are depicted as mean ± SD. Asterisk (*) denotes statistical significant differences compared with baseline levels at PIH 0 ($P < 0.0063$). Please note that clinical and hematological responses for the horses with systemic inflammation have also been presented in the article Vinther et al. [28]. PIH: post-injection hour

(high expression) were more pronounced and fluctuating for the SI group. Genes clustered in 4 groups with similar expression patterns (Cluster A-D) across study groups. The genes *IL8* and *IL1RN* and to a lesser degree *IL1B* in Cluster C were transcribed to different levels for SI and LI horses independent of time since injection of LPS. Expression levels of all other genes only differed between SI and LI horses in certain time intervals during the two disease courses. In the SI group, gene expression levels in Cluster A (*IL18* to *TIMP1*) increased from low to high over time, having a markedly different level than those of the LI horses at PIH 1-2 and PIH 5-8, but not at PIH 3-4. Genes in Cluster B (*IL15, CD14, MAPK14, ITGAM*) showed different expression levels between the LI group and the first 3 h post-injection for the SI group. Genes in Cluster D (*IL10, TNF, IL17, IL6, CCL5*) were most highly expressed during the first hours in the SI group but otherwise showed heterogeneous transcription patterns.

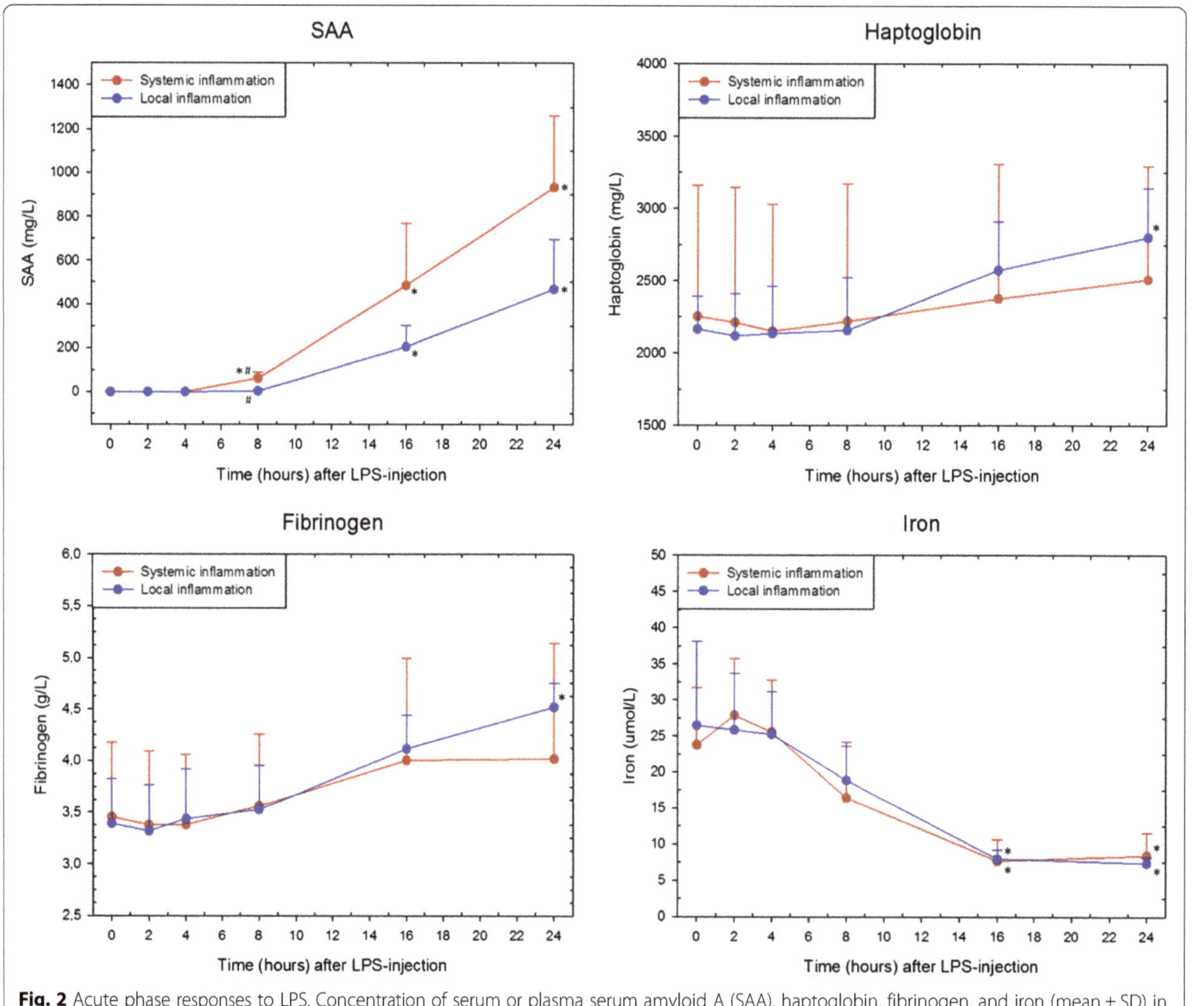

Fig. 2 Acute phase responses to LPS. Concentration of serum or plasma serum amyloid A (SAA), haptoglobin, fibrinogen, and iron (mean ± SD) in horses with induced systemic (*red*) and local (*blue*) inflammation, respectively. Asterisk (*) denotes statistical significant differences compared with baseline levels at PIH 0, while hash tag (#) denotes statistical significant differences between SI and LI horses ($P < 0.0063$). PIH: post-injection hour

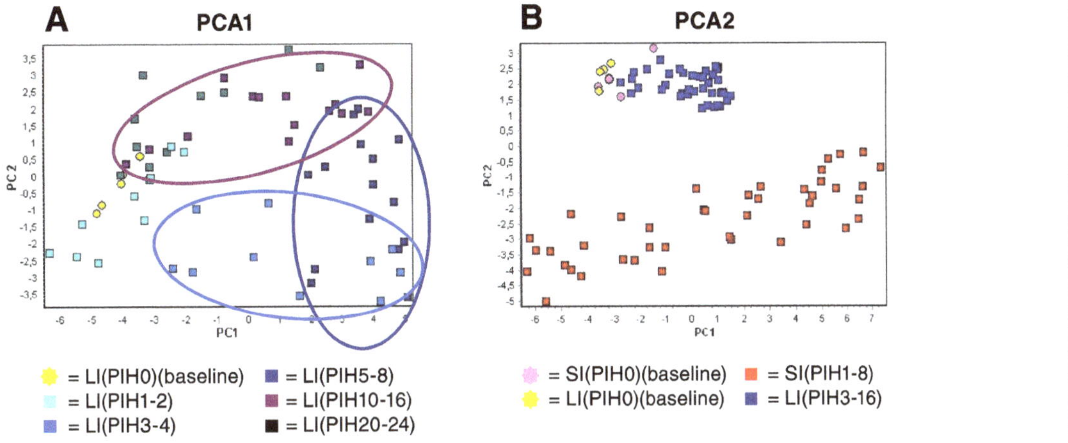

Fig. 3 Clustering of samples from systemically and locally inflamed horses. Clustering of samples based on the co-expression patterns of 22 innate genes measured in PBLs. Clustering was performed by use of principal component analysis (PCA) in which individual samples were plotted by their two first principal components. **a** Expression data of all LI samples was divided into the following subgroups: PIH 0 (baselines), PIH 1-2, PIH 3-4, PIH 5-8, PIH 10-16, and PIH 20-24. Ellipses are drawn to highlight the sample subgroups at PIH 3-4, PIH 5-8, and PIH 10-16. **b** Expression data of SI (PIH 1-8) samples, LI (PIH 3-16) samples, and baseline samples for both study groups. PIH: post-injection hour

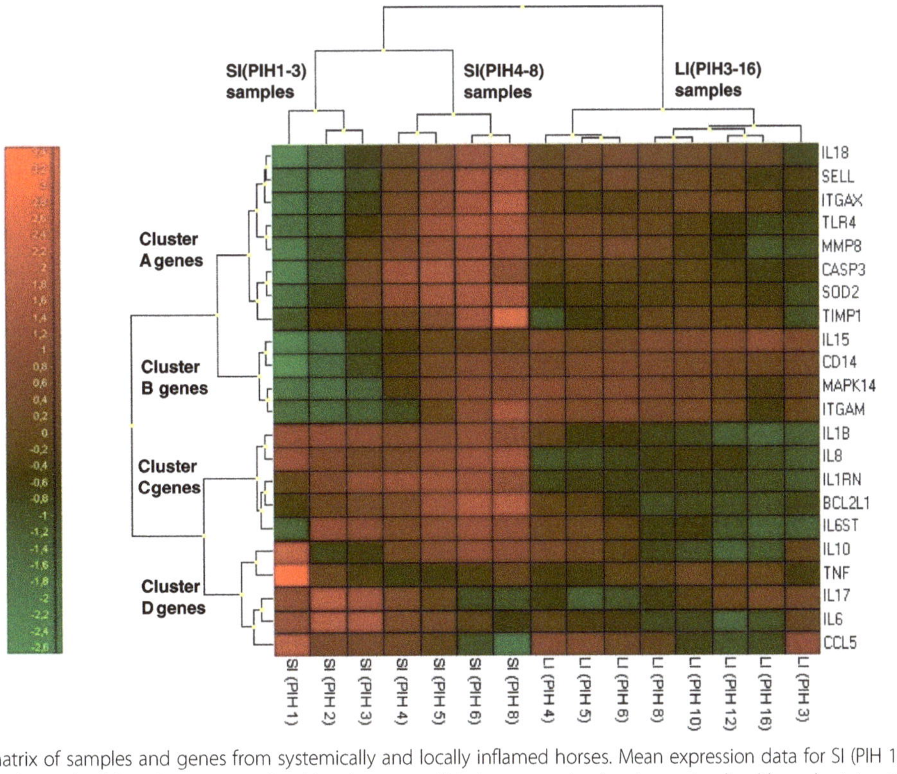

Fig. 4 Expression matrix of samples and genes from systemically and locally inflamed horses. Mean expression data for SI (PIH 1-8) samples and LI (PIH 3-16) samples for each of the 22 genes visualized in a heat map. Relative expression levels are visualized by color intensities where red corresponds to high expression and green corresponds to low expression. Please note that baseline samples are not included to indicate baseline levels. The upper dendrogram shows clustering of samples across genes. The dendrogram to the left indicates clustering of genes across samples (time points and study groups). Clustering is based on similarities between samples/genes indicated by the distance at which they are joined. Short distance indicates high similarity. PIH: post-injection hour

Discussion

All horses in the SI and LI group developed classic signs of mild endotoxemia and moderate aseptic synovitis, respectively, and both clinical disease courses were self-limiting and lasted approximately for the expected 24 h. In addition to local inflammation and pain, LI-horses showed the systemic response of fever, increased peripheral blood WBC counts, and alterations in serum acute phase reactants previously described in other studies of equine induced local inflammation [9–11], clinically indicating the systemic presence of pro-inflammatory cytokines [14–17]. A hallmark of acute systemic inflammation is the activation of the endothelium, leading to expression of adhesion molecules and leukocyte margination [5]. This is clinically reflected by a rapidly initiated marked decrease in circulating leukocytes, which could be appreciated in the SI horses as early as one hour after induction of inflammation. In contrast, this hallmark of early systemic inflammation was not detected at any time point in the LI horses (Fig. 1). As all other vital clinical and hematological disease parameters, except for RR, in the LI horses only changed mildly, there is a strong indication that intraarticular LPS injection in these horses had not elicited secondary systemic inflammation.

Figure 3a reveals that samples LI (PIH 3-4) and LI (PIH 5-8) clustered distinctly from baseline samples, while LI (PIH 10-16) samples clustered with a partial overlap to baseline samples. This indicates changes in transcription of innate immune genes in circulating leukocytes approximately 3–16 h after intraarticular LPS injection. To our knowledge, this study is the first in horses to show that LPS-induced local inflammation can initiate a complex transcriptional response of innate immune genes in circulating leukocytes. As the horses by all appearances did not experience actual systemic inflammation, this further implies that the detection of transcription activity of innate inflammatory mediators in circulating leukocytes not necessarily equals a systemic inflammatory process in an individual. Whether the systemic response of fever, increased WBC count, and changes in serum APR concentration was due to the release of pro-inflammatory cytokines from the inflamed joint, or whether it was mediated by cytokines produces by activated PBLs, is unknown. However, the increase in WBC count and rectal temperature was not detectable until after the transcriptional onset of genes in PBLs, indicating a relationship between these signs of disease and activation of PBLs. In contrast to this, there was no PBL transcriptional response in LI horses at PIH 2, where horses were almost non-weight-bearing lame, had mild tachycardia and tachypnea, and had signs of acute joint inflammation such as local swelling and heat and elevated synovial fluid total protein and WBC count. This clinical presentation thus reflected a purely local inflammatory process where tachycardia and tachypnea may have been responses to pain from the inflamed joint [39]. Our results thus suggest that elevated body temperature and an increased WBC count may be more indicative of the presence of innate inflammatory mediators in circulation than tachycardia and tachypnea. However, WBC was still significantly elevated at PIH 20 where the systemic expression response had abated.

The presence of inflammatory mediators in circulation in response to local inflammation has only been sparsely investigated in horses. In Cook et al. [40], increased serum levels of the lipid mediators prostaglandin E_2 and thromboxane B_2 were found after ischemia-induced local inflammation of the jejunum, and in a study by Campebell et al. [11], no difference in serum TNF-*a* concentration could be detected in LPS-induced synovitis despite significant changes in body temperature and circulating white blood cell composition in response to LPS. In induced endometritis in mares, transcription of the cytokines *IL1B, IL6, IL8, IL10, TNF,* and *IL1RN* were investigated in serial samples after inflammatory induction, but no differences were found when compared with pre-induction levels [41]. In dogs, mild changes in WBC counts and in markers of coagulation (activated partial thromboplastin time, fibrinogen, protein-C, and -S) were seen in peripheral blood in response to experimentally induced localized soft tissue inflammation [42]. Increased levels of for example serum IL-6, IL-10, and HMGB-1 were reported in pigs and mice in more severe models of local inflammation such as multiple trauma [43, 44] and acute kidney injury [45]. Even though the exact mechanisms determining whether local inflammation is strictly contained, mediates a systemic response, or initiates actual systemic inflammation are unknown, the severity of the initial insult is probably an important factor. A general dose dependency of the *in vivo* response to LPS has previously been shown in an experimental study in cattle [46]. Furthermore, a study on equine LPS-induced synovitis by Palmer and Bertone [10] specifically supports a dose dependency in the transition of local inflammation to a systemic state, as signs of systemic involvement (fever, altered leukocyte differential count) were only observed after high intraarticular injection doses of LPS. Factors such as pathogen virulence [47, 48] and immune competence of the host [49, 50] most likely contribute to determine the outcome of local inflammation.

In the present study, the PBL transcriptional response to local inflammation at PIH 3-16 was compared with the PBL transcriptional response during the early phase of acute systemic inflammation. The clear separation of all LI samples from all SI samples (Fig. 3b) indicates that expression profiles in the two study groups were substantially different independent of time since LPS injection. In this study, expression profiles of innate immune genes in circulating leukocytes could thus differentiate horses with systemic inflammation from horses with local inflammation and a secondary systemic transcriptional response, during both onset and progression of inflammation. For comparison, pronounced leukopenia, which is pathognomonic for systemic inflammation, was only seen in SI horses for approximately 4 h (Fig. 1). Like WBC, the APRs SAA, haptoglobin, fibrinogen, and iron are also widely used in the diagnosis of inflammation in both veterinary and human medicine [17, 51–53]. While transcription of leukocyte genes in the present study could be detected within the first few hours after LPS injection in both study groups, the acute phase reactants did not significantly change until PIH 8 at the earliest. A delay in acute phase responses compared with cytokine responses is not surprising as cytokines are well known mediators of APRs [17]. It was however remarkable that the 24-h acute phase responses were similar between SI and LI horses despite the two distinctly different gene expression profiles.

To further characterize differences in the PBL transcriptional responses in the two study groups, a heat map combining two dendrograms was constructed using mean expression values for horses for SI (PIH 1-8) and LI (PIH 3-16) samples (Fig. 4). Color intensities in the heat map suggest that the expression response for the SI group in general was more pronounced and with greater fluctuations than for the LI group. This may also be appreciated by the scattered cluster of individual SI samples and the much tighter cluster of individual LI samples in Fig. 3b. The clustering of study groups seen in Fig. 3b is supported by the upper dendrogram in Fig. 4, which furthermore suggests subgrouping of SI samples in PIH 1-3 and PIH 4-8. The subgrouping of SI samples indicates a marked change in the expression profile for SI horses as the disease course progressed, which was likely due to the transition from relative lymphocytosis to relative neutrophilia. The dependence of gene expression results upon the composition of the various types of white blood cells is supported by both equine [54] and human [55, 56] studies and is for the present experiment discussed in detail in Vinther et al. [28]. The heat map suggests that especially IL6, IL17, and most of the genes in Cluster A contributed to subgrouping of the inflammatory response in the SI group.

The 22 genes included in the study clustered in four main groups based on their temporal expression patterns (Cluster A-D, Fig. 4). Grouping of the genes *IL8*, *IL1RN*, and *IL1B* (Cluster C) was also seen in a study by Prabhakar et al. [57], in which 16 different genes were clustered by similarities in time profiles during a 24-h course of human LPS-induced systemic inflammation. In the present study, genes within especially Cluster A and B tended to follow the same expression pattern in the SI group, which indicates that the panel of genes could probably be reduced within these clusters without affecting the overall clustering analyses. Assessed visually, only few of the specific genes displayed distinctly different expression levels between study groups, irrespective of time since LPS injection. Except for *IL8*, *IL1RN*, and *IL1B* which seemed to be consistently higher expressed in SI-horses, single gene expressions thus only discriminated LI samples from SI samples in certain time intervals. *IL1RN* has not previously been investigated in systemic inflammation in the horse, and while *IL8* and *IL1B* seem to be differentially expressed in equine experimental systemic inflammation [25, 27, 58], the transcriptional response of these genes in equine septic patients is more unclear [54, 59–61].

By evaluating expression profiles across the 22 genes instead of expression levels of each gene one by one, a more detailed and comprehensive picture of the complex immunological states of the cells was obtained. Multivariate methods for analysis such as PCA and hierarchical clustering are not frequently used in studies of innate immune gene expressions in large animals even though the well described interactions of inflammatory mediators [23, 62] make it a biologically relevant approach. Theoretically, the optimal experimental model to study innate immune processes during the transition of local inflammation into a systemic state would be a model in which systemic inflammation was obtained by inducing local inflammation. To our knowledge, no such experimental model is however well characterized in the horse and we therefore used the traditional intra-articular LPS model as a model for a localized inflammation, even though this may not be the most frequent source of a systemic inflammation. Another limitation of the study includes the fact that the group of horses was quite different according to gender, age, and origin. Furthermore, the complete clinical history of the horses prior to the inclusion examinations was not known, leaving questions on possible endotoxin tolerance due to previous endotoxin related diseases. However, none of the experimental horses showed a decreased pyrogenic response or lack of illness behavior, which are considered clinical hallmarks of tolerance [30, 63]. Additionally, induced in vivo endotoxin tolerance in horses subsides within 2-3 weeks [30], meaning that the risk of endotoxin tolerance in the experimental horses was low. The 4 week wash out period between the two challenges reduced the risk of induction of endotoxin tolerance, and the cross over study minimized effects of the large individual variation known to characterize the innate immune response [55, 64, 65]. The clear clustering of genes described here therefore represent robust findings, despite the low number of experimental horses and their different backgrounds. In the present study, by restricting samples in the SI group to the first 8 h after LPS injection, this model of systemic inflammation was representative of the expression profiles during the onset and progression of early systemic inflammation. Since there is very little knowledge of the possible PBL response to local inflammation, samples from LI horses were included covering the complete course of disease from clinical onset to approximate recovery (PIH 1-24).

To our knowledge, this is the first study in horses that specifically assesses the PBL transcriptional response to local inflammation and describes it in the context of onset and early development of systemic inflammation. New insights into the molecular mechanisms involved in the transition of local inflammation into a systemic state is needed for future advances in the diagnostic opportunities of early systemic inflammation in horses. A recent study by Hooijberg et al. [66] highlights the need for this to supplement standard clinical and paraclinical parameters in the differentiation of equine local and systemic inflammation. The study retrospectively investigated the diagnostic efficacy and combined predictive capability of the myeloperoxidase index (MPXI), and plasma fibrinogen, iron, and SAA concentrations for the diagnosis of systemic inflammation, local inflammation, or non-inflammatory disease in hospitalized horses. Based on quantitative results of these markers, diagnostic guidelines could not be formulated by Hooijberg and colleagues. The group of systemically inflamed horses in the study did not specifically represent early systemic inflammation, but given the results of the APRs in the present study this might not have changed the conclusion. In the present study, an important finding with direct clinical implication is the fact that self-limiting, moderate, local inflammation of brief duration mediated transcription of inflammation-related genes in PBLs. This suggests that a substantial part of equine patients with local inflammatory conditions are exposed to inflammatory mediators on a systemic level, emphasizing the relevance of close monitoring of this patient group.

Conclusions

The results in this study showed that LPS-induced local inflammation in horses could initiate a transcriptional response of inflammation-related genes in circulating leukocytes, and that such a response not necessarily equals a systemic inflammatory process in the individual. The co-expression patterns of multiple genes differentiated horses with a systemic transcriptional response to local inflammation from horses with early systemic inflammation, independent of the progression of inflammatory states. Only few of the individual genes displayed distinctly different expression levels between study groups irrespective of time since LPS injection. This study may provide new insights into the immunobiology of PBLs during the transition of local inflammation into a systemic state and encourage further research in the subject.

Endnotes

[1]Clinical Refractometer T2, Atago.

[2]Marienfeld, Th. Geyer.

[3]LZ test SAA, EIKEN Chemocal Co., Tokyo, Japan.

[4]Phase Range Hp Assay, Tridelta Development Ltd., Kildare, Ireland.

[5]ADVIA 1650, Siemens, Lyngby, Denmark.

[6]ACL 9000, Instrumentation Laboratory, Barcelona, Spain.

[7]NanoDrop ND-1000 spectrophotometer, Saveen and Werner AB, Limhamn, Sweden.

[8]SPD111V SpeedVac Concentrator, Thermo Scientific, Slangerup, Denmark.

[9]Agilent 2100 Bioanalyzer, Agilent Technologies, Naerum, Denmark.

[10]Tprofessional TRIO 3x48, Fisher Scientific Denmark, Slangerup, Denmark

[11]48.48 Dynamic Array Integrated Fluidic Circuits and BioMark Fluidigm thermocycler, Fluidigm, California, USA.

[12]GenEx5, MultiD, Göteborg, Sweden.

Additional files

Additional file 1: Time-table of clinical examinations performed during the 24-h experimental period. All procedures were carried out in a similar manner for both systemically and locally injected (LI) horses, except only LI horses underwent lameness evaluation. PIH: post-injection hour. (DOCX 14 kb)

Additional file 2: Primers and reaction conditions. *PCR efficiency and correlation coefficient for IL6 is calculated based on a specific high-responder standard curve made from dilution series of a pool of cDNA samples showing high expression levels of the implicated gene. r^2 = correlation coefficient. (DOCX 28 kb)

Abbreviations

APR, acute phase reactant; CRT, capillary refill time; DAMP, damage associated molecular patterns; HR, heart rate; LI, locally inflamed; LPS, lipopolysaccharide; PAMP, pathogen-associated molecular pattern; PCA, principal component analysis; PIH, post-induction hour; RIN, RNA integrity number; RR, respiratory rate; RT, rectal temperature; RT-qPCR, reverse transcription quantitative real-time polymerase chain reaction; SAA, serum amyloid A; SI, systemically inflamed; WBC, white blood cell.

Acknowledgements

The authors gratefully acknowledge laboratory technician Karin Tarp for exceptional help with RT-qPCR analysis, and Claus Thorn Ekstrøm for performing statistical analyses of clinical and paraclinical data.

Funding

The experimental and laboratory procedures for the study presented in this paper were funded by Hesteafgiftsfonden, Kongeriget Danmarks Hesteforsikring, Foreningen Kustos af 1881, and University of Copenhagen.

Availability of data and materials
Data and materials supporting the results of this article are available upon request. Transcript IDs, primer sequences, amplicon lengths, reaction efficiencies, and correlation coefficients for gene specific primer pairs are shown in Additional file 2.

Authors' contributions
AMLV conceived the study, designed and carried out experimental and laboratory procedures, performed the statistical analyses of expression data, and drafted the manuscript. PMHH and RB participated in the interpretation of analytical data and in drafting the manuscript. KS participated in designing and performing laboratory procedures and carried out data pre-processing. SMA participated in the design of the study and carried out experimental procedures. PHA conceived the study, participated in its design and coordination and in drafting the manuscript. All authors have revised the manuscript critically for important intellectual content and have read and approved the final manuscript.

Competing interests
The authors declare that they have no competing interests.

Consent to publish
Not applicable.

Author details
[1]Department of Large Animal Sciences, Faculty of Health and Medical Sciences, University of Copenhagen, Taastrup, Denmark. [2]Innate Immunology Group, Section for Immunology and Vaccinology, National Veterinary Institute, Technical University of Denmark, Frederiksberg, Denmark. [3]Department of Clinical Sciences, Faculty of Veterinary Medicine and Animal Science, Swedish University of Agricultural Sciences, Uppsala, Sweden.

References
1. Fry DE. Sepsis, systemic inflammatory response, and multiple organ dysfunction: the mystery continues. Am Surg. 2012;78(1):1–8.
2. Peters K, Unger RE, Brunner J, Kirkpatrick CJ. Molecular basis of endothelial dysfunction in sepsis. Cardiovasc Res. 2003;60(1):49–57.
3. Zhang Q, Raoof M, Chen Y, Sumi Y, Sursal T, Junger W, Brohi K, Itagaki K, Hauser CJ. Circulating mitochondrial DAMPs cause inflammatory responses to injury. Nature. 2010;464(7285):104–7.
4. Seeley EJ, Matthay MA, Wolters PJ. Inflection points in sepsis biology: from local defense to systemic organ injury. Am J Physiol Lung Cell Mol Physiol. 2012;303(5):L355–363.
5. Hack CE, Zeerleder S. The endothelium in sepsis: source of and a target for inflammation. Crit Care Med. 2001;29(7 Suppl):S21–27.
6. Janeway Jr CA, Travers P, Walport M, Shlomchik MJ. Innate Immunity. In: Austin P, Lawrance E, editors. Immunobiology. 5th ed. New York City: New York Garland Publishing; 2001. p. 35–91.
7. MacKay RJ. Endotoxemia. In: Smith BP, editor. Large Animal Internal Medicine. 4th ed. St. Louis, Missouri: Mosby Elsevier; 2009. p. 711–23.
8. Dellinger RP, Levy MM, Rhodes A, Annane D, Gerlach H, Opal SM, Sevransky JE, Sprung CL, Douglas IS, Jaeschke R et al. Surviving Sepsis Campaign: international guidelines for management of severe sepsis and septic shock, 2012. Intensive Care Med. 2013;39(2):165–228.
9. Jacobsen S, Niewold TA, Halling-Thomsen M, Nanni S, Olsen E, Lindegaard C, Andersen PH. Serum amyloid A isoforms in serum and synovial fluid in horses with lipopolysaccharide-induced arthritis. Vet Immunol Immunopathol. 2006;110(3-4):325–30.
10. Palmer JL, Bertone AL. Experimentally-induced synovitis as a model for acute synovitis in the horse. Equine Vet J. 1994;26(6):492–5.
11. Campebell RC, Peiró JR, Valadão CAA, Santana AE, Cunha FQ. Effects of lidocaine on lipopolysaccharide-induced synovitis in horses. Arq Bras Med Vet Zootec. 2004;56:281–91.
12. Hulten C, Gronlund U, Hirvonen J, Tulamo RM, Suominen MM, Marhaug G, Forsberg M. Dynamics in serum of the inflammatory markers serum amyloid A (SAA), haptoglobin, fibrinogen and alpha2-globulins during induced noninfectious arthritis in the horse. Equine Vet J. 2002;34(7):699–704.
13. Christoffersen M, Baagoe CD, Jacobsen S, Bojesen AM, Petersen MR, Lehn-Jensen H. Evaluation of the systemic acute phase response and endometrial gene expression of serum amyloid A and pro- and anti-inflammatory cytokines in mares with experimentally induced endometritis. Vet Immunol Immunopathol. 2010;138(1-2):95–105.
14. Netea MG, Kullberg BJ, Van der Meer JW. Circulating cytokines as mediators of fever. Clin Infect Dis. 2000;31 Suppl 5:S178–184.
15. von Vietinghoff S, Ley K. Homeostatic regulation of blood neutrophil counts. J Immunol. 2008;181(8):5183–8.
16. Christopher MJ, Link DC. Regulation of neutrophil homeostasis. Curr Opin Hematol. 2007;14(1):3–8.
17. Gabay C, Kushner I. Acute-phase proteins and other systemic responses to inflammation. N Engl J Med. 1999;340(6):448–54.
18. Lohmann KL, Barton MH. Endotoxemia. In: Reed SM, Bayly WM, Sellom DC, editors. Equine Internal Medicine. 3rd ed. St. Louis, Missouri: Saunders Elsevier; 2010. p. 807–23.
19. Andonegui G, Goyert SM, Kubes P. Lipopolysaccharide-induced leukocyte-endothelial cell interactions: a role for CD14 versus toll-like receptor 4 within microvessels. J Immunol. 2002;169(4):2111–9.
20. Stockham SL, Scott MA. Fundamentals of veterinary clinical pathology. 2nd ed. Chichester: John Wiley and Sons Ltd; 2010.
21. Henriquez-Camacho C, Losa J. Biomarkers for sepsis. BioMed Res Int. 2014; 2014:547818.
22. Lichtenstern C, Brenner T, Bardenheuer HJ, Weigand MA. Predictors of survival in sepsis: what is the best inflammatory marker to measure? Curr Opin Infect Dis. 2012;25(3):328–36.
23. Calvano SE, Xiao W, Richards DR, Felciano RM, Baker HV, Cho RJ, Chen RO, Brownstein BH, Cobb JP, Tschoeke SK et al. A network-based analysis of systemic inflammation in humans. Nature. 2005;437(7061):1032–7.
24. Talwar S, Munson PJ, Barb J, Fiuza C, Cintron AP, Logun C, Tropea M, Khan S, Reda D, Shelhamer JH et al. Gene expression profiles of peripheral blood leukocytes after endotoxin challenge in humans. Physiol Genomics. 2006; 25(2):203–15.
25. Nieto JE, MacDonald MH, Braim AE, Aleman M. Effect of lipopolysaccharide infusion on gene expression of inflammatory cytokines in normal horses in vivo. Equine Vet J. 2009;41(7):717–9.
26. Fossum C, Hjertner B, Olofsson KM, Lindberg R, Ahooghalandari P, Camargo MM, Brojer J, Edner A, Nostell K. Expression of tlr4, md2 and cd14 in equine blood leukocytes during endotoxin infusion and in intestinal tissues from healthy horses. Vet Immunol Immunopathol. 2012;150(3-4):141–8.
27. Tadros EM, Frank N. Effects of continuous or intermittent lipopolysaccharide administration for 48 hours on the systemic inflammatory response in horses. Am J Vet Res. 2012;73(9):1394–402.
28. Vinther AM, Skovgaard K, Heegaard PM, Andersen PH. Dynamic expression of leukocyte innate immune genes in whole blood from horses with lipopolysaccharide-induced acute systemic inflammation. BMC Vet Res. 2015;11:134.
29. Aiello SEE. The Merck veterinary manual. 8th ed. Whitehouse Station: Merck & Co., Inc; 1998.
30. Holcombe SJ, Jacobs CC, Cook VL, Gandy JC, Hauptman JG, Sordillo LM. Duration of *in vivo* endotoxin tolerance in horses. Vet Immunol Immunopathol. 2016;173:10–6.
31. Anonymous. Definition and classification of lameness. In: Guide for Veterinary Service and Judging of Equestrian Events. 4th ed. Lexington: American Association of Equine Practitioners; 1991. p. 19.
32. Schroeder A, Mueller O, Stocker S, Salowsky R, Leiber M, Gassmann M, Lightfoot S, Menzel W, Granzow M, Ragg T. The RIN: an RNA integrity number for assigning integrity values to RNA measurements. BMC Mol Biol. 2006;7:3.
33. Skovgaard K, Cirera S, Vasby D, Podolska A, Breum SO, Durrwald R, Schlegel M, Heegaard PM. Expression of innate immune genes, proteins and microRNAs in lung tissue of pigs infected experimentally with influenza virus (H1N2). Innate Immun. 2013;19(5):531–44.
34. Andersen CL, Jensen JL, Orntoft TF. Normalization of real-time quantitative reverse transcription-PCR data: a model-based variance estimation approach

Background

Osteochondrosis (OC) is a clinically important joint disorder that occurs in many animal species, including dogs, horses, pigs, and humans. It is defined as a focal disturbance of endochondral ossification and is believed to be a multifactorial disease. The most frequently described etiologic factors are heredity, rapid growth, trauma, and dietary imbalances, although its etiology remains incompletely understood [1–3]. Recently, it was shown in foals [4], pigs [5] and horses [6] that biomechanical damage to the juvenile blood supply plays an important role in the pathogenesis of OC [7]. Consistently, mRNA expression was demonstrated to differ in bone and cartilage after experimental mechanical loading between OC-negative and -positive foals [8]. Early OC lesions in pigs and foals appear to be very similar. In foals, OC is associated with the necrosis of cartilage vessels at between 3 weeks and 5 months of age. Cartilage canals are absent in foals aged over 7 months and, subsequently, nutrition is supplied to the cartilage through diffusion [4]. A recent study [9] showed that vessel failure is associated with ischemic necrosis in growing cartilage. However, OC is a dynamic disorder: during the first few months of life, lesions occur and repair spontaneously [10, 11]. This implies a "window of susceptibility" in which OC development and cartilage repair vary between joints [12, 13]. Dik et al. [10] showed that OC lesions become permanent from approximately 5 months of age in the tarsocrural (TC) joint, although disease stability is reached at a later stage; similarly, OC lesions become permanent from the age of approximately 8 months in the femoropatellar (FP) joint.

Exercise is an important environmental factor for musculoskeletal development in foals [1, 14], but is thought to have only an indirect effect on OC development [13]. Permanent box rest combined with exercise bouts every other day from 1 month of age did not significantly influence the number of OC lesions, but after 5 months of age, there was a tendency towards other, severe degenerative musculoskeletal lesions [13]. Recently, Van Grevenhof et al. [15] showed that rapidly growing pigs housed on conventional, slippery flooring developed OC significantly more frequently than those housed on a deep litter floor with restricted feeding. On standing up, one or more limbs can slip sideways as a result of weaker, less effective muscle function, a slippery floor, or a combination of these factors. Aberrant standing-up behavior may cause excessive changes in peak pressure on joint surfaces, damaging the vasculature of juvenile cartilage and resulting in avascular necrosis. Indeed, lateral wedge-induced, out-of-plane hindlimb movement has been shown to cause considerable asymmetrical loading of the fetlock, hock, and stifle joints in ponies [16, 17].

Acute joint overloading resulting from incidental, out-of-plane limb movements may cause the fragmentation of OC-prone cartilage. Lepeule et al. [18] reported that increased growth and reduced or irregular physical activity during the first weeks of life are responsible for lesions being more severe, whereas Praud et al. [19] found that "mixed housing" during winter and pastures with rough ground were significantly associated with OC with a poor radiologic score.

Based upon these epidemiologic findings, we hypothesized that farms in which foals experienced a higher sliding frequency, resulting in a greater number of incidental, out-of-plane limb movements, would have a higher prevalence of OC.

Therefore, in this study, we examined three questions:

1) Is there a difference between the standing-up behavior of foals between different farms during the first 12 months of life?
2) Is there a difference in OC development in foals between different farms during the first 12 months of life?
3) Is standing-up behavior linked to OC development in these foals during the first 12 months of life?

Methods

Foals

Five farms in The Netherlands were selected to participate in this pilot study, from which a total of 50 Dutch Warmblood foals were chosen for their immediate availability. Two foals died before the completion of the observations, so the data reported relate to 48 foals. The participating farms differed with respect to management factors, including their housing and exercise systems, and were ranked according to their sliding ratio. On average, 10 foals (range: 6–18 foals) per farm participated in the study. These foals were aged 6–12 months and their height at the withers ranged from 1.36–1.44 m.

Observations of standing-up behavior

Standing-up behavior was observed in weaned, group-housed foals aged 6–12 months at each farm; the sliding ratio (%) was calculated from the recorded number of incidences of sliding of the total number of recorded incidences of standing up. Standard black-and-white surveillance cameras (Mitsubishi) and video recorders (Sanyo) were used to record each group of foals for 3–4 h per session up to a total of 24 h per group, and the images were stored on videotapes. A video splitter device (EQM 100 Monoquad) was used to create one image containing different views of the stable. Each foal was scored using a predefined ethogram (Additional file 1) containing definitions of all the behaviors observed.

Using the all-occurrences method [20], each standing-up maneuvre (when fully visible) was evaluated. Several characteristics were scored: baseline behavior (rolling; hindlimbs lying laterally left or right; or forelimbs lying sternally left or right); provocation (following another foal; external disturbance; panic; another cause; or cause not visible); standing-up behavior (forelimb or hindlimb first [Fig. 1]; direction of movement in relation to the center of gravity; normal or sliding [and which limb was sliding]; and the extent of sliding [short, medium, or long]).

Scoring osteochondrosis

Standardized radiologic examinations (fetlock: LM, hock: AP/LM/APLO, and stifle: LM) were routinely performed at 6 and 12 months of age using standard mobile digital equipment. OC was scored by a board-certified radiologist from A–E using a categorical scale [10] as follows: A, no abnormalities; B, smooth flattening of the bony contour delineating the cartilage; C, slight irregular flattening of the bony contour delineating the cartilage; D, minor fragmentation; and E, a severely irregular ossification front with pronounced fragmentation. The OC ratio (%) was calculated from the recorded number of OC-positive joints of the total number of evaluated joints. Four joints were scored bilaterally for OC: the metacarpo- and metatarsophalangeal (fore and hind "fetlock"), TC ("hock"), and FP ("stifle") joints. The original scores were subsequently converted to scores of 0–4 on a quantitative scale to emphasize the quantitative continuous characteristics of the trait, enabling summation at an animal level (quantitatively explained as follows: 1, a regular subchondral ossification front [SOF]; 2, focal smooth flattening of the SOF; 3, irregular flattening of the SOF; 4, minor [and what that was, such as fragments of 1–2 mm or <3 mm] fragmentation of the SOF; and 5, an irregular SOF with osteochondral fragments >3 mm). The degree or severity of OC at the animal level was obtained by the summation of the scores for all eight scored joints. For example, eight joints with the scores "0, 0, 3, 0, 1, 1, 0, 0", would score "5" at the animal level. Only 39 radiographs at both 6 and 12 months were available for follow-up.

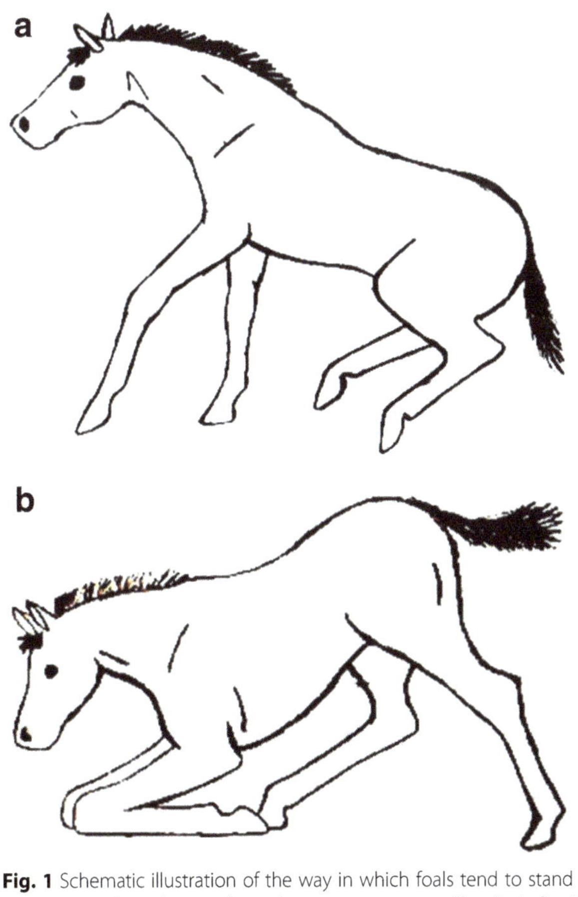

Fig. 1 Schematic illustration of the way in which foals tend to stand up, as scored in this study: **a** the more common "forelimb-first" (equine) method, and **b** the less common "hindlimb-first" (bovine) method

Parameter estimation

In the univariate general linear models (Models 1 and 3, presented below), analyses and assessments of correlation (Pearson correlation coefficient) were performed using PROC GLM in SAS 9.2 (SAS Institute Inc., Cary, NC, USA; [21]). In the bivariate model (Model 2), parameter estimations and correlations were performed via the residual maximum-likelihood method using ASReml [22].

For the estimation of parameters, the following linear univariate general linear model was used:

$$Y_{ijkl} = \mu + farm_i + OC12_j + reason_k + e_{ijkl} \quad (1)$$

in which Y_{ijkl} is the frequency ratio of sliding versus normal standing-up behavior of an individual; μ is the mean frequency ratio; $farm_i$ is the fixed-class effect of farm (i = 1, 2, ..., 5); $OC12_j$ is the continuous effect of OC at the animal level at 12 months of age (j = 0, 1, 2, ..., 8); $reason_k$ is the fixed-class effect of the reason for standing up, if any (k = 0, 1); and e_{ijkl} is the residual. All fixed effects in the model had a significance level of $P < 0.05$.

Additionally, the following bivariate model was used, in which we assumed the two processes of OC at the animal level and standing-up behavior were co-occurring:

$$Y_{tr1}\ Y_{tr2} = \mu_{tr1} + \mu_{tr2} + farm_{tr1} + reason_{tr1} + e_{tr1} + e_{tr2} \quad (2)$$

in which Y_{tr1} is the frequency ratio of sliding versus normal standing-up behavior; Y_{tr2} is the prevalence of OC at the animal level at 12 months of age; μ_{tr1} is the mean of Y_{tr1}; μ_{tr2} is the mean of Y_{tr2}; $farm_{tr1}$ is the fixed-class effect of farm (i = 1, ..., 5) on Y_{tr1}; $reason_{tr1}$ is the fixed-class effect of the reason for standing, if any (k = 0, 1) on Y_{tr1}; e_{tr1} is the residual of Y_{tr1}; and e_{tr2} is the residual of Y_{tr2}. All fixed effects in the model had a significance level of P < 0.05. No traits significantly influenced Y_{tr2} (OC at 12 months of age) in the bivariate model.

Standing-up behavior with respect to whether foals rose "forelimbs first" or "hind limbs first" (Fig. 1) and the effect of this on OC prevalence were also analyzed. The following univariate model was used:

$$Y_{ijkl} = \mu + \text{hindlimbs first}_i + e_{ijkl} \quad (3)$$

in which Y_{ijkl} represents OC in the hock joint at 6 months of age; μ is the mean prevalence; hindlimbs $first_i$ is the fixed-class effect of the number of standing-up maneuvres performed with the hindlimbs first (i = 0, 1, 2); and e_{ijkl} is the residual. All fixed effects in the model had a significance level of P < 0.05. No other traits affected Y in Model 3. In general, all traits were tested in the model. Those traits that did not significantly influence the model were not taken into account in the models described above.

Results

The mean ratio of sliding versus normal standing-up behavior was 29% (range: 0–50%), i.e., foals experienced limb sliding during around 29% of standing-up maneuvres, based on a mean of 46 observations of standing up per foal. Per foal, on average, there was a particular reason for the foal to stand up 43% of the time; the reason was most often the approach of another foal. The frequency ratio of sliding versus normal standing-up behavior and that of an identifiable reason for standing (following another foal; external disturbance; panic; or another cause) versus no visible reason were not significantly correlated. The sliding ratio was significantly determined by farm (r^2 = 0.323, P = 0.0072, n = 5), and was also significantly different between farms (one-factor analysis of variance F = 4.455, P < 0.005).

The prevalence of OC at the animal level varied from zero to eight joints affected and 25% of foals were free of OC at 6 months of age (i.e., 75% OC). During the following 6 months, 23% of foals with OC recovered (i.e., 52% OC) and 10% developed OC (i.e., 62% OC), resulting in 38% of foals free of OC at 12 months of age (Fig. 2, Table 1). Around 52% of foals had OC at both 6 and 12 months of age. Table 1 presents the total OC scores. For example, when a foal scored 0 in the fetlock, 4 in the hock, and 2 in the stifle joint, the total OC score representing OC severity at the animal level was 6.

In the stifle joint, the total OC severity varied from 0 to 4 at 12 months of age (Table 1; warmblood studbook score: A-E), whereas the OC prevalence decreased from 38% at 6 months to 26% at 12 months of age. In the hock joint, the total OC severity varied from 0 to 8 at 12 months of age (Table 1), whereas the OC prevalence decreased from 41% at 6 months to 28% at 12 months of age. In the fetlock joint, the total OC severity varied from 0 to 6 at 12 months of age (Table 1), whereas the OC prevalence decreased from 28% at 6 months to 18% at 12 months of age.

In general, the OC prevalence at 6 months showed a significant correlation with that at 12 months of age for all joints (r = 0.46, P = 0.003), but also for each individual joint, with r varying per joint from 0.6–0.67

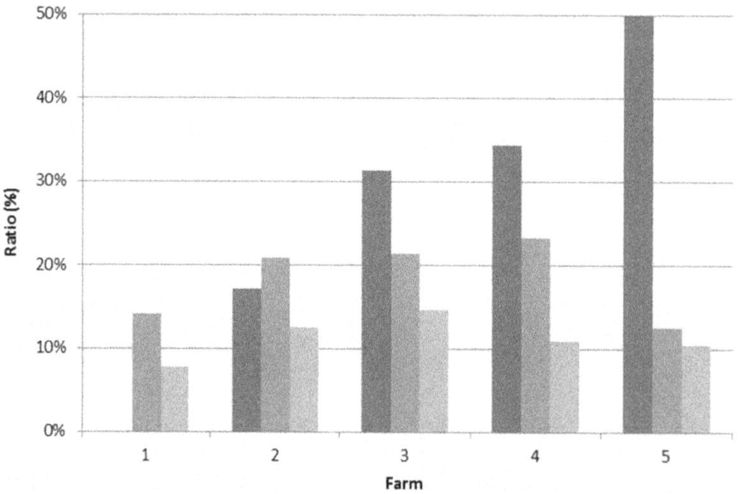

Fig. 2 Bar chart displaying the sliding ratio (dark bar) and osteochondrosis ratio at 6 (medium-dark bar) and 12 (light bar) months of age for the farms listed on the x-axis in the order of their sliding ratio, with Farm 1 showing the lowest sliding ratio (0%) and Farm 5 the highest sliding ratio (50%)

Table 1 The prevalence of osteochondrosis at the animal and joint levels among 48 foals aged 6 and 12 months

Variable	Age (months)	*n* (animals)	Mean OC score	SD	Minimum OC score	Maximum OC score
OC animal	6	40	2.7	2.7	0	12
OC hock	6	41	1.2	1.7	0	6
OC stifle	6	41	0.9	1.4	0	5
OC fetlock	6	40	0.6	1.3	0	6
OC animal	12	48	1.9	2.3	0	8
OC hock	12	48	0.9	1.7	0	8
OC stifle	12	48	0.4	0.9	0	4
OC fetlock	12	48	0.7	1.5	0	6

n number of animals available for each trait, *SD* standard deviation, *OC* osteochondrosis

($P < 0.01$) at 6 months and 0.35–0.67 ($P < 0.02$) at 12 months of age. Moreover, individual joints showed a significant within-joint correlation at both ages in OC prevalence, with r varying from 0.41–0.43 ($P < 0.01$).

The univariate Model 1 demonstrated that "farm", "foal OC score at 12 months of age", and "reason to stand" were significantly and positively associated with "sliding ratio" ($r^2 = +0.52$, $P < 0.05$; Tables 2 and 3). The bivariate Model 2 demonstrated that "foal OC score at 12 months of age" and "sliding ratio" were significantly but negatively correlated ($r^2 = -0.36$, $P < 0.05$). Both "farm" (positively) and "reason to stand" (negatively) significantly influenced "sliding ratio" ($P < 0.05$). None of the variables significantly influenced OC at the animal level at 12 months of age. Finally, the univariate Model 3 examined the method of standing-up (whether foals rose using either their forelimbs or hindlimbs first [Fig. 1])

Table 2 Model parameters (univariate) affecting the group of 48 foals for the farms listed in order of sliding ratio, with Farm 1 showing the lowest sliding ratio (0%) and Farm 5 showing the highest sliding ratio (50%) The farms (n = 5) are listed 1–5 (detailed information provided in Table 3)

Parameter	Subcategory of OC	Estimate/LSM	*P*
Intercept		+0.07	<0.001*
Farm			<0.001*
	1	−0.16	0.099
	2	+0.38	<0.001*
	3	+0.27	0.003*
	4	+0.35	<0.001*
	5	+0.41	<0.001*
OC at 12 months		−0.03	0.018*
Specific reason for standing-up versus no reason		−0.41	0.005*

Least squares means are presented at the farm level; standard deviation varied from 0.05–0.1 (*$P < 0.05$). The most frequent reason for standing up was the approach of another foal; OC = osteochondrosis; LSM = least squares mean

Table 3 Schematic tabular overview of the differences between Farms 1–5 listed in order of sliding ratio, with Farm 1 showing the lowest sliding ratio (0%) and Farm 5 showing the highest sliding ratio (50%)

Farms	1 (n = 8)	2 (n = 6)	3 (n = 12)	4 (n = 18)	5 (n = 6)
Average birth month	April	April	April	April	June
Average withers height	1.4 m	1.37 m	1.36 m	1.44 m	1.4 m
Stable size (avg/foal)	1250 m^2	18.3 m^2	8 m^2	5 m^2	7.5 m^2
Cleaning frequency	1/wk	1/wk	1/wk	1/wk	2–3/wk
Feeding amount	2 kg/d	1 kg/d	2 kg/d	n.a.	2–4 kg/d
Roughage	Silage	Silage	Hay	Silage	Silage
Ground flooring	Bricks/sand	Straw bed	Straw bed	Straw bed	Straw bed
Exercise area	Pasture	Arena	Paddock	Paddock	Arena
Exercise duration/d	24 h/d	0.5–1 h/d	1–2 h/d	1–1.5 h/d	3–4 h/d
Hoof trimming	No	Yes	Yes	Yes	Yes

n.a. not available, *d* day, *wk.* week, *avg.* average

with respect to its effect on OC development. Use of the hindlimbs first (Fig. 1) showed only a slight tendency to be related to OC prevalence in the hock joint at 6 months of age ($r^2 = 0.094$, $P = 0.057$).

Discussion

We hypothesized that farms in which foals showed a higher sliding frequency would house animals that experienced more incidental, out-of-plane limb movements. We also speculated that farms with a higher sliding frequency would exhibit a higher prevalence of OC. Each farm used a different type of housing and this was associated with the ratio of sliding versus normal standing-up behavior shown by the foals. The management variations between the horse farms evaluated in this study were similar to those found in pig farming, in which both deep litter and conventional flooring systems are used: the frequency of renewal ranged from 1 to 3/wk., and the type of bedding provided ranged from straw to no bedding at all. The exercise area and duration/day differed between the farms (Table 3): Farm 5 gave the longest "active" exercise duration to their foals (3–4 h/day) in contrast to the longer "passive" pasture exercise given at Farm 1. Generally, moderate exercise improves the circulation and development of cartilage and bone, thereby influencing the risk of OC development [13]. Differences in management at individual farms were not identified as causal factors for sliding behavior by our statistical analyses, but only five farms participated in this study.

The prevalence of OC changed significantly over time at the animal level, and also within each joint ($P < 0.05$). The process of OC development and recovery varied over time and differed depending on the joint affected [23]. Both the development of new lesions and fragments and the repair of OC lesions were observed between the ages of 6 and 12 months, confirming early reports [13]. OC was found most frequently and most severely (scores D and E) in the hock joints (not shown), followed by the stifle joints, and finally the fetlock joints.

The prevalence of OC at the animal level at 12 months of age was somewhat negatively correlated with the sliding ratio; possibly, pain may have contributed to some animals showing sliding during standing up less frequently. This implies that, in foals suffering from OC, the resulting pain caused relatively careful standing-up behavior, as significantly less sliding behavior was evident.

The development of OC in joints before 6 months of age depends on many different factors, of which standing-up behavior is just one (Fig. 2, Table 3). After 6 months, it remains possible for OC to develop and repair. The process of repair is also multifactorial and the time frame (window of susceptibility) differs between joints [23]. A relatively high, significant correlation was found between the prevalences of OC at 6 and 12 months of age (Pearson's $r = 0.36$, $P < 0.05$), but the OC prevalence at 6 months (19.1%) was significantly higher than that at 12 months (10%; paired t test; $P < 0.05$), in accordance with earlier studies on OC prevalence [13].

Only the number of times a foal stood up with its hindlimbs first showed a tendency ($P = 0.057$) to be related to OC in the hock joint at 6 months of age. From a practical perspective, the method of standing up can be expected to have an effect on OC development in foals. Lifting the body by raising the hindlimbs first is usually seen in calves, in contrast to the forelimb-first approach more commonly used by foals. A tendency towards a correlation between sliding during standing up and OC was evident in the hock joint alone and only at 6 months of age. The association with OC in the hock joint at 12 months was not significant. In this respect, we assume that foals who often stand up hindlimbs first aged 6–12 months will also have done so aged 0–6 months, and anticipate a stronger statistical effect with a larger sample size. It can also be expected that the differences in OC prevalence between the housing systems and specific management factors evaluated would become significant were more farms to participate in this type of study.

Conclusions

The ratio of sliding versus normal standing-up behavior differed significantly between farms and depended on the reason for standing up, such as the approach of another foal. The prevalence of OC at the animal level at 12 months of age was negatively correlated with the ratio of sliding during standing up, but it is possible that pain might have prevented the animals from showing frequent sliding. There are small but significant differences between farms in sliding frequency and the total incidence of OC in their Warmblood foals, but whether environmental factors are causally related to these differences requires further elucidation.

Abbreviations

AP: Anterior-posterior; APMO: Anterior-posterior, medial oblique; FP: Femoropatellar; LM: Lateromedial; OC: Osteochondrosis; SOF: Subchondral ossification front; TC: Tarsocrural

Acknowledgments

This work has been published as an MSc thesis by Aukje Gezelle Meerburg for Utrecht University in collaboration with GD-Animal Health Service, Deventer, The Netherlands. The authors would like to thank GD Deventer for their technical support, Mirjam Nielen and Jan van den Broek for their statistical support, and Cambridge Language Consultants (CLC) for English language correction.

Funding

This study was done in close collaboration with the GD-Animal Health Service, Deventer, The Netherlands, and they allowed us to collect and use these data for publication, while permission had been gained from the horse owners and farmers.

Authors' contributions

MD, BS, PM, and WB conceived and planned the study and participated in the decision to publish. AG collected data on farms, IG performed statistical analyses and prepared the manuscript, and AB scored all osteochondric lesions. AG, IG, and MD reviewed the literature. All authors read and approved the manuscript.

Consent for publication

This study was done in close collaboration with the GD-Animal Health Service, Deventer, The Netherlands, and they provided us with Consent to Publish.

Competing interests

The authors declare that they have no competing interests.

Author details

[1]Animal Breeding and Genomics Centre, Wageningen UR Livestock Research, Wageningen, The Netherlands. [2]Department of Equine Sciences, Utrecht University, Utrecht, The Netherlands. [3]Behavioral Biology Group, Utrecht University, Utrecht, The Netherlands. [4]Department of Companion Animal Sciences, Division of Diagnostic Imaging, Utrecht University, Utrecht, The Netherlands. [5]GD-Animal Health Service, Deventer, The Netherlands. [6]Ridderkerk Equine Clinic, Ridderkerk, The Netherlands. [7]Department of Surgery and Anaesthesiology of Domestic Animals, Ghent University, Merelbeke, Belgium.

References

1. Jeffcott LB. Osteochondrosis in the horse-searching for the key to pathogenesis. Equine Vet J. 1991;23:331–8.
2. Philipsson J, Andreasson E, Sandgren B, Dalin G, Carlsten J. Osteochondrosis in the tarsocrural joint and osteochondral fragments in the fetlock joints in Standardbred trotters. II. Heritability. Equine Vet J. 1993;S16:38–41.
3. Wittwer C, Hamann H, Rosenberger E, Distl O. Prevalence of osteochondrosis in the limb joints of south German Coldblood horses. J Vet Med. 2006;A53:531–9.
4. Carlson CS, Cullins LD, Meuten DJ. Osteochondrosis of the articular-epiphyseal cartilage complex in young horses: evidence for a defect in cartilage canal blood supply. Vet Pathol. 1995;32:641–7.
5. Ytrehus B, Carlson CS, Ekman S. Etiology and pathogenesis of osteochondrosis. Vet Pathol. 2007;44:429–48.
6. Hendrickson EH, Olstad K, Nødtvedt A, Pauwels E, van Hoorebeke L, Dolvik NI. Comparison of the blood supply to the articularepiphyseal growth complex in horse vs. pony foals. Equine Vet J. 2015;47:326–32.
7. Olstad K, Hendrickson EH, Carlson CS, Ekman S, Dolvik NI. Transection of vessels in epiphyseal cartilage canals leads to osteochondrosis and osteochondrosis dissecans in the femoro-patellar joint of foals; a potential model of juvenile osteochondritis dissecans. Osteoarthr Cartil. 2013;21:730–8.
8. Desjardin C, Vaiman A, Mata X, Legendre R, Laubier J, Kennedy SP, Laloe D, Barrey E, Jacques C, Cribiu EP, Schibler L. Next-generation sequencing identifies equine cartilage and subchondral bone miRNAs and suggests their involvement in osteochondrosis physiopathology. BMC Genomics. 2014;15:798.
9. Olstad K, Ytrehus B, Ekman S, Carlson CS, Dolvik NI. Early lesions of osteochondrosis in the distal tibia of foals. J Orthop Res. 2007;25:1094–105.
10. Dik KJ, Enzerink E, Van Weeren PR. Radiographic development of osteochondral abnormalities, in the hock and stifle of dutch warmblood foals, from age 1 to 11 months. Equine Vet J. 1999;S31:9–15.
11. Van Weeren PR. Etiology, diagnosis, and treatment of OC(D). Clin Techn Equine Pract. 2006;5:248–58.
12. Van Grevenhof EM, Ducro BJ, Van Weeren PR, Van Tartwijk JMFM, Van den Belt AJM, Bijma P. Prevalence of various radiographic manifestations of osteochondrosis and their correlations between and within joints in Dutch Warmblood horses (KWPN). Equine Vet J. 2009;41:11–6.
13. Van Weeren PR, Barneveld A. The effect of exercise on the distribution and manifestation of osteochondrotic lesions in the warmblood foal. Equine Vet J. 1999;S31:16–25.
14. Rogers CW, Firth EC, McIlwraith CW, Barneveld A, Goodship AE, Kawcak CE, Smith RK, Van Weeren PR. Evaluation of a new strategy to modulate skeletal development in thoroughbred performance horses by imposing track-based exercise during growth. Equine Vet J. 2008;40:111–8.
15. Van Grevenhof EM, Ott S, Hazeleger W, Van Weeren PR, Bijma P, Kemp B. The effects of housing system and feeding level on the joint-specific prevalence of osteochondrosis in fattening pigs. Livest Sci. 2011;135:53–61.
16. Back W, Schamhardt HC, Barneveld A, Van Weeren PR. Longitudinal development of kinematics in shetland ponies and the influence of feeding and training regimes. Equine Vet J. 2002;34:609–14.
17. Back W, Remmen JL, Knaap J, De Koning JJ. Effect of lateral heel wedges on sagittal and transverse plane kinematics of trotting Shetland ponies and the influence of feeding and training regimes. Equine Vet J. 2003;35:606–12.
18. Lepeule J, Bareille N, Robert C, Valette J-P, Jacquet S, Blanchard G, Denoix JM, Seegers H. Association of growth, feeding practices, and exercise conditions with the severity of the osteoarticular status of limbs in French foals. Vet J. 2013;197:65–71.
19. Praud A, Dufour B, Robert C, Valette J-P, Denoix J-M, Crevier-Denoix N. Effects of management practices as risk factors for juvenile osteochondral conditions in 259 French yearlings. Vet J. 2013;197:72–6.
20. Lehner PN. Handbook of ethological methods. 2nd ed. Cambridge: Cambridge University Press (CUP); 1996.
21. SAS. Qualification tools user's guide. Version 9.2. Cary: Statistical Analysis Systems (SAS) Institute Inc; 2010.
22. Gilmour AR, Gogel BJ, Cullis BR, Thompson R. ASReml User Guide. Release 2.0. Hemel Hempstead: VSN International Ltd; 2006.
23. Van Weeren PR, Sloet van Oldruitenborgh-Oosterbaan MM, Barneveld A. The influence of birth weight, rate of weight gain and final achieved height and seks on the development of osteochondrotic lesions in a population of genetically predisposed Warmblood foals. Equine Vet J. 1999;S31:26–30.

16

Visual evoked potentials in the horse

L. Ström* and B. Ekesten

Abstract

Background: Electrical potentials generated in the central nervous system in response to brief visual stimuli, flash visual evoked potentials (FVEPs), can be recorded non-invasively over the occipital cortex. FVEPs are used clinically in human medicine and also experimentally in a number of animal species, but the method has not yet been evaluated in the horse. The method would potentially allow the ophthalmologist and equine clinician to evaluate visual impairment caused by disorders affecting post-retinal visual pathways. The aim was to establish a method for recording of FVEPs in horses in a clinical setting and to evaluate the waveform morphology in the normal horse.

Methods: Ten horses were sedated with a continuous detomidine infusion. Responses were recorded from electrodes placed on the scalp. Several positions were evaluated to determine suitable electrode placement. Flash electroretinograms (FERGs) were recorded simultaneously. To evaluate potential contamination of the FVEP from retinal potentials, a retrobulbar nerve block was performed in two horses and transection of the optic nerve was performed in one horse as a terminal procedure.

Results: A series of positive (P) and negative (N) peaks in response to light stimuli was recorded in all horses. Reproducible wavelets with mean times-to-peaks of 26 (N1), 55 (P2), 141 (N2) and 216 ms (P4) were seen in all horses in all recordings. Reproducible results were obtained when the active electrode was placed in the midline rostral to the nuchal crest. Recording at lateral positions gave more variable results, possibly due to ear muscle artifacts. Averaging ≥100 responses reduced the impact of noise and artifacts. FVEPs were reproducible in the same horse during the same recording session and between sessions, but were more variable between horses. Retrobulbar nerve block caused a transient loss of the VEP whereas transection of the optic nerve caused an irreversible loss.

Conclusions: We describe the waveform of the equine FVEP and our results show that it is possible to record FVEPs in sedated horses in a clinical setting. The potentials recorded were shown to be of post-retinal origin. Further studies are needed to provide normative data and assess potential clinical use.

Keywords: Visual evoked potentials, Visual cortical evoked potentials, Flash visual evoked potentials, VEP, FVEP, Horse, Equine, Visual impairment

Background

Techniques currently available to evaluate visual impairment in the horse are limited to reflex testing, falling cotton-ball testing, ability to navigate an obstacle course and evaluation of findings detected during a routine ophthalmic examination. The results from these tests can be difficult to interpret and visual dysfunction may have to be severe before it can be diagnosed. Objective methods for evaluation of the function of visual pathways, such as visual evoked potentials (VEPs), would be a useful addition to the armamentarium of the veterinary ophthalmologist and equine clinician, to detect and evaluate disorders affecting post-retinal pathways in this species.

In 1954, Adrian & Matthews [1] reported that electrical potentials generated by neurons in the central nervous system in response to visual stimuli, VEPs, can be recorded from electrodes placed over the human occipital cortex. Dawson [2] developed a superimposition technique, called signal averaging, to extract the VEP waveform from the electroencephalogram (EEG) and subsequently, Cigánek [3] described the waveform morphology of the human VEP. In the years following further work was performed to explore the VEP in

* Correspondence: lena.strom@slu.se
Department of Clinical Sciences, Swedish University of Agricultural Sciences, PO Box 7054, SE-750 07 Uppsala, Sweden

human beings, improve the technique and implement recording of VEPs as a clinical, diagnostic aid.

Diseases or trauma affecting neurons in the retina or the post-retinal visual pathways can alter the waveform of the VEP. Therefore this technique can be used to investigate the functional integrity of post-retinal visual pathways, including the optic nerves, optic tract and optic radiation to the visual cortex. The method is used as a complement to the electroretinogram (ERG) and allows functional assessment of the visual pathways, while imaging techniques, such as magnetic resonance imaging (MRI) and computer tomography (CT), mainly are used to obtain information about structural integrity in patients. VEPs are used in human medicine to evaluate visual impairment due to several pathological conditions, such as cortical trauma or anoxia, meningitis, optic neuritis, optic neuropathies, demyelinating diseases and compression of post-retinal pathways due to tumors [4].

The International Society for Clinical Electrophysiology of Vision (ISCEV) regularly updates the human standard protocols for clinical electrophysiological examinations [5]. Guidelines are also issued by the American Clinical Neurophysiology Society [6]. In human patients, VEPs are usually recorded non-invasively using skin electrodes on the scalp overlying the visual cortex. In routine testing the VEP scalp electrodes are normally placed according to either the International 10/20 system [7] or the Queen Square System [8]. Midline electrodes over the visual cortex are used to assess prechiasmal function, while additional lateral electrode positions are used to assess chiasmal and postchiasmal dysfunction.

Usually one eye at a time is stimulated by flashes of white light (flash-VEP; FVEP) or a reversing, isoluminant pattern (pattern-VEP; PVEP). In human subjects, PVEPs are preferred in most clinical situations due to less variability than the FVEPs and PVEPs will also enable a better quantitative assessment of visual function. However, when using a pattern stimulus it is critical that the subject focuses on the pattern and in infants and patients with poor cooperation, poor fixation or poor acuity, the FVEP can be more useful [5].

While VEPs have been evaluated in a number of animal species such as the dog [9, 10], cat [11–13], cow [14], pig [15], sheep [16] and a large number of laboratory animal species [17], the use of VEP in the horse has only been very scantily described [18]. A simple protocol for recording the equine VEP in a clinical setting would be a useful tool to assess function of post-retinal pathways in this species and also of use in the investigation of neurological disorders.

Methods

The aim of this study was to establish a method for recording FVEP in horses in a clinical setting and to evaluate the waveform morphology in the normal horse. Electrode positions and the number of averaged responses needed to produce robust and reproducible VEP waveforms were also assessed. Additionally, recording of FVEPs after retrobulbar nerve block and transection of the optic nerve was performed to evaluate possible contamination of the FVEP by retinal potentials (ERGs).

Horses and procedures

Ten Swedish Standardbred Trotters, seven mares and three geldings, aged 3 to 19 years (median age: 11 years) were included in the study (Table 1). All horses were deemed to be healthy on the basis of a physical examination prior to the recordings. The regional Ethical Committee (Uppsala Djurförsöksetiska nämnd, Sweden, dnr C254/10 and dnr C39/12) approved the use of the horses in the study and the experiments were carried out following the guidelines of the ARVO guidelines for animal experimentation.

Table 1 FVEPs and FERGs were performed in ten healthy Swedish Standardbred Trotters

Horse	Age at examination (years)	Gender	Eye(s)	ERG	Number of averages	Electrode positions	Lateral positions	Transection	Retro bulbar nerve block	Reproducibility
A	19	M	OD	X	4–196	P_{z-0} P_{z-15} P_{z-30} P_{z-45} P_{z-60}	P_1 P_2 P_3 P_4	-	-	X
B	10, 11	G	OS	X	144	P_{z-0} P_{z-15} P_{z-30} P_{z-45} P_{z-60}	P_1 P_2 P_3 P_4	-	-	X
C	16, 17	M	OD	X	144	P_{z-0} P_{z-15} P_{z-30} P_{z-45} P_{z-60}	P_1 P_2 P_3 P_4	-	-	-
D	16, 17	M	OS (OD)	X	144	P_{z-0} P_{z-15} P_{z-30} P_{z-45} P_{z-60}	P_1 P_2 P_3 P_4	-	-	X
E	10, 11	M	OD	X	4–196	P_{z-0} P_{z-15} P_{z-30} P_{z-45} P_{z-60}	P_1 P_2	-	-	X
F	5	G	OS	X	4–196	P_{z-45}	P_1 P_2 P_3 P_4	-	-	X
G	17	M	OS	X	144	P_{z-45}	P_1 P_2	-	-	-
H	3	M	OS (OD)	X	144	P_{z-45}	P_1 P_2	-	-	-
I	4	M	OS (OD)	X	144	P_{z-45}	P_1 P_2	X	X	-
J	5	G	OD (OS)	X	144	P_{z-45}	P_1 P_2	-	X	-

Recordings from one, randomly selected eye were performed in six horses and from both eyes in four horses. "X" indicates that a certain procedure was performed. There were no statistical differences comparing eyes in this study

In five horses (horses A-E) five electrode positions in the midline, and two to four lateral positions were evaluated (Table 1). In the remaining five horses (horses F-J), one position in the midline and two to four lateral positions were evaluated. FVEPs were recorded at two separate sessions in five horses (horses A-B and D-F). The minimum number of responses averaged to obtain a waveform with clearly distinguishable wavelets was evaluated in three horses (horses A, E and F). For comparison between eyes in the same horse, both eyes were evaluated in horses D and H-J. Unilateral retrobulbar nerve blocks were performed in two horses (horses I and J) and transection of the optic nerve was performed in one horse (horse I). This was performed as a terminal procedure.

Ophthalmic examination

An ophthalmic examination including a cotton-ball test, reflex testing, slit-lamp biomicroscopy (Kowa SL-15; Kowa Co.Ltd., Tokyo, Japan), direct and indirect ophthalmoscopy (Heine Beta 200 and Heine Omega 500; Heine Optotechnik, Herrsching, Germany) and tonometry (TonoVet®, Icare Finland Oy, Helsinki, Finland) was performed in all horses. Furthermore, all horses were evaluated in a simple obstacle course, with five obstacles, under room-light conditions to assess vision prior to the recordings. No findings indicating visual impairment, ocular or neurological disease were observed in any of the horses.

Preparations

Pupils were dilated with 1–2 drops of tropicamide (Mydriacyl, 0.5 %, Alcon, Stockholm, Sweden). Recordings were started when pupils were fully dilated. The size of the pupils was monitored repeatedly during the recording sessions. The cornea was kept moist throughout the experiment by repeated instillation of artificial tears (ZilkEye, Evolan Pharma AB, Stockholm, Sweden).

The horses were sedated with a bolus injection of detomidine, 0.01 mg/kg (Domosedan vet, 10 mg/ml, Orion Pharma Animal Health, Sollentuna, Sweden) intravenously, followed by maintenance of sedation throughout the recording sessions using a continuous intravenous infusion of 2 % detomidine in physiologic saline solution (Natriumklorid, 7 mg/ml, Fresenius Kabi, Uppsala, Sweden) (2.0–2.6 mg detomidine hydrocloride/ 100 kg*h). The depth of sedation was monitored continuously and the infusion rate was adjusted to maintain a stable level of sedation where the horses kept their heads resting heavily on a padded headstand without becoming unsteady.

Akinesia and partial analgesia of the eyelids was induced with subcutaneous injections of 2.5 ml mepivacaine (Carbocain, 20 mg/ml, AstraZeneca, Södertälje, Sweden) over the auriculopalpebral and supraorbital nerves. A local infiltration of mepivacaine (Carbocain, 20 mg/ml, AstraZeneca, Södertälje, Sweden) was performed at each electrode position to reduce muscle contractions that potentially could give rise to artifacts in the FVEP. The cornea was anaesthetized with topical application of 2–3 drops of tetracaine (Tetrakain Chauvin, 1 %, Bausch & Lomb Nordic AB, Stockholm, Sweden).

The contralateral eye was covered by a black, opaque plastic eye shield during recordings (Fig. 1). Wads of cotton were stuffed in the ear canals to reduce noise-induced artifacts.

Retrobulbar nerve block and transection of the optic nerve

Horses I and J were prepared as for a transpalpebral enucleation in the standing, sedated horse [19, 20]. A single dose of flunixin meglumine (Flunixin N-vet, 50 mg/ml, N-vet, Uppsala, Sweden), 1.1 mg/kg, was administered intravenously pre-operatively. Akinesia of the eyelids, anesthesia at electrode positions and topical

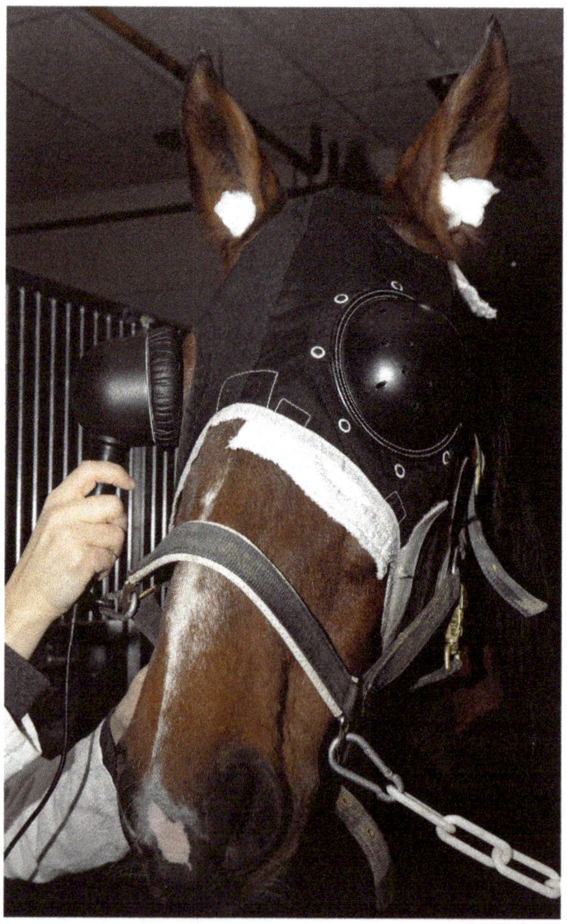

Fig. 1 FVEP in a horse using a handheld flash photostimulator. An opaque plastic eye shield covers the fellow eye and cotton wads in the ears are used to reduce auditory stimuli that could influence the level of consciousness

anesthesia were performed and sedation maintained as previously described. FERGs and FVEPs were recorded. A retrobulbar nerve block was performed as has been described by Gilger & Davidson [21] and the surgical site was infiltrated with mepivacain (Carbocain, 20 mg/ml, AstraZeneca, Södertälje, Sweden). Transection of the optic nerve was performed (horse I) 45 min after the retrobulbar nerve block. FERGs and FVEPs were recorded and rebound tonometry performed before and after the retrobulbar block in both horses, as well as before and after transection of the optic nerve in horse I. Both horses (I and J) were euthanized after the recordings in accordance with the ethical approval.

Electrophysiological testing

Light-adapted FERGs and FVEPs were recorded simultaneously. The recordings were performed in an examination room at the horse clinic of the University Animal Hospital, SLU, Uppsala. All electronic equipment not required for the study was disconnected to avoid mains current interference, but no other adjustments were made to the room.

A handheld dome-shaped photostimulator (Retiport mini-Ganzfeld, Acrivet GmbH, Hennigsdorf, Germany) with a background light intensity of 25 cd/m^2 and a flash intensity of 3 cd/m^2/s was used to light-adapt and stimulate the retina (Fig. 1). The stimuli were presented at a frequency of 1.09 Hz. The sweep duration was 500 ms post-stimulus. The room was kept dark during recordings to avoid interference by stray light.

Responses were amplified (amplifier set to x10^4), band-pass filtered (0.1–300 Hz for FERGs and 1–100 Hz for FVEPs), stored and analyzed using an Acrivet RETI-*port* (Acrivet GmbH, Hennigsdorf, Germany, software version 5.9.7.0) and a laptop computer.

FERGs

Light-adapted FERGs were recorded using corneal electrodes (Gold-foil corneal electrodes, CH electronics, Bromley, UK). Metal corkscrew electrodes (Stainless Steel Disposable Corkscrew Electrode, Cephalon A/S, Nørresundby, Denmark) served as ground and reference electrodes and were placed at the forehead and approximately 3 cm caudal to the lateral canthus, respectively. Electrode impedance was kept below 3 kΩin all FERG recordings.

FVEPs

Electrode positions were essentially selected based on a combination of the International 10–20 system [7] and Queen Square System [8] for human beings. Figure 2 shows the position of evaluated electrode positions in the midline in relation to the equine brain and the primary visual cortex heralded by the stria of Gennari in our postmortem specimens. The electrodes were placed at distances relative to bony landmarks on the head to adjust for head size. The distance between the nuchal crest and a line connecting the supraorbital foramina was measured and the electrodes were positioned at preselected percentages along this distance (Fig. 3), where the nuchal crest was the starting point (0 %) and the level of the supraorbital foramina was at 100 %. The first active electrode was placed on the nuchal crest in the midline (P_{z-0}; z = in the sagittal midline, 0 = 0 % rostral

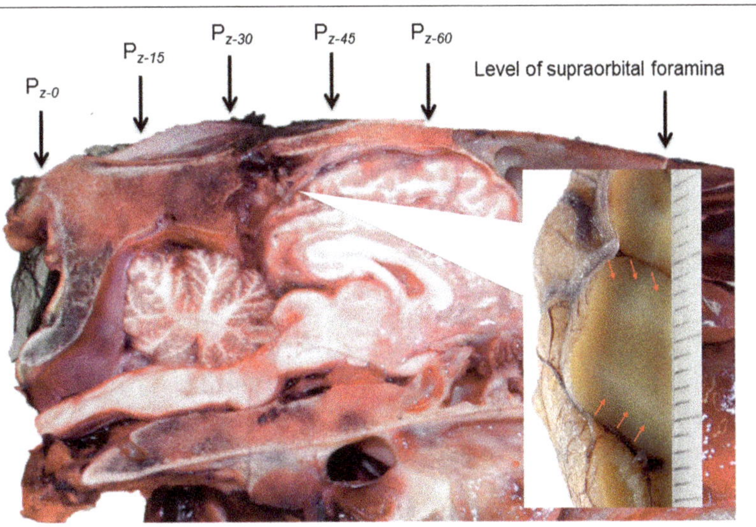

Fig. 2 The evaluated electrode positions along the midline shown relative to the equine brain. The subscript. "z" denotes a position in the sagittal midline and the numerical subscript reflects a percentage of the distance between the nuchal crest and the connecting line between the supraorbital foramina. The insert shows the approximate position of the striate cortex, or primary visual cortex. Red arrows indicate the stria of Gennari, which represents myelinated axons from the lateral geniculate body terminating in the gray matter in the primary visual cortex

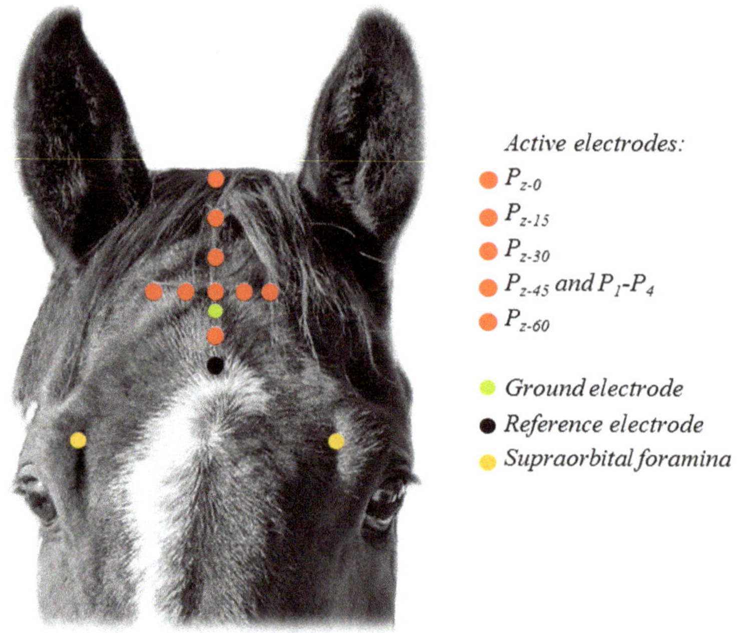

Fig. 3 The first active electrode was placed on the nuchal crest in the midline (P_{z-0}). The following positions along the midline were placed in front of the nuchal crest at regular intervals (P_{z-0} to P_{z-60}). Lateral positions, P_1 to P_4, were evaluated with electrodes placed two and four centimeters lateral to the midline electrode at P_{z-45}. Photo: Dr Måns Michanek

to the nuchal crest). The following positions of active electrodes along the midline were placed in front of the nuchal crest at intervals of 15 % (P_{z-0} to P_{z-60}). Lateral positions, P_1 to P_4, were evaluated with electrodes placed two and four centimeters lateral to the midline electrode at P_{z-45}. The ground and reference electrodes were positioned in the midline at 50 % and 70 %, respectively, of the distance from the nuchal crest and a line connecting the supraorbital foramina. Electrode impedance was kept below 2 kΩ in all FVEP recordings.

Averaged responses of 4, 16, 32, 64, 100, 144 and 196 sweeps were evaluated in horse A, E and F. The duration of a recording was therefore 3.7, 15, 29, 59, 92, 132 and 180 s respectively (frequency of 1.09 Hz). Averaged responses of 144 sweeps were evaluated in the remaining horses (horses B-D and G-J) and the duration of each averaged response was 2.2 min (132 s).

Monocular recordings were obtained from one randomly selected eye in all horses. In addition to this, recordings from the fellow eye were obtained in four horses (horses D and H-J to be able to enable comparison of recordings between eyes of the same horse.

Evaluation of ERGs and VEPs

FERG a- and b-wave amplitudes and times-to-peak (latencies) were measured according to convention [22]. The wavelets in the recorded FVEP waveform were named according to the nomenclature by Harding [23], a classification that was adopted in veterinary neurophysiology by Strain et al. in 1986 [14]. FVEP wavelet amplitudes and times-to-peak were measured as shown in Fig. 4. Results from recordings before and after retrobulbar nerve block (horse I-J) and transection of the optic nerve (horse I) were compared to evaluate if retinal potentials were recorded at the electrode positions on the scalp, thus contaminating the waveform, or if the recorded FVEP only represented post-retinal potentials.

Data analyses

Descriptive data; means, standard deviations (SDs) and ranges are reported for FVEP wavelet peak-to-peak amplitudes (μV) and times-to-peaks (ms). Data were analyzed using the non-parametric Kruskal-Wallis one-way analysis of variance by ranks and by regression analysis. Levene´s test was used to assess the equality of variances between groups. P-values ≤0.05 were considered statistically significant. Bonferroni correction was used to adjust the p-values when multiple comparisons were performed. Statistical analysis was performed with JMP® Pro 11 (Version 11.0.0, SAS Institute Inc., Cary, NC).

Results

Reproducible FVEPs and light-adapted FERGs were readily recorded from all ten, sedated horses. It was not possible to obtain reproducible recordings with acceptable quality from non-sedated horses due to movement and muscle artifacts. The waveform of the equine FVEP consisted of three major positive, P2-P4, and two major

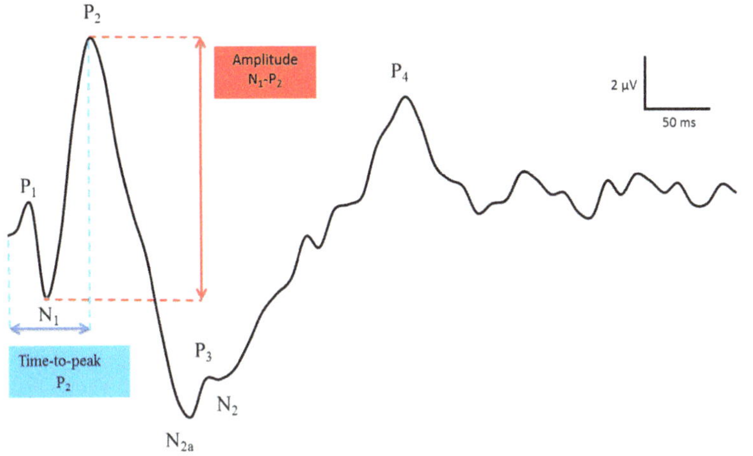

Fig. 4 Measurements of amplitudes and times-to-peak in the equine FVEP. Amplitudes were measured from the peak or trough of wavelet preceding the trough or peak of interest. Time-to-peak measurements were made from the onset of the stimulus to the maximum amplitude of the peak or trough

negative wavelets, N1-N2 (Fig. 5). The first negative wavelet, N1 and the most prominent positive wavelet, P2, was present in all recordings in all horses, as well as the large negative N2-complex and subsequent positive P4-wavelet. An early positive peak, P1, with a time-to-peak between 12–20 ms, was present in all horses, but only in approximately 50 % of the recordings in each horse. A small positive wavelet, P3, splitting the N2-complex into two parts, N2a and N2, was observed in all horses and in 82 % of recordings on average. A late positive wavelet, P5, with a time-to-peak between 275–375 ms was present in 74 % of recordings in nine out of the ten horses. Times-to-peak and peak-to-peak amplitudes are shown in Table 2.

The waveform was reproducibly recorded and no significant differences in amplitudes or times-to-peaks were found when comparing recordings from the same recording session and at the following session ($p > 0.05$; Fig. 6). There were no significant differences in times-to-peaks or amplitudes regardless of eye stimulated.

The overall waveform morphology did not differ between the electrode positions in the midline ($P_{z\text{-}0}$, $P_{z\text{-}15}$, $P_{z\text{-}30}$, $P_{z\text{-}45}$ and $P_{z\text{-}60}$) but the FVEP waveform was somewhat difficult to evaluate at the most caudal positions (Fig. 7). Average times-to-peak did not differ significantly between electrode positions but there were significant differences comparing amplitudes N1P2, P2N2 and N2P4 from recordings at $P_{z\text{-}0}$ and $P_{z\text{-}60}$ to recordings at $P_{z\text{-}30}$

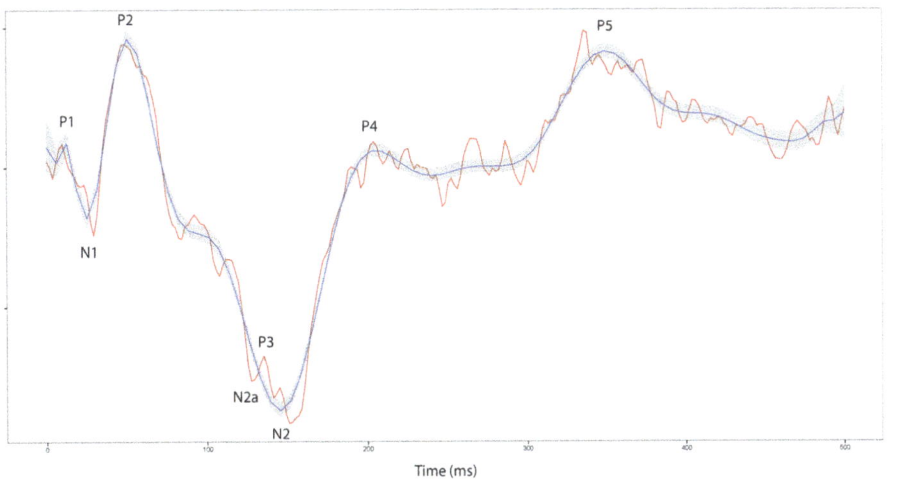

Fig. 5 Equine FVEP waveform. An equine FVEP waveform with typical peaks and troughs indicated (from horse B; position $P_{z\text{-}30}$). The red curve shows one recording (144 averaged responses). Data from 40 recordings (with 144 responses each) are averaged into the blue curve and shown with a 95 % confidence interval. Vertical division = 5 µV

Table 2 FVEP amplitudes and times-to-peaks in ten horses with the active electrode at $P_{z\text{-}45}$

		Time-to-peak (ms)								Peak-to-peak amplitude (μV)			
Horse	Eye	P1	N1	P2	N2a	P3	N2	P4	P5	N1-P2	P2-N2	N2a-P3	N2-P4
A	OD	13 ± 1.4	28 ± 1.2	59 ± 2.0	124 ± 3.5	137 ± 8.7	150 ± 4.6	273 ± 8.6	Missing	9.6 ± 1.0	11.9 ± 1.8	2.5 ± 1.2	8.4 ± 2.9
B	OS	12 ± 3.5	26 ± 1.1	54 ± 2.3	113 ± 4.2	123 ± 3.5	145 ± 8.8	218 ± 13.8	373 ± 31.4	7.4 ± 0.7	15.7 ± 3.6	0.4 ± 0.1	13.5 ± 4.9
C	OD	14 ± 0.1	28 ± 1.0	62 ± 2.5	104 ± 4.1	125 ± 5.6	153 ± 7.5	207 ± 13.8	354 ± 15.8	6.1 ± 1.2	5.9 ± 1.4	1.5 ± 1.0	3.7 ± 1.1
D	OS	18 ± 1.0	25 ± 2.0	47 ± 1.5	87 ± 12.0	107 ± 10.0	138 ± 8.8	200 ± 15.3	260 ± 7.2	2.2 ± 0.8	4.5 ± 1.0	1.5 ± 1.0	4.5 ± 1.1
E	OD	16 ± 3.5	29 ± 0.1	55 ± 4.0	99 ± 1.4	116 ± 2.8	153 ± 6.8	234 ± 24.6	375 ± 10.2	3.8 ± 1.5	4.6 ± 1.4	2.9 ± 1.1	4.5 ± 1.5
F	OS	12 ± 2.3	27 ± 1.4	54 ± 4.6	105 ± 4.6	122 ± 7.6	148 ± 6.4	213 ± 18.2	361 ± 26.2	7.3 ± 1.8	13.8 ± 3.9	0.4 ± 0.3	11.1 ± 3.7
G	OS	17 ± 1.2	30 ± 2.7	58 ± 3.0	104 ± 22.2	111 ± 18.9	137 ± 5.5	213 ± 13.3	355 ± 30.3	2.6 ± 0.6	6.2 ± 1.2	2.8 ± 4.7	8.3 ± 2.5
H	OD	20 ± 1.1	32 ± 5.9	57 ± 4.0	84 ± 28.9	95 ± 27.0	133 ± 7.9	202 ± 13.7	337 ± 32.7	2.5 ± 0.5	5.7 ± 1.5	0.9 ± 0.4	8.0 ± 2.6
I	OS	12 ± 3.1	17 ± 3.5	48 ± 2.2	103 ± 6.2	110 ± 8.2	130 ± 3.0	193 ± 7.1	305 ± 12.1	1.8 ± 0.2	3.5 ± 0.7	0.2 ± 0.1	4.5 ± 0.7
J	OD	6 ± 2.6	22 ± 5.9	52 ± 4.8	94 ± 12.5	107 ± 12.9	140 ± 6.8	206 ± 20.0	347 ± 19.8	2.2 ± 0.6	5.0 ± 1.5	4.8 ± 1.5	4.8 ± 1.7
95 % CI		12–14	26–28	55–57	99–105	113–120	137–142	213–224	324–345	4.3–5.5	6.1–7.8	1.3–2.3	5.9–7.4

A minimum of 5 recordings (144 averaged responses each) are included from each horse
Data are shown as means ± SDs

and $P_{z\text{-}45}$ ($p < 0.001$). Amplitudes increased when moving the electrode position rostrally from the most caudal position at the nuchal crest, reaching maximum amplitudes at the positions $P_{z\text{-}30}$ and $P_{z\text{-}45}$ (Fig. 8). Moving further rostrally decreased the amplitudes.

At lateral positions (P_1 to P_4), both amplitudes and times-to-peak showed a greater variability compared to recordings in the midline but there were no significant differences between means of times-to-peaks and amplitudes in our material.

When averaging 4 and 16 responses, the waveform was very difficult to evaluate and most recordings were discarded. Times-to-peaks and amplitudes from recordings with 32 and 64 averaged responses did not differ statistically compared to recordings with 100, 144 and 196 averages, but the waveform was more challenging for the examiner to evaluate due to artifacts. By averaging no less than 100 responses, the background noise was reduced, which resulted in more easily discernible peaks and reduced number of peaks split by large artifacts.

The FVEP was affected by the retrobulbar nerve block (Figs. 9 and 10). FVEP amplitudes were gradually reduced in both horses (I and J) and after 25 min, only a minuscule remnant of P2 was discernible, despite that FERG a- and b-wave times-to-peaks were within normal limits and b-wave amplitudes were only marginally reduced. Intraocular pressures (IOPs) in the examined eye were 42 mmHg (horse J) and 56 mmHg (horse I) 25 min after the block (IOPs before retrobulbar block; 19 and 22 mmHg respectively). In horse J, the peaks and troughs of the equine FVEP were clearly visible again approximately 90 min after the nerve block, although FVEP amplitudes and times-to-peaks were not yet within normal limits. In horse I, the FVEP disappeared when the optic nerve was transected (Fig. 10). FERG a- and b-wave times-to-peaks were within normal limits after the transection, but amplitudes were somewhat

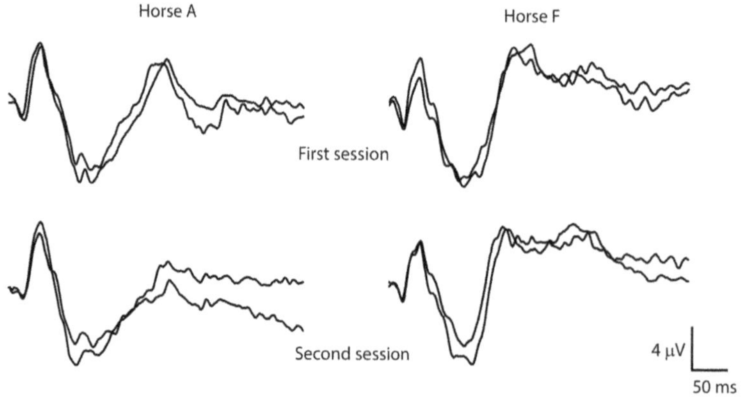

Fig. 6 Reproducibility of equine FVEPs. FVEPs recorded from two horses (horses A and F) with the active electrode at $P_{z\text{-}45}$. Two recordings from the first (top row) and second sessions several weeks later (bottom row) are shown

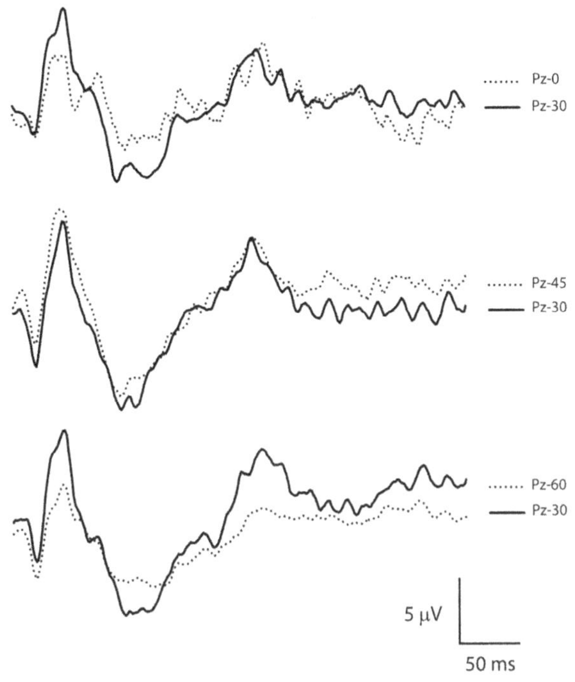

Fig. 7 FVEPs from different electrode positions. The overall waveform morphology did not differ between electrode positions in the midline but amplitudes were significantly reduced at the most caudal ($P_{z\text{-}0}$) and rostral ($P_{z\text{-}60}$) positions and it was often more difficult to obtain a reasonably smooth waveform due to artifacts. Recordings from horse A

reduced. The intraocular pressure was measured to 57 mmHg at this point. When the IOP rose above 70 mmHg due to retrobulbar hemorrhage and oedema (65 min after the nerve block), FERG a- and b-waves were also abolished.

Light-adapted FERGs were recorded from all eyes included in the study (Table 3). Times-to-peaks of the a- and b-waves were highly consistent both within and between horses while peak-to-peak amplitudes were more variable. Means for times-to-peaks of the a- and b-waves were 13 ± 1 ms and 30 ± 1 ms respectively. The mean a- and b-wave amplitudes were 10 ± 3 and 63 ± 15 μV respectively.

Discussion

VEPs were easily recorded in sedated horses under standard clinical settings and the overall waveform resembled those recorded in other mammalian species [5, 9–11, 13, 14, 16]. It has been shown in dogs that the signal recorded from VEP scalp electrodes can include a far-field conducted ERG [24, 25]. On the other hand, Sims et al. [9] have shown that transection of the optic nerve completely abolishes the light-adapted canine FVEP and that the recorded potentials are of post-retinal origin under light-adapted conditions. The horse has a large retina and the potentials generated are sizable. Thus it was important to evaluate if the equine FVEP recorded over the scalp consisted of post-retinal potentials only, or if a far-field conducted FERG was included. It is known in humans that a retrobulbar nerve block can cause temporary blindness due to a complete block of transmission in the optic nerve [26, 27] and we used retrobulbar anaesthesia and optic nerve transection to show that our recordings of the equine FVEP must be of post-retinal origin. Although the intraocular pressures were increased and FERG amplitudes were somewhat affected during the experiments, it has been shown in other species that increased intraocular pressures will result in reduced b-wave amplitudes [28–30], but simultaneous FVEP recordings will be more resistant to the pressure change [31]. Thus, it is highly unlikely that the increased intraocular pressure was causing the loss of the FVEPs in our recordings.

By using the FVEP we have also verified that the retrobulbar nerve block causes a transient loss of vision in the horse which should be considered when this method of anesthesia is employed for performing standing ocular surgery in this species.

Waveform morphology, times-to-peaks and peak-to-peak amplitudes were reproducible between recording sessions in the same horse and also similar between horses, although some differences in times-to-peaks and especially amplitudes were observed. It is well known from human beings [5] and other species [11] that the FVEP is highly variable between individuals. Four wavelets (N1, P2, N2 and P4) were observed in all FVEPs from all horses whereas additional wavelets were present in some recordings in all horses (P1, N2a and P3). A late, positive component (P5) was present in the majority of recordings in most horses. The differences in waveform morphology might represent the variability of the FVEP between individuals. However, in our study, these differences might also be due to varying depth of sedation, or that the individual horse might react differently to external disturbances in the clinical setting. The most reliable wavelets in the equine FVEP were the N1, P2, N2 and P4, which indicate that these peaks could be of use when the FVEP is employed as a diagnostic tool to diagnose and localize abnormal function in post-retinal visual pathways.

Choosing the suitable electrode position enhanced detection of the individual wavelets composing the equine FVEP. The wavelets became more obvious and distinct when the active electrodes were positioned well in front of the nuchal crest at electrode positions $P_{z\text{-}30}$ and $P_{z\text{-}45}$. These positions correspond well to the position of the occipital cortex and presumed anatomical location of the visual cortex in the horse [32], which we confirmed using the stria of Gennari as our landmark in postmortem specimens (Fig. 2). As expected, the electrode

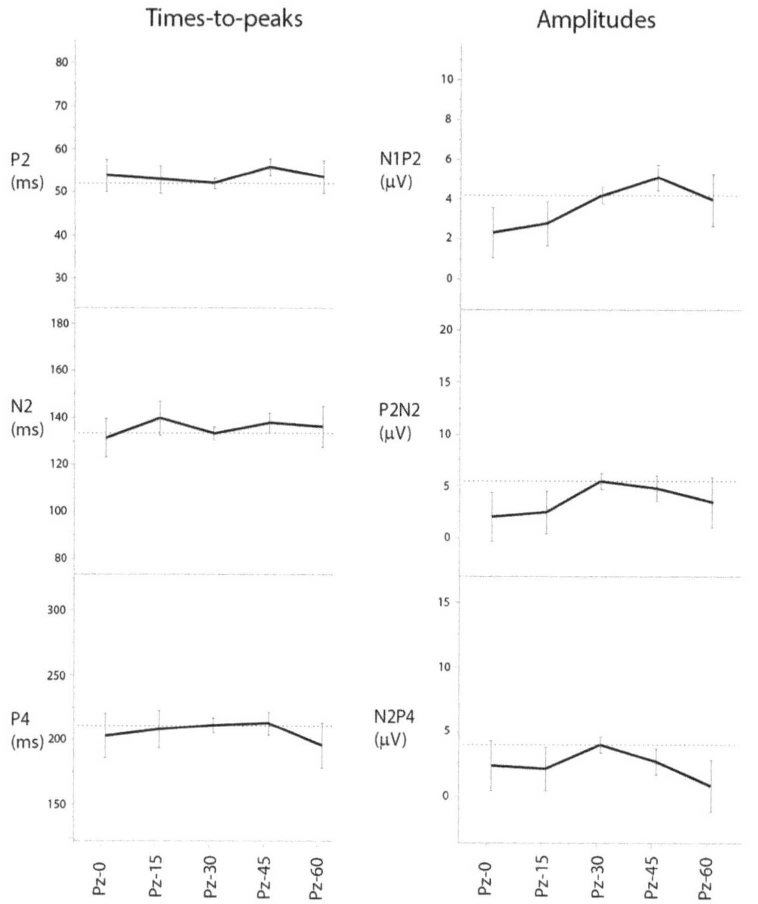

Fig. 8 Amplitudes and times-to-peaks at different electrode positions. Amplitudes increased when moving the electrode position rostrally from the most caudal position at the nuchal crest, reaching maximum amplitudes at the positions P_{z-30} and P_{z-45}. Moving further rostrally decreased the amplitudes. Times-to-peaks did not differ significantly between any of the positions tested in this material

positions providing the largest amplitudes in the horse differ considerably from the positions recommended in the guidelines for clinical VEPs in human beings [5], reflecting the substantial anatomical differences between man and horse. An asymmetry was expected when recording the FVEP at ipsi- and contralateral positions due to the high degree of decussation of optic fibers in the horse of more than 80 % [33]. It was thus somewhat surprising that results from lateral positioning of the active electrodes were inconclusive, with regards to the separation of signals from the right and left hemispheres. The reason for this may be due to the close proximity to the ears of the horse. The lateral electrode positions were close to the base of the pinnae and the large muscles controlling ear movements in this species [34]. However, while muscle artifacts due to ear movements may have masked more subtle differences between lateral and midline positions, a method for obtaining FVEPs from the more lateral electrode positions without picking up artifacts from the ear muscles would be useful, because these positions may be important for differing between pre- and postchiasmal lesions in certain horses with visual impairment.

Sufficient and stable depth of sedation enhanced recording of FVEPs and FERGs in our study, which is in accordance with previous studies when FERGs were recorded in the standing, sedated horse [35–37]. Movement artifacts did occasionally overshadow a response, but the stable level of sedation obtained by continuous infusion of detomidine kept the horses sufficiently immobile to allow recording of reproducible FERGs and FVEPs. It was obvious that the depth of sedation influenced the amount of artifacts and the quality of recordings. When only light sedation was used, ear movements and head movements in response to sudden sounds caused muscle movement artifacts, which could wipe out the normal waveform. The use of earplugs reduced the episodes of increased alertness due to undesired auditory stimuli affecting the level of sedation. In summary, we conclude that a sufficiently stable level and depth of sedation can be accomplished by a bolus dose followed by a continuous infusion of detomidine.

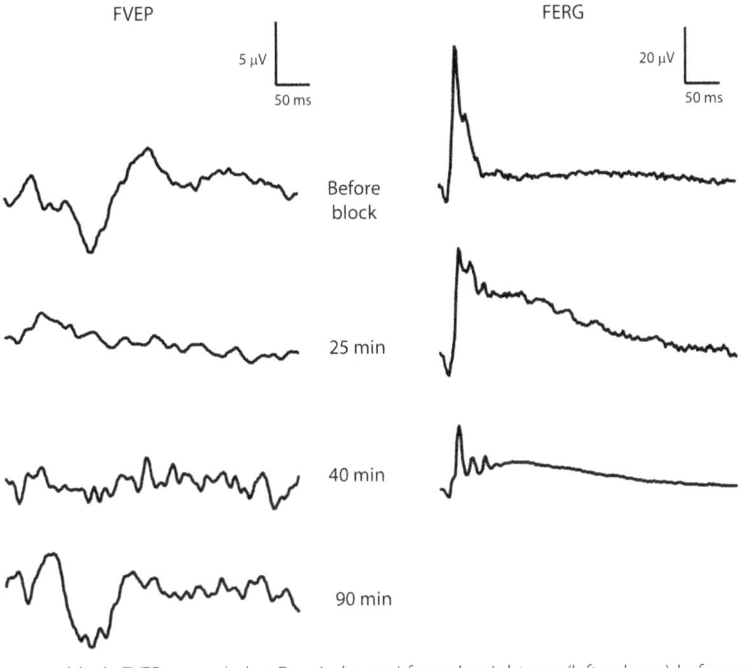

Fig. 9 FVEPs after retrobulbar nerve block. FVEPs recorded at $P_{z\text{-}45}$ in horse J from the right eye (left column) before retrobulbar nerve block (uppermost tracing), 25, 40 and 90 min after retrobulbar nerve block (lower tracings). Simultaneous FERG recordings from the right eye are shown in the right column

However, it is important to note that in human beings and other species, some sedatives have been shown to affect the VEP [38], causing reduced amplitudes and somewhat prolonged times-to-peak. To our knowledge, the specific effects of detomidine on the VEP has not yet been evaluated. It is reasonable to believe that amplitudes and times-to-peaks may be affected by sedation with alfa-2-agonists, but we were unable to obtain recordings with acceptable quality from non-sedated horses. Therefore, we recommend performing recordings under sedation. To be able to compare results between horses, it is important to continuously monitor the level of sedation during the recordings.

According to our results, at least 100 sweeps are needed in sedated horses to suppress electrical noise and obtain curves with easily discernible, reproducible peaks and troughs. This might be especially important for the inexperienced examiner. Overall, the equine FVEP appears to require more averaged responses than suggested by the ISCEV guidelines, which is likely to reflect that both the amount of potentially disturbing auditory stimuli in a busy equine clinic is higher than in a human ophthalmology or neurology clinic, and that our equine patients may be less cooperative and docile than the average adult human patient. However, it is important to note that 32–64 averaged responses in the horse may also produce acceptable results. Although there might be more noise and artifacts affecting the quality of the recordings, our results indicate that times-to-peaks and amplitudes will not be significantly affected when measured by an experienced examiner. The advantage of averaging fewer responses is that the time to perform the examination will be reduced, which might be important when recording the VEP in a critically ill patient.

The corkscrew electrodes used were applied without difficulties and were well tolerated by the horses and maintained good contact between electrodes and horse (impedance <2 kΩ) throughout the recordings. Other technical considerations include keeping the connecting wires as short as possible to avoid mains interference from electric equipment in the examination room and ensuring equipment not necessary for the procedure are unplugged to reduce sources of electromagnetic noise.

Although this work confirmed that the potentials in the equine FVEP are of post-retinal origin, the exact sources of the potentials producing each of the wavelets in the equine VEP are still unknown. The FVEP has been a diagnostic aid in human medicine for decades, but the origins of some wavelets seen in the human FVEP are still not clearly identified. Further studies are needed in the horse to evaluate the components of the equine VEP with the aim of establishing the anatomical location of lesions causing certain aberrations in the VEP waveform.

The first study of the human VEP in different age groups was performed by Dustman and Beck in 1969

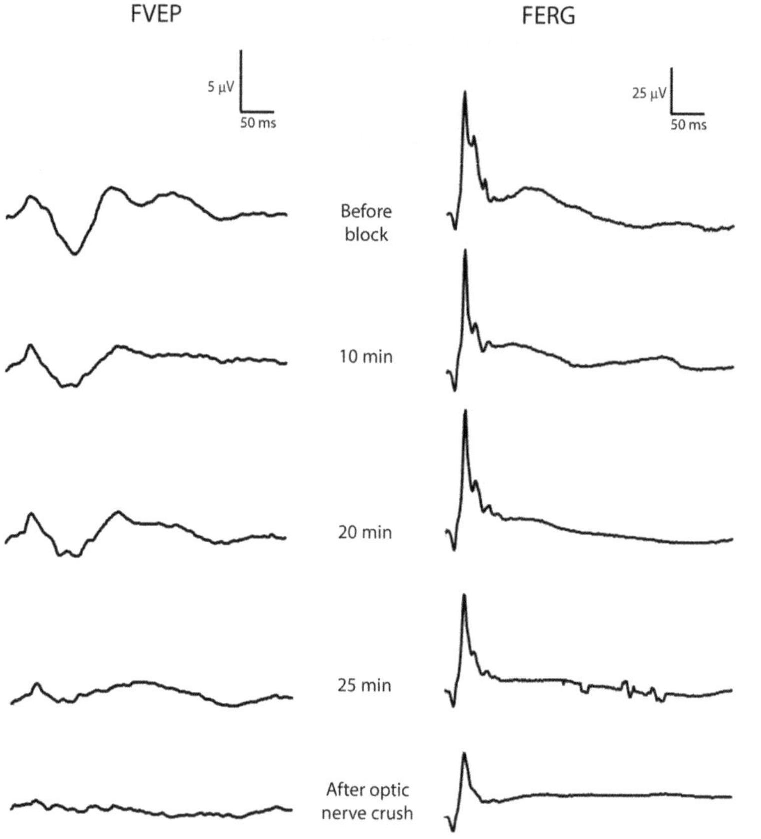

Fig. 10 FVEPs after retrobulbar nerve block and transection of the optic nerve. FVEPs recorded at $P_{z\text{-}45}$ in horse I from the right eye (left column) before retrobulbar nerve block (uppermost tracing), 10, 20 and 25 min after retrobulbar nerve block (middle rows) and after transection of the optic nerve (lowermost tracing). Simultaneous FERG recordings from the right eye are shown in the right column

[39] who showed that with increasing age, the human VEP alters in regards to times-to-peak and amplitudes, although the overall morphology stays the same throughout the adult life-span. Studies on the age-related development of VEPs have been performed in several animal species such as for example in cats [13], dogs [40], ruminants [16, 41, 42] and pigs [15]. As we have only few horses at each age in this study, further studies are needed to evaluate the normal variation of the equine VEP related to increasing age.

The FVEP in combination with concurrent FERGs provide an opportunity for objective, functional assessment of the post-retinal visual pathways in the standing, sedated equine patient. Equine FVEPs would be valuable for evaluating similar types of pathological conditions affecting the visual system as in human patients, such as cortical trauma, optic neuritis, tumors and demyelinating diseases. Additionally, FVEPs offer an opportunity to detect persistent visual impairment or blindness after retrobulbar nerve block, which has been reported as a rare complication in human patients [43]. As well the technique could be an important additional diagnostic tool to rule out visual impairment in cases of abnormal behavior.

Table 3 Flash electroretinograms

	a-wave	b-wave
Implicit time (ms)	13 ± 1	30 ± 1
Amplitude (μV)	10 ± 3	63 ± 15

Light-adapted FERG amplitudes and times-to-peak in ten healthy horses. Data are shown as means ± SDs

Conclusions

We have recorded FVEPs in healthy, sedated horses in a clinical setting. Active electrodes positioned at $P_{z\text{-}30}$ to $P_{z\text{-}45}$ provided a good signal-to-noise ratio and reproducible waveforms. Results from recordings after retrobulbar anaesthesia and transection of the optic nerve showed that the recorded responses were of post-retinal origin. Our technique for FVEP in the horse is relatively easily mastered and opens up for objective, functional assessment of the post-retinal visual pathways in this species.

Acknowledgements
We wish to thank Dr. John Pringle, Dr. Katarina Nostell, Dr. Marie Rhodin and Dr. Pia Haubro-Andersen for comments on the draft of the manuscript. We also wish to thank Dr. Ove Wattle for performing retrobulbar nerve blocks and transection of the optic nerve. We are grateful to the Mari Wallbring, Kenneth Larsson, Dennis Larsson and Carola Jansson in the Department for Clinical Sciences, SLU, for their excellent assistance during all the examinations of the horses.

Authors' contributions
LS participated in the design of the study, performed ophthalmic and electrophysiological examinations and drafted the manuscript. BE proposed the study, assisted in ophthalmic and electrophysiological examinations and in the drafting of the manuscript. Both authors read and approved the final manuscript.

Competing interests
The authors declare that they have no competing interests.

Consent for publication
Not applicable.

References

1. Adrian ED, Matthews BH. The interpretation of potential waves in the cortex. J Physiol. 1934;81(4):440–71.
2. Dawson GD. A summation technique for the detection of small evoked potentials. Electroencephalogr Clin Neurophysiol. 1954;6(1):65–84.
3. Ciganek L. The EEG response (evoked potential) to light stimulus in man. Electroencephalogr Clin Neurophysiol. 1961;13:165–72.
4. Young B, Eggenberger E, Kaufman D. Current electrophysiology in ophthalmology: a review. Curr Opin Ophthalmol. 2012;23(6):497–505.
5. Odom JV, Bach M, Brigell M, Holder GE, McCulloch DL, Tormene AP, Vaegan. ISCEV standard for clinical visual evoked potentials (2009 update). Doc Ophthalmol. 2010;120(1):111–9.
6. American Clinical Neurophysiology S. Guideline 9B: guidelines on visual evoked potentials. Am J Electroneurodiagnostic Technol. 2006;46(3):254–74.
7. Jasper HH. The ten-twenty electrode system of the International Federation. Electroencephalogr Clin Neurophysiol. 1958;10:371–5.
8. Blumhardt LD, Barrett G, Halliday AM. The asymmetrical visual evoked potential to pattern reversal in one half field and its significance for the analysis of visual field defects. Br J Ophthalmol. 1977;61(7):454–61.
9. Sims MH, Laratta LJ, Bubb WJ, Morgan RV. Waveform analysis and reproducibility of visual-evoked potentials in dogs. Am J Vet Res. 1989; 50(11):1823–8.
10. Strain GM, Jackson RM, Tedford BL. Visual evoked potentials in the clinically normal dog. J Vet Intern Med. 1990;4(4):222–5.
11. Creel DJ, Dustman RE, Beck EC. Visually evoked responses in the rat, guinea pig, cat, monkey, and man. Exp Neurol. 1973;40(2):351–66.
12. Padnick LB, Linsenmeier RA. Properties of the flash visual evoked potential recorded in the cat primary visual cortex. Vision Res. 1999;39(17):2833–40.
13. Rose GH, Gruenau SP, Spencer JW. Maturation of visual electrocortical responses in unanesthetized kittens: effects of barbiturate anesthesia. Electroencephalogr Clin Neurophysiol. 1972;33(2):141–58.
14. Strain GM, Olcott BM, Hokett LD. Electroretinogram and visual-evoked potential measurements in Holstein cows. Am J Vet Res. 1986;47(5):1079–81.
15. Strain GM, Tedford BL, Gill MS. Brainstem auditory evoked potentials and flash visual evoked potentials in Vietnamese miniature pot-bellied pigs. Res Vet Sci. 2006;80(1):91–5.
16. Strain GM, Claxton MS, Prescott-Mathews JS, LaPhand DJ. Electroretinogram and visual-evoked potential measurements in sheep. Can J Vet Res. 1991; 55(1):1–4.
17. Ridder W, J.R H, G.B A. Visual Evoked Potentials in Animals. In: Principles and Practice of Clinical Electrophysiology of Vision. 2nd ed. USA: MIT Press; 2006. p. 935–47.
18. Brooks DE. Equine Ophthalmology. In: Gelatt KN, editor. Veterinary Ophthalmology, vol. II. 3rd ed.: Philadelphia: Lippincott Williams & Wilkins; 1999. p. 1108.
19. Hewes CA, Keoughan GC, Gutierrez-Nibeyro S. Standing enucleation in the horse: a report of 5 cases. Can Vet J. 2007;48(5):512–4.
20. Pollock PJ, Russell T, Hughes TK, Archer MR, Perkins JD. Transpalpebral eye enucleation in 40 standing horses. Vet Surg. 2008;37(3):306–9.
21. Gilger BC, Davidson MG. How to prepare for ocular surgery in the standing horse. In: Annual Convention of the AAEP. Orlando: IVIS; 2002.
22. Ekesten B, Komaromy AM, Ofri R, Petersen-Jones SM, Narfstrom K. Guidelines for clinical electroretinography in the dog: 2012 update. Doc Ophthalmol. 2013;127(2):79–87.
23. Harding GFA. The visual evoked response. In: Roper-Hall MJea, editor. Advances in Ophthalmology. Basel: S Karger AG; 1974. p. 2–28.
24. Bichsel P, Oliver Jr JE, Coulter DB, Brown J. Recording of visual-evoked potentials in dogs with scalp electrodes. J Vet Intern Med. 1988;2(3):145–9.
25. Malnati GA, Marshall AE, Coulter DB. Electroretinographic components of the canine visual evoked response. Am J Vet Res. 1981;42(1):159–63.
26. Atkinson WS. Local anesthesia in ophthalmology. Am J Ophthalmol. 1948; 31(12):1607–18.
27. Cyriac IC, Pineda 2nd R. Postoperative complications of periocular anesthesia. Int Ophthalmol Clin. 2000;40(1):85–91.
28. Uenoyama K, McDonald JS, Drance SM. The effect of intraocular pressure on visual electrical responses. Arch Ophthalmol. 1969;81(5):722–9.
29. Uenoyama K, McDonald JS, Drance SM. The effect of artificially elevated intraocular pressure on the electroretinogram of the cat. Can J Ophthalmol. 1968;3(2):154–8.
30. Howard DR, Sawyer DC. Electroretinography of acute hypoxic and increased intraocular pressure status in the dog. Am J Vet Res. 1975;36(1):81–4.
31. Tsai TI, Bui BV, Vingrys AJ. Effect of acute intraocular pressure challenge on rat retinal and cortical function. Invest Ophthalmol Vis Sci. 2014;55(2):1067–77.
32. Takeuchi T, Sugita S. Cytoarchitectonic study of the cerebral cortex in the horse brain–classification of the cortex area. Kaibogaku Zasshi. 2001;76(2): 213–21.
33. Duke-Elder S. The eye in evolution. In: System of Ophthalmology, vol. I. London: Henry Kimpton; 1958. p. 446–504.
34. Budras KD. Anatomy of the Horse. 6th ed. Hannover: Schluetersche; 2012.
35. Ben-Shlomo G, Plummer C, Barrie K, Brooks D. Characterization of the normal dark adaptation curve of the horse. Vet Ophthalmol. 2012;15(1):42–5.
36. Church ML, Norman JC. Electroretinogram responses of the normal thoroughbred horse sedated with detomidine hydrochloride. Vet Ophthalmol. 2012;15 Suppl 2:77–83.
37. Komaromy AM, Andrew SE, Sapp Jr HL, Brooks DE, Dawson WW. Flash electroretinography in standing horses using the DTL microfiber electrode. Vet Ophthalmol. 2003;6(1):27–33.
38. Sannita WG. Neuropsychiatric drug effects on the visual nervous system. In: Heckenlively JR, Arden GB, editors. Principles and practice of clinical electrophysiology of vision. USA: Mosby Year Book; 1991. p. 167–73.
39. Dustman RE, Beck EC. The effects of maturation and aging on the wave form of visually evoked potentials. Electroencephalogr Clin Neurophysiol. 1969;26(1):2–11.
40. Strain GM, Jackson RM, Tedford BL. Postnatal development of the visual-evoked potential in dogs. Am J Vet Res. 1991;52(2):231–5.
41. Strain GM, Graham MC, Claxton MS, Olcott BM. Postnatal development of brainstem auditory-evoked potentials, electroretinograms, and visual-evoked potentials in the calf. J Vet Intern Med. 1989;3(4):231–7.
42. Takeuchi T, Suzuki M, Sitizyo K, Saito T. Postnatal development of visual evoked potentials in Japanese black calves. Jpn J Physiol. 1993;43(6):809–15.
43. Kumar CM, Dowd TC. Complications of ophthalmic regional blocks: their treatment and prevention. Ophthalmologica. 2006;220(2):73–82.

17

Elevated urinary excretion of free pyridinoline in Friesian horses suggests a breed-specific increase in collagen degradation

Veronique Saey[1*], Jonathan Tang[2], Richard Ducatelle[1], Siska Croubels[3], Siegrid De Baere[3], Stijn Schauvliege[4], Gunther van Loon[5] and Koen Chiers[1]

Abstract

Background: Friesian horses are known for their high inbreeding rate resulting in several genetic diseases such as hydrocephaly and dwarfism. This last decade, several studies focused on two other presumed hereditary traits in Friesian horses: megaoesophagus and aortic rupture. The pathogenesis of these diseases remains obscure but an important role of collagen has been hypothesized. The purpose of this study was to examine possible breed-related differences in collagen catabolism. Urinary specimens from Friesian ($n = 17$, median age 10 years old) and Warmblood horses ($n = 17$, median age 10 years old) were assessed for mature collagen cross-links, i.e. pyridinoline (PYD) (=hydroxylysylpyridinoline/HP) and deoxypyridinoline (DPD) (lysylpyridinoline /LP). Solid-phase extraction was performed, followed by reversed-phase ion-paired liquid chromatography prior to tandem mass spectrometry (MS/MS) detection.

Results: Mean urinary concentrations of free PYD, expressed as fPYD/creatinine ratio, were significantly higher in Friesian horses compared to Warmblood horses (28.5 ± 5.2 versus 22.2 ± 9.6 nmol/mmol, $p = 0.02$) while mean fDPD/creatinine ratios were similar in both horse breeds (3.0 ± 0.7 versus 4.6 ± 3.7 nmol/mmol, $p = 0.09$).

Conclusions: Since DPD is considered a specific bone degradation marker and PYD is more widely distributed in connective tissues, the significant elevation in the mean PYD/DPD ratio in Friesian versus Warmblood horses (9.6 ± 1.6 versus 5.7 ± 1.8, $p < 0.0001$) suggests a soft tissue origin for the increased fPYD levels. Considering that a previous study found no differences in total collagen content between Friesian and Warmblood horses for tendon and aortic tissue, this indicates a higher rate of collagen degradation. The latter might, at least in part, explain the predisposition of Friesians to connective tissue disorders.

Keywords: Horse, Aortic rupture, Megaoesophagus, Mass spectrometry, Collagen, Cross-links

Background

The Friesian horse breed is challenged by several traits in which an important role of collagen has been suggested. In some of these, such as in dwarfism and hydrocephaly, the genetic background has been elucidated. In both diseases, a mutation in a gene involved in protein glycosylation was detected. Dwarfism in Friesian foals, which are also affected by tendon laxity [1], has been attributed to a splice site mutation in B4GALT7 [2]. B4GALT7 has an important role in collagen fibrillinogenesis [3]. A nonsense mutation in B3GALNT2 has been shown to be involved in hydrocephaly in Friesian foals [4]. In humans, mutations in B3GALNT2 are associated with both muscular and brain anomalies [5].

In aortic rupture and megaoesophagus, both presumed to be hereditary traits in Friesians, an increased deposition of clumped collagen at the site of the aortic rupture and in both the dilated and non-dilated part of the

* Correspondence: veronique.saey@ugent.be
[1]Laboratory of Veterinary Pathology, Department of Pathology, Bacteriology and Avian Diseases, Faculty of Veterinary Medicine, Ghent University, Merelbeke, Belgium

oesophagus has been reported [6, 7]. Friesian horses seem to be predisposed for primary gastric rupture as well. It has been suggested that this could also be a manifestation of a presumed underlying soft tissue disorder [8].

Research on collagen metabolism in Friesians is scarce. Our research group has demonstrated an increased area percentage of collagen type I in the thoracic aortic media of Friesians compared to Warmblood horses. This finding suggests the presence of a primary collagen disorder [9]. However, no differences were found in the aortic matrix between Friesian and Warmblood horses using high performance liquid chromatography. In contrast, within the deep digital flexor tendon, Friesians had less lysine hydroxylation and pyrrole cross-linking than Warmblood horses [10].

Mature, trivalent cross-links, called pyridinoline (PYD) or hydroxylysylpyridinoline (HP) and deoxypyridinoline (DPD) or lysylpyridinoline (LP), bind adjacent collagen chains in a triple-helix structure. During tissue degradation, these cross-links are released into the circulation, followed by urinary excretion [11]. In human medicine, excretion of PYD and DPD is mainly monitored to diagnose diseases associated with excessive bone resorption [12, 13].

Not much is known regarding urinary excretion of collagen cross-links in horses. An age-related and diurnal variation, similar as in humans, has been reported [14].

To demonstrate a breed-related change in collagen metabolism, urinary excretion of collagen degradation products (fPYD and fDPD) was measured in Friesian horses and compared to Warmblood horses.

Methods

Animals and urine collection

All information regarding the sampled horses is summarized in Table 1. Age, gender, physical exercise, reason for surgery and time of sampling are displayed.

Seventeen Warmblood horses (6 months-20 years old; median age: 10 years old; 5 mares, 3 stallions and 9 geldings) and 17 Friesian horses (2–19 years old; median age: 10 years old; 15 mares, 1 gelding and 1 stallion) were used in this study. Sixteen Friesian and 8 Warmblood horses were sampled once at home during spontaneous urination. The remaining 10 horses (1 Friesian and 9 Warmblood horses) were sampled once during general anesthesia for surgical procedures at the Faculty of Veterinary Medicine of Ghent University, Belgium.

The specimens were immediately frozen. A written consent was obtained from the owner/keeper of each animal. All Friesian horses were purebred and had a pedigree.

The urine was collected in a plastic container and stored at − 20 °C until further analysis (maximum storage time was 9 months). Prior to analysis, samples were defrosted, gently stirred and transferred to 1.5 mL eppendorf tubes.

It is likely that training could have an influence on the pyridinoline cross-linking in bone and possibly in muscle in horses. Both groups of horses however were used for pleasure riding or low level competition.

All samples were collected between September–April, stored and analysed in May of the same year. Time intervals between harvesting and analysis were similar in both horse groups.

Urine analysis for free pyridinoline (fPYD), deoxypyridinoline (fDPD) and creatinine

Analysis for urinary fPYD and fDPD were conducted following the method described by Tang et al. [15]. Human urine pyridinium cross-links calibrator was used from Immundiagnostik (Bensheim, Germany). In short, 0.5 mL of urine samples pretreated with 0.5 mL hydrochloric acid (40% concentrate) was extracted using solid phase extraction (SPE) column packed with cellulose slurry. Pyridinium cross-links were eluted from the SPE columns and the extracts were transferred into microcentrifuge vials. The eluents were then centrifuged at 10,000 g for 5 min, followed by liquid chromatography and mass spectrometric analysis (LC–MS/MS). The lower limit of quantification was 6 nmol/L for fPYD and 2.5 nmol/L for fDPD.

Urine creatinine was measured to obtain fPYD and fDPD to urine creatinine ratio. A second generation kinetic colorimetric assay was performed, based on the Jaffe method on the COBAS® C501 analyser.[1] Final results were expressed as nmol/L of fPYD or fDPD per mmol/L creatinine or thus nmol/mmol fPYD or fDPD/creatinine ratio.

Statistical data analysis

T-tests were performed to compare fPYD/creatinine and fDPD/creatinine urinary ratios between Friesian and Warmblood horses. Significance was set at $p < 0.05$. Commercial statistical software was used (GraphPad Prism, USA).

Results

The mean urinary fPYD/creatinine ratio was 28.5 (19.6–39.9) ± 5.2 nmol/mmol and 22.2 (7.1–47.4) ± 9.6 nmol/mmol in Friesian and Warmblood horses, respectively (Fig. 1). The mean urinary fDPD/creatinine ratio was 3.0 (2.2–4.4) ± 0.7 and 4.6 (1.2–17.1) ± 3.7 nmol/mmol in Friesian and Warmblood horses, respectively (Fig. 2). The mean urinary fPYD/fDPD ratio was 9.6 (6.9–13.7) ± 1.6 in Friesian and 5.7 (2.4–9.1) ± 1.8 in Warmblood horses (Fig. 3).

There was a significantly higher mean urinary fPYD/creatinine ratio ($p = 0.02$) and fPYD/DPD ratio ($p < 0.0001$) in Friesian versus Warmblood horses. No significant differences were found for mean fDPD/creatinine ratio ($p = 0.09$).

Table 1 Overview of animals sampled with information regarding age of horses, gender, regular physical exercise, reason for surgery, time of sampling and measured results

Horse	Age yrs	Gender	Exercise	Surgery	Time hrs	fPYD/creatinine nmol/mmol	fDPD/creatinine nmol/mmol	PYD/DPD	Horse	Age yrs	Gender	Exercise	Surgery	Time hrs	fPYD/creatinine nmol/mmol	fDPD/creatinine nmol/mmol	PYD/DPD
Friesian	2	stallion		castration	10	34.8	4.4	8.0	warmbl	6 m	stallion		tendon retraction	9	41.1	17.1	2.4
Friesian	3	mare	yes		17	39.9	4.4	9.0	warmbl	7 m	mare		cecocolic invagination	10	47.4	8.3	5.7
Friesian	4	mare	yes		20	23.4	2.3	10.0	warmbl	3	stallion	yes	castration	9	19.0	4.7	4.0
Friesian	4	mare			17	31.5	2.3	13.7	warmbl	4	mare	yes	sarcoid	10	25.4	4.6	5.5
Friesian	5	mare			18	28.4	3.0	9.6	warmbl	4	mare	yes	Laryngeal hemiplegia	12	20.3	3.5	5.8
Friesian	8	mare	yes		16	26.9	2.7	10.0	warmbl	6	gelding	yes	sarcoid	10	17.1	4.7	3.6
Friesian	8	mare	yes		16	22.9	2.7	8.5	warmbl	6	stallion	yes	castration sarcoid	10	14.1	2.1	6.7
Friesian	9	mare	yes		19	24.9	2.4	10.3	warmbl	7	gelding	yes		13	18.1	2.4	7.4
Friesian	10	mare	yes		11	32.2	3.4	9.4	warmbl	10	gelding	yes	cornea ulcer	11	16.0	2.4	6.7
Friesian	10	mare	yes		16	32.2	3.2	9.9	warmbl	10	gelding	yes		12	16.7	3.0	5.6
Friesian	10	mare	yes		11	32.8	3.1	10.7	warmbl	12	mare		subepiglottis cyst	16	7.1	1.2	6.0
Friesian	11	mare	yes		18	26.0	2.3	11.3	warmbl	12	gelding	yes		12	26.8	2.9	9.1
Friesian	12	mare	yes		11	27.9	2.6	10.9	warmbl	13	gelding	yes		14	19.1	4.2	4.6
Friesian	13	mare	yes		17	26.4	3.6	7.3	warmbl	17	gelding	yes		13	24.0	2.9	8.4
Friesian	15	mare	yes		17	22.5	2.5	8.9	warmbl	18	gelding	yes		15	25.0	7.7	3.3
Friesian	16	mare			17	31.7	3.6	8.8	warmbl	20	mare		urinary bladder tumor	10	18.9	2.9	6.4
Friesian	19	gelding	yes		17	19.6	2.8	6.9	warmbl	20	gelding	yes		14	20.7	3.6	5.8
Median	10				17	27.9	2.8	9.6		10				12	19.1	3.5	5.8

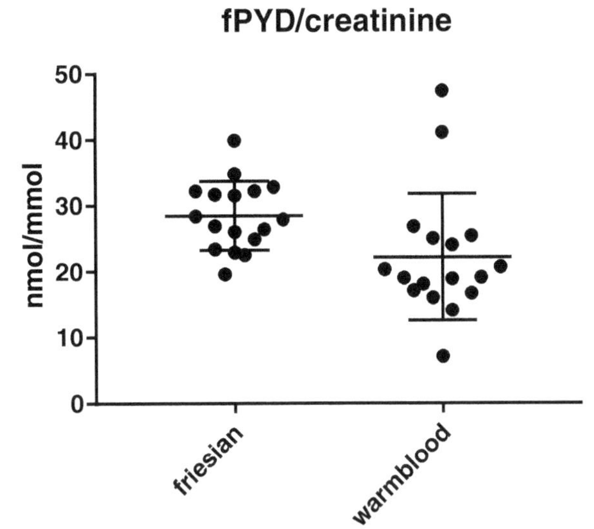

Fig. 1 Friesian horses have a significantly higher mean urinary fPYD/creatinine concentration (28.5 ± 5.2 nmol/mmol) (, $n = 17$) in the urine compared to Warmblood horses (22.2 ± 9.6, $n = 17$) ($p = 0.02$)

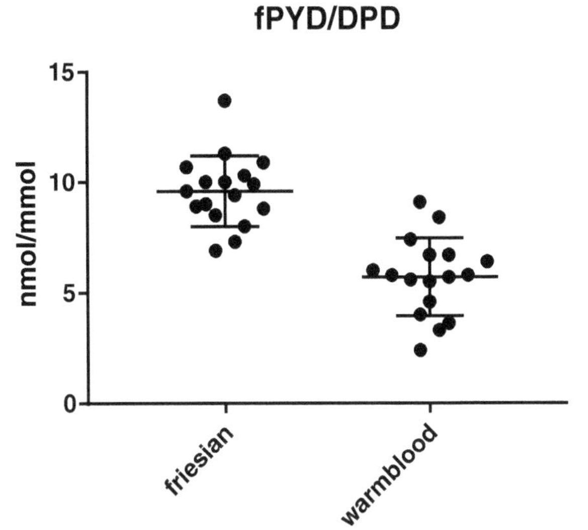

Fig. 3 Mean fPYD/fDPD ratios in the urine from Friesian horses (9.6 ± 1.6, $n = 17$) were significantly higher compared to Warmblood horses (5.7 ± 1.8, $n = 17$) ($p < 0.0001$)

Discussion

In several mature fibrillar collagen types (I, II, III, V and XI) both PYD and DPD have been detected and their concentration varies considerably between tissues [14]. DPD is mainly present in bone and dentine of type I collagen, while PYD is more widespread and is found in several connective tissues [16]. In the early days, pyridinium cross-links were considered as specific urinary indicators of bone resorption. However, later on, their specificity to bone was questioned. Not only bone and cartilage, but also blood vessels, fascia, muscles and ligaments contain pyridinolines [17].

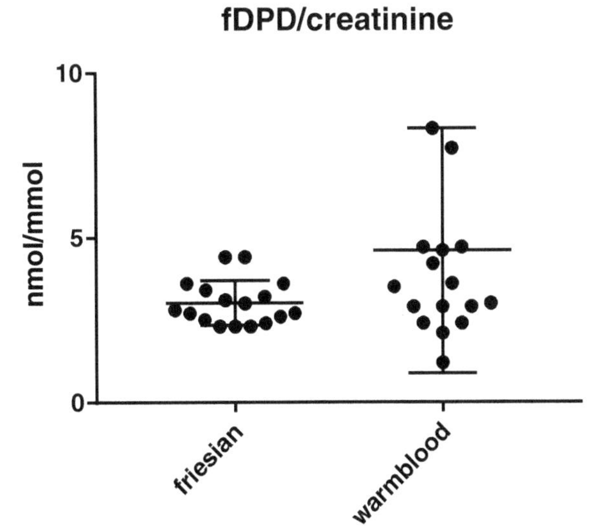

Fig. 2 Mean urinary fDPD/creatinine concentrations in Friesian horses (3.0 ± 0.7 nmol/mmol, $n = 17$) showed no significant differences to the values in Warmblood horses (4.6 ± 3.7 nmol/mmol, $n = 17$) ($p = 0.1$)

Urine analysis does not allow to determine the origin of the collagen cross-links. However, considering the low PYD/DPD ratio of bone (3.8 in monkeys) [18] compared to soft tissues, and the mean PYD/DPD ratio of 9.6 in the Friesian horses in this study being very high compared to the result in Warmblood horses (5.8), it is likely that these cross-links originate from soft tissue catabolism. This difference in collagen metabolism could explain the known soft tissue abnormalities in Friesian horses, such as megaoesophagus and aortic rupture. Indeed, a similar elevation in the PYD/DPD ratio has been observed in human patients with scleroderma, a disease characterized by systemic excessive collagen deposition [19] and is attributed to increased collagen catabolism [20]. Furthermore, it was recently shown that deep flexor tendon tissue from Friesian horses contains less PYD cross-links compared to Warmblood horses. Increased collagen catabolism in the tendons from Friesian horses could explain the lower amount of PYD in the tendons [10] and might be related to the increased urinary levels seen in this study. Other connective tissues, such as ligaments, vessels and internal organs, might also have contributed to the high urinary PYD/DPD ratios in Friesian horses.

Besides the origin of collagen cross-links in urine, several other factors might influence their urinary excretion. An age-related variation in PYD and DPD excretion has been reported. The highest amounts of collagen cross-links in urine are found in children during the first

27. Naylor KE, Iqbal P, Fledelius C, Fraser RB, Eastell R. The effect of pregnancy on bone density and bone turnover. J Bone Miner Res. 2000;15(1):129–37.
28. Yoshida K, Jiang H, Kim M, Vink J, Cremers S, Paik D, Wapner R, Mahendroo M, Myers K. Quantitative evaluation of collagen cross-links and corresponding tensile mechanical properties in mouse cervical tissue during normal pregnancy. PLoS One. 2014; https://doi.org/10.1371/journal.pone.0112391.
29. Stone PJ, Franzblau C. Increase in urinary desmosine and pyridinoline during postpartum involution of the uterus in humans. Proc Soc Exp Biol Med. 1995;210(1):39–42.
30. Belstra BA, Flowers WL, Croom WJ, DeGroot J, See MT. Urinary excretion of collagen degradation markers by sows during postpartum uterine involution. Anim Reprod Sci. 2005;85(1–2):131–45.
31. Müller A, Jakob K, Hein GE. Evaluation of free and peptide bound collagen crosslink excretion in different skeletal diseases. Ann Rheum Dis. 2003;62:65–7.

18

Conjunctival aerobic bacterial flora in healthy Silesian foals and adult horses in Poland

A. Zak[1*], N. Siwinska[1], M. Slowikowska[1], H. Borowicz[1], K. Ploneczka – Janeczko[2], P. Chorbinski[2] and A. Niedzwiedz[1]

Abstract

Background: Commensal bacterial and fungal flora of the conjunctival sac has been described in horses and other animals. The identification of commensal flora of the conjunctival sac may aid in the diagnosis of ocular inflammatory diseases, such as conjunctivitis or more severe ulcerative keratitis, common in horses. Moreover, damage of ocular protective barriers may lead to an opportunistic infection. The study was carried out in Silesian horses kept at a single breeding center in South-western Poland, in order to limit any breed-dependant and climate-dependant variables affecting the results. Following an ophthalmic examination that revealed no abnormalities, sterile swabs were collected from conjunctival sac in 26 adult horses and 11 foals. The obtained swabs were subjected to bacterial culture testing. In case of *Staphylococcus* spp. isolation, susceptibility to methicillin was evaluated.

Results: Forty- three bacterial isolates, representing eleven genera of bacteria were cultured from 30 (81%) horses. Gram-positive bacteria were the dominant isolates (72%) ($p < 0.001$). The most commonly isolated Gram-positive bacteria were *Staphylococcus* spp., while *Moraxella* spp. were the most frequently isolated Gram-negative bacteria. There was no significant influence of sex and age on the frequency and type of microbial isolates.

Conclusions: Commensal flora is present in the conjunctival sac of healthy horses in Poland. Age does not affect the abundance and type of microbial isolates.

Keywords: Horse, Conjunctival sac, Bacteriology

Background

Under physiological conditions, the surface of the eye is protected by a number of innate and adaptive immune systems, such as mucin and epithelium barrier functions, tear film that enables flushing of the ocular surface, antimicrobial tear components (lysozyme, beta-lysin, lactoferrin, blood cells, IgA), antigen presenting cells and special antigen-recognition receptors called Toll-like receptors [1–3]. Commensal flora also protects against microbial pathogens by producing antibacterial substances, limiting the surface of growth of other microorganisms and decreasing the nutrient content on the surface of the corneal and conjunctival epithelium [3–5]. Commensal bacterial and fungal flora of the conjunctival sac is commonly reported in horses and other animals [3, 5–14]. The most common Gram-positive bacterial isolates from the conjunctival sac in horses include *Staphylococcus* spp., *Streptococcus* spp., *Corynebacterium* spp. and *Bacillus* spp. [3, 5–9]. Gram-negative bacteria, such as *Pseudomononas* spp., *Moraxella* spp. and *Acinetobacter* spp., are much less common [6–8]. Studies on the commensal flora of the conjunctival sac report a mixed growth of at least two bacterial species, characterised by meagre growth often accompanied by fungi (i.e. *Aspergillus* spp., *Penicillium* spp., *Fusarium* spp., *Cladosporium* spp.) [6–8, 10, 11]. The study by Moore et al. 1988 did not find anaerobic bacteria in conjunctival swab cultures in healthy horses [5]. Damage of the protective barriers of the ocular

* Correspondence: agnieszka.zak@upwr.edu.pl
[1]Department of Internal Medicine and Clinic of Diseases of Horses, Dogs and Cats, Faculty of Veterinary Medicine, Wroclaw University of Environmental and Life Sciences, pl. Grunwaldzki 47, 50-366, Wroclaw, Poland

surface, including injury to the cornea, may lead to the formation of pathological bacterial and fungal flora [1, 9, 15]. Under pathological conditions, commensal flora may become opportunistic (eg. *Pseudomonas aeruginosa*) [3, 7, 15, 16]. The most commonly isolated bacteria in the case of equine ulcerative keratitis include: *Pseudomonas aeruginosa, Streptococcus equi* subspecies *zooepidemicus* and *Staphylococcus* spp. [1, 16–18]. According to Moore et al., Gram-negative bacteria are isolated more commonly than Gram-positive bacteria in the course of ulcerative keratitis [5]. The identification of the physiological commensal flora in the conjunctival sac is essential in order to diagnose inflammatory ocular disorders and to choose the most effective antibacterial treatment prior to obtaining antibiogram results.

To date, no studies have been carried out in East-Central Europe assessing the commensal microbial flora of the conjunctival sac in horses. According to several studies, the geographical location, the climate zone and the age and breed of the horses may affect the type of bacterial flora in the conjunctival sac [5, 7]. The presented study was carried out in one breed of horse, in animals stabled in a single facility in South-Western Poland, in order to eliminate any possible effects of breed and climate on the results. The aim of the study was to assess the presence of commensal flora in the conjunctival sac of healthy adult horses and foals and to determine the effect of sex, age and housing type on the amount of this flora. To the authors' best knowledge, there are no reports assessing commensal flora in the conjunctival sac of foals or the effect of age on the frequency and type of microbial isolates.

Methods

The study samples were collected from the "Książ Stallion Stud Farm" (coordinates: 50°50′35.6"N 16°17′58.4″E) on a single day in July 2017. Consent by the owner was obtained prior to sample collection. According to the present law in Poland (Experiments on Animals Act from January 15th 2015, Journal of Laws of the Republic of Poland from 2015, item. 266), the study did not require the approval of the Ethics Committee. The highest ambient temperature during the day was 18.5 °C during sample collection. According to the information provided by the owner of the horses, none of the animals had any ocular disorders and no systemic or topical ophthalmic drugs were used. A screening ophthalmic examination including the assessment of the eye accessory organs, the pupillary light reflex and the eye was carried out using a source of light (Ri-light, Riester, Jungingen, Germany). An assessment of the corneal surface, the anterior chamber, lens, the vitreous chamber and the ocular fundus with imaging of the optic disc was carried out using a direct ophthalmoscope (Ri-scope L2, Riester, Jungingen, Germany). The examination was performed by a qualified veterinary surgeon in a darkened environment. The cornea was not stained with fluorescein due to its potential bactericidal effect on the culture [19]. The clinical examination was normal. Thirty-seven horses were included in the study, which were divided into three groups: I. adult stallions (15 horses), with a mean age of 6.6 years (range: 3–16); II. adult mares (11 horses), with a mean age of 11.1 years (range: 5–18); III. healthy foals of both sexes up to 6 months old from mares from group II (11 horses), with a mean age of 4.1 months (range: 2–6 months). The stallions in group I were kept in uniform conditions in separate stalls in one stable with one to two hours of training/lunging per day. Groups II and III were stabled together. Groups II and III had a 9–10 h access to a pasture and spent the remaining time in a common stable without boxes. Straw (from the same source) was used as bedding in both stables, and the horses were fed hay ad libitum and oat according to their requirements. The swabs were collected from the right eye of each horse without sedation or local anaesthesia. The globe was retropulsed in order to visualise the conjunctiva of the lower eyelid and third eyelid, which was protruded. The sample was collected from the external surface of the third eyelid using a sterile swab (Sarstedt, Copan, Italia) covered by sterile saline. The swab was moved toward the conjunctival sac avoiding contact with the palpebral margin, eyelashes and vibrissae. The swabs were placed directly on an Amies (Sarstedt, Copan, Italia) transport medium and transported for analysis within an hour in temperature 4 °C. The microbiological analysis was carried out at Epi-Vet, a veterinary diagnostic laboratory operating at the Department of Epizootiology with Clinic for Birds and Exotic Animals of the Wroclaw University of Environmental and Life Sciences, in accordance with the prevailing microbiological regulations (according ISO 9001:2008) [20].

Obtaining isolates

The samples were placed in a liquid medium containing broth and glucose (Oxoid, USA) and were then incubated at 37 °C for 24 h under aerobic conditions. Following incubation, the samples were cultured on a solid medium containing Columbia agar with 5% sheep blood (Oxoid, USA), Mannitol Salt Agar (Oxoid, USA), MacConkey Agar with Crystal violet (Oxoid, USA). The plates were incubated at 37 °C for another 24 h under aerobic conditions. The colonial morphology was assessed using commercial biochemical tests, described below.

Species strain identification

In order to identify the species, the following tests were carried out: a. the oxidase test (with the use of a 1% solution of tetramethyl-p-phenylenediamine); b. the catalase test was performed with 3% hydrogen peroxide; c.

detection of the Clumping Factor (CF) of Staphylococcus (using undiluted rabbit plasma). The species identification of the *Staphylococcus* isolates was determined using the RapID STAPH Plus (Remel Inc., USA) test. The species identification of non-fermenting Gram-negative bacterial isolates was determined using the RapID NF Plus test (RemelInc, USA).*The assessment of Staphylococcus susceptibility to methicillin:*

The susceptibility of the *Staphylococcus* spp. isolates to methicillin was examined using Mueller-Hinton agar (Oxoid, USA) and 30 μg cefoxitin discs (Oxoid, USA), with incubation at 37 °C for 24 h [21].

Statistical analysis

The Chi-squared test was used to assess the differences in proportions between the groups, and due to the small sample size, the Yate's correction was applied. All the analyses were performed at a 5% significance level (p = 0,05) using the PQStat for Windows 1.6.2 software.

Results

Positive bacterial cultures were observed in 30 of the 37 healthy horses (81%), including 12 of the 15 stallions (80%), eight of the 11 mares (72%) and 10 of the 11 healthy foals (91%). No statistically significant relationship between the bacteriological result, the sex and the age of the horses was observed. Forty-three isolates were identified from the samples. The details of the bacterial isolates are presented in Table 1. Growth of a single species of bacteria was noted in 17 healthy horses (56.7% positive bacteriological pathogen detection), including five stallions, six mares and six healthy foals. Mixed bacterial growth was observed in 13 healthy horses (43.3% positive bacteriological pathogen detection), including in seven stallions, two mares and four foals. The complete results, including the age and sex of the animals, is presented in Table 2. There was no statistically significant difference between the type of bacterial growth, the sex and the age of the animals. There were significantly more Gram-positive bacteria (31 isolates - 72%) than Gram-negative bacteria (12 isolates - 28%) (p < 0.001). No statistically significant difference between the growth of Gram-positive and Gram-negative bacteria and the age and sex of the horses was found. The same bacterial isolates were found in two out of 11 mare-foal pairs. The detailed results are presented in Table 3. Most of the *Staphylococcus* bacteria were coagulase-negative (three isolates were of undetermined morphology using commercial biochemical tests) and no methicillin-resistant *Staphylococcus* spp. strains were observed.

Discussion

The Silesian horse is a breed of warmblood horse native to Upper and Lower Silesia in Poland and is part of the Conservation Program of Genetic Diversity in Farm Animals in Poland. The obtained results are consistent with similar studies in horses in other geographical regions. Positive bacteriological pathogen detection and bacterial culture was present in 81% of the healthy horses. Other studies often reported less frequent bacterial growth in samples from conjunctival sac, occurring in 30 to 52% of the healthy horses [6, 8]. In the study by Gemensky-Metzler et al. 2005 bacterial isolates were presented more frequent: in 78% of the corneal swabs and 90% of the conjunctival sac swabs [3].

This study confirmed the tendency toward a more frequent Gram-positive bacterial culture than a Gram-negative bacterial culture of conjunctival sac swabs from healthy horses [3, 5–9]. In the present study, 66.67% of the isolates from adult horses and 71.43% of the isolates from foals were

Table 1 Bacterial isolates from normal eyes

Isolate	Gram positive/ Gram negative	Total number of isolates	Percent of total isolates
Staphylococcus sciuri[a]	Positive	16	37.2%
Staphylococcus xylosus[a]	Positive	9	21.0%
Staphylococcus spp.*[b]	Positive	4	9.3%
Staphylococcus simulans[a]	Positive	1	2.3%
Brevibacterium diminuta	Positive	1	2.3%
Moraxella lacunata	Negative	5	11.6%
Pseudomonas aeruginosa	Negative	1	2.3%
Sphingomonas paucimobilis	Negative	2	4.7%
Burkholderia cepacia	Negative	2	4.7%
Comamonas acidovorans	Negative	1	2.3%
Chryseobacterium meningosepticum	Negative	1	2.3%
Total		43	100%

(*not subjected to further strain identification)
[a]MSCNS– methicillin-susceptible coagulase-negative *Staphylococcus*)
[b]MSS – methicillin-susceptible *Staphylococcus*

Table 2 Bacterial isolates from normal eyes

Groups	Group I - stallions	Group II - mares	Group I and II – adult horses	Group III foals
Number of horses	15	11	26	11
No growth	3	3	6	1
Number of isolates	19	10	29	14
Isolates:	Number of isolates/% isolates	Number of isolates/% isolates	Number of isolates/% isolates	Number of isolates/% isolates
Staphylococcus sciuri[a]	8 / 42.1%	4 / 40%	12 / 41.4%	4 / 28.6%
Staphylococcus xylosus[a]	2 / 10.5%	4 / 40%	6 / 20.7%	3 / 21.45%
Staphylococcus spp.[*b]	2 / 10.5%	0	2 / 6.9%	2 / 14.3%
Staphylococcus simulans[a]	0	0	0	1 / 7.1%
Brevibacterium diminuta	0	0	0	1 / 7.1%
Moraxella lacunata	3 / 15.7%	0	3 / 10.3%	2 / 14.3%
Pseudomonas aeruginosa	1 / 5.3%	0	1 / 3.45%	0
Sphingomonas paucimobilis	1 / 5.3%	1 / 10%	2 / 6.9%	0
Burkholderia cepacia	1/ 5.3%	0	1 / 3.45%	1 / 7.1%
Comamonas acidovorans	1 / 5.3%	0	1 / 3.45%	0
Chryseobacterium meningosepticum	0	1 /10%	1 / 3.45%	0

(*not subjected to further strain identification)
[a]MSCNS– methicillin-susceptible coagulase-negative *Staphylococcus*)
[b]MSS – methicillin-susceptible *Staphylococcus*

from the *Staphylococcus* spp. genus. This finding was confirmed in previous studies, where *Staphylococcus* spp. was reported to be the most prevalent or the second most prevalent type of isolated bacteria [3, 5–8]. However, this microbial species was much more common in the present study than in previous studies [3, 5–8]. The cause of varying dominant isolates between studies is not clear, although hypothetically it may be associated with different maintenance conditions, such as different bedding, hay, weather and climate. The majority of the isolated *Staphylococcus* spp. bacteria were coagulase-negative. Coagulase-negative Staphylococci (CNS) are considered non-pathogenic commensal bacteria, although it is currently believed that they may be a cause of opportunistic infections [22]. All of the isolates were methicillin- sensitive using the cefoxitin disc susceptibility test [21]. According to literature, methicillin-resistant *Staphylococcus aureus* (MRSA) was isolated in one case of ulcerative keratitis in a horse in Japan [23]. Methicillin-resistant Staphylococci (MRS) were also isolated from a horse with a surgical wound complication as well as from the nares and skin of healthy horses [24–26]. MRS are also a cause of ocular disease in humans [27, 28].

The second most common bacterial isolate was *Moraxella* spp. (11.6%), all of which are identified as *Moraxella lacunata*, which are gram-negative coccobacilli responsible for opportunistic infections. This finding has been confirmed in literature, where *Moraxella* spp. was reported to be the most prevalent or the second most prevalent type of isolated gram-negative bacteria [3, 5, 7]. According to Moore et al. 1983, the majority of commensal flora is formed by Gram-positive bacteria, while Gram-negative microorganisms usually cause septic infections [16]. In the study by Sauer et al. 2003 *Pseudomonas aeruginosa* was the most frequently isolated

Table 3 Bacterial isolates from healthy mares and their folas

pair nr	mare isolates	foal isolates
1.	*S. sciuri*[a]	*S. simulans*[b]
2.	*S.sciuri*[a]	*S. xylosus*[c] *B. cepacia*[d]
3.	No growth	*S. sciuri*[a] *M. lacunata*[e]
4.	No growth	*S. xylosus*[c]
5.	*S. xylosus*[c] *S. paucimobilis*[f]	*S. sciuri*[a]
6.	*S. xylosus*[c]	*Staphylococcus spp.*[i]
7.	*S. xylosus*[c]	no growth
8.	***S. sciuri*[a]**	***S. sciuri*[a]**
9.	*S. sciuri*[a]	*Staphylococcus spp.*[i]
10.	no growth	*S. sciuri*[a] *B. diminuta*[g]
11.	***S. xylosus*[c]** *Ch. meningosepticum*[h]	***S. xylosus*[c]** *M. lacunata*[e]

Bacterial isolates from healthy mares and their foals: the overlapping results are presented in bold (legend: [a]*Staphylococcus sciuri*, [b]*Staphylococcus simulans*, [c]*Stapylococcus xylosus*, [d]*Burkholderia cepacia*, [e]*Moraxella lacunata* [f]*Sphingomonas paucimobilis*, [g]*Brevibacterium diminuta*, [h]*Chryseobacterium meningosepticum*, [i]not subjected to further analysis) pathogen in the case of equine bacterial ulcerative keratitis [17]. *Pseudomonas aeruginosa* is an opportunistic bacterium considered to be the most virulent corneal pathogen, with the capacity to produce proteases and exotoxins and causing serious damage of the corneal epithelium and endothelium [1, 15]. In the present study, *P. aeruginosa* was isolated from one healthy horse. *P. aeruginosa* showed mixed growth with *Staphylococcus sciuri*, suggestive of a commensal infection. However, other authors reported that Gram-positive bacteria, such as *Streptococcus equi* subsp. *zooepidemicus* and *Staphylococcus aureus*, were isolated more frequently from cases of ulcerative keratitis than the Gram-negative *P. aeruginosa* [1, 15, 18, 29, 30]. The commensal flora isolated from the conjunctival sac in the present study was also isolated in cases of ulcerative keratitis, i.e. *Staphylococcus xylosus*, *Staphylococcus sciuri*, *Sphingomonas paucimobilis* (isolated by Hidaka et al. [18]) and *Burkholderia cepacia* (isolated by Sauer et al. 2003) [17, 18]. This confirmed the hypothesis that commensal bacteria may act as pathogens in case of damage of the ocular barrier. One must also keep in mind that bacterial ulcerative keratitis should be diagnosed based on a swab obtained directly from the ulcer, as the conjunctival flora is not directly related to the corneal flora.

There were no statistically significant differences between the number of isolates, the age and the similarity of the isolates between the mother-foal pairs. This suggests that foals above two months of age develop their own commensal flora, independent of their mothers' flora. Mycological research carried out by Sgorbini et al. 2008 on newborn foals reported a commensal flora immediately after delivery [10]. Those authors found this flora to be diverse in the first month of life, which may be associated with later bacterial colonisation of the conjunctival sac [10]. In order to understand the detailed mechanisms of conjunctival sac bacterial colonisation, further studies assessing conjunctival swabs immediately after birth and during the first days of life are needed. Studies carried out on newborn children immediately after delivery revealed more prevalent conjunctival sac bacterial isolates in newborns born vaginally compared with those delivered by caesarean [31]. However, the number of bacterial isolates from the conjunctival sac was similar in both groups in the second day of life [31]. This indicates contamination of the newborn conjunctival sac with vaginal commensal bacteria, as well as subsequent colonisation with bacteria from the children's carers and from their surroundings [31].

The presence of insects in stables, which often have contact with horse eyes and mechanically transmit microorganisms, may be an important factor contributing to bacterial colonisation of the equine conjunctival sac. In the study by Butler et al. 2010, *Staphylococcus sciuri* and *Staphylococcus xylosus*, which were isolated from the horses in the present study, were isolated from the surface of the housefly (*Musca domestica* L) [32]. The study by Andrew et al. 2003 and Moore et al. 1988 revealed no impact of the season and surroundings on the conjunctival sac bacterial flora [5, 7]. In the presented study, the horses from group I spent less time on pasture than horses from group II and III. However, this did not affect the number of microbiological isolates.

Conclusions

The commensal flora in the equine conjunctival sac of horses in Poland may vary despite the horses living in uniform environment conditions. In addition, foals older than two months of age have fully developed bacterial commensal flora, independent of the mothers' flora.

Abbreviations

CNS : coagulase-negative Staphylococcus; MRS : methicillin-resistance Staphylococcus; MSCNS : methicillin-susceptible coagulase-negative Staphylococcus

Acknowledgements

We are grateful to Angelika Bryla and Kinga Malewska, students of Faculty of Veterinary Medicine, Wroclaw University of Environmental and Life Sciences for help during sample collecting.

Funding

The research was financed from statutory research no B010/0017/17 at Department of Epizootiology with Clinic for Birds and Exotic Animals, Wroclaw University of Environmental and Life Sciences.

Authors' contributions

AŻ participated in the design of the study and was a major contributor in writing the manuscript. NS, MS, HB participated in the design of the study, in the collection of the samples and in the analysis and the interpretation of data. KP-J, PC and AN participated in the design of the study and performed the statistical analysis. KP-J, PC and AN conceived the study, participated in its design and coordination and helped to draft the manuscript. All authors were involved in a continuous discussion regarding the manuscript. All authors read and approved the final manuscript.

Consent for publication

Not applicable.

Competing interests

The authors declare that they have no competing interests. None of the aforementioned companies had any direct or indirect involvement in the development of the manuscript and the works reflects the opinions of the authors.

Author details

[1]Department of Internal Medicine and Clinic of Diseases of Horses, Dogs and Cats, Faculty of Veterinary Medicine, Wroclaw University of Environmental and Life Sciences, pl. Grunwaldzki 47, 50-366, Wroclaw, Poland. [2]Department of Epizootiology with Clinic for Birds and Exotic Animals, Wroclaw University of Environmental and Life Sciences, pl. Grunwaldzki 45, 50-366, Wroclaw, Poland.

References

1. Keller RL, Hendrix DVH. Bacterial isolates and antimicrobial susceptibilities in equine bacterial ulcerative keratitis (1993-2004). Equine Vet J. 2005;37(3): 207–11.
2. Gilger BC. Immunology of the ocular surface. Vet Clin Small Anim. 2008;38: 223–31.
3. Gemensky-Metzler AJ, Wilkie DA, Kowalski JJ, Schmall LM, Willis A, Yamagata M. Changes in bacterial and fungal ocular flora of clinically normal horses following experimental application of topical antimicrobial or antimicrobial-corticosteroid ophthalmic preparations. Am J Vet Res. 2005;66:800–11.
4. Halbert SP. Inhibitory properties of the ocular flora. In: Lacatcher-Khorazo D, Seegal BC, editors. Microbiology of the eye. St Louis: Mosby; 1972. p. 24–40.
5. Moore CP, Heller N, Majors LJ, Whitley RD, Burgees EC, Weber J. Prevalence of ocular microorganisms in hospitalized and stabled horses. Am J Vet Res. 1988;49:773–9.
6. Whitley RD, Burgess EC. Moore CP. Microbial isolates of the normal equine eye. Equine vet J. Suppl. 1983;2:138–40.
7. Andrew SE, Nguyen A, Jones GL, Brooks DE. Seasonal effects on the aerobic bacterial and fungal conjunctival flora of normal thoroughbred brood mares in Florida. Vet Ophthalmol. 2003;6:45–50.
8. Johns IC, Baxter K, Booler H, Hicks C, Menzies-Gow N. Conjunctival bacterial and fungal flora in healthy horses in the UK. Vet Ophthalmol. 2011;14(3):195–9.
9. Whitley RD, Moore CP. Microbiology of the equine eye in health and disease. Vet Clin N Am-Large. 1984;6(3):451–66.
10. Sgorbini M, Barsotti G, Nardoni S, Mancianti F, Rossi S, Corazza M. Fungal Flora of normal eyes in healthy newborn foals living in the same stud farm in Italy. J Equine Vet Sci. 2008;28(9):540–3.
11. Rosa M, Cardozo LM, Pereira J, Brooks DE, Martins ALB, Florido PSS, Stussi JSP. Fungal flora of normal eyes of healthy horses from the state of Rio de Janeiro, Brazil. Vet Ophthalmol. 2003;6(1):51–5.
12. Kiełbowicz Z, Płoneczka-Janeczko K, Bania J, Bierowiec K, Kiełbowicz M. Characteristics of the bacterial flora in the conjunctival sac of cats from Poland. J Small Anim Pract. 2015;56:203–6.
13. Urban M, Wyman M, Rheins M, Marraro RV. Conjunctival flora of clinically normal dogs. JAVMA. 1972;161:201–6.
14. Wilcox GE. Bacterial flora of the bovine eye with special reference to the *Moraxella* and *Neisseria*. Aust Vet J. 1970;46:253–7.
15. Leigue L, Montiani-Ferreirm F, Moore BA. Antimicrobial susceptibility and minimal inhibitory concentration of Pseudomonas aeruginosa isolated from septic ocular surface disease in different animal species. Open Veterinary Journal. 2016;6(3):215–22.
16. Moore CP, Fales WH, Whittington P, Bauer L. Bacterial and fungal isolates from Equidae with ulcerative keratitis. JAVMA. 1983;182(6):600–3.
17. Sauer P, Andrew SE, Lassaline M, Gelatt KN, Denis HM. Changes in antibiotic resistance in equine bacterial ulcerative keratitis (1991–2000): 65 horses. Vet Ophthalmol. 2003;6(4):309–13.
18. Hidaka S, Kobayashi M, Ando K, Fujji Y. Efficacy and safety of lomefloxacin on bacterial extraocular disease in the horse. J Vet Med Sci. 2015;77(7):829–35.
19. Malhotra S, Kaur N, Mehta DK. Bactericidal effect of trypan blue and fluorescein sodium in ophthalmic practice. Nepal J Ophthalmol. 2012;4(1):80–3.
20. Szewczyk ME. Diagnostyka bakteriologiczna. Warszawa: Wydawnictwo Naukowe PWN; 2013.
21. Pourmand MR, Hassanzedeh S, Mashhadi R, Askari E. Comparison of four diagnostic methods for detection of methicillin resistant Staphylococcus aureus. Iranian Journal of Microbiology. 2014;6(5):341–4.
22. Kloos WE, Bannerman TL. Update of clinical significance of coagulaso-negative Staphylococci. Clinical Microbiology Reviews. 1994;7(1):177–14023.
23. Kuroda T, Kinoshita Y, Niwa H, Mizobe F, Ueno T, Kuwano A, Hatazoe T, Hobo S. Methicillin-resistant Staphylococcus aureus ulcerative keratitis in a thoroughbred racehorse. J Equine Sci. 2015;26(3):95–8.
24. Karakulska J, Fijałkowski K, Nawrotek P, Pobucewicz A, Poszumski F, Identification C-FD. Methicillin resistance of coagulase-negative staphylococci isolated from nasal cavity of healthy horses. J Microbiol. 2012; 50(3):444–51.
25. Yasuda R, Kawano J, Onda H, Takagi M, Shimizu A, Anzai T. Methicillin-resistant coagulase-negative staphylococci isolated from healthy horses in Japan. AJVR. 2000;61(11):1451–5.
26. Hartmann FA, Troslte SS, Klohnen AA. Isolation of methicillin-resistant Staphylococcus aureus from a postoperative wound infection in a horse. J Am Med Assoc. 1997;211:590–2.
27. Shanmuganathan VA, Armstrong M, Buller A, Tullo AB. External ocular infections due to methicillin-resistant Staphylococcus aureus (MRSA). Eye. 2005;19:284–91.
28. Blomquist PH. Methicillin-resistant Staphylococcus aureus infections of the eye and orbit. Trans Am Ophthalmol Soc. 2006;104:322–45.
29. Brooks DE, Andrew SE, Biros DJ, Denis HM, Cutler TJ, Strubbe DT, Gelatt KN. Ulcerative keratitis caused by beta-hemolytic *Streptococcus equi* in 11 horses. Vet Ophthalmol. 2000;3:121–5.
30. McLaughlin SA, Brightman AH, Helper LC, Manning JP, Tomes JE. Pathogenic bacteria and fungi associated with extraocular disease in the horse. J Am Vet Med Assoc. 1983;182(3):241–2.
31. Lee PW, Jun AK, Cho BC. A study of microbial flora of conjunctival sac in newborns. Korean J Ophthalmol. 1989;3(1):33–7.
32. Butler JF, Garcia-Maruniak A, Meek F, Maruniak JE. Wild Florida house flies (Musca Domestica) as carriers of pathogenic bacteria. Fla Entomol. 2010; 93(2):218–23.

The detection of capsaicin and dihydrocapsaicin in horse serum following long-term local administration

A. Zak[1*], N. Siwinska[1], M. Slowikowska[1], H. Borowicz[1], P. Szpot[2], M. Zawadzki[2] and A. Niedzwiedz[1]

Abstract

Background: Capsaicin and dihydrocapsaicin are alkaloids with analgesic effects in humans and animals. When used locally, both of them minimalise pain sensation by defunctionalising nerve endings. According to the Federation Equestrian International Prohibited Substances List, these are substance banned in horse competitions. The aim of the study was to determine the detection time of capsaicin in both plasma and serum after long-term use of a gel recommended for commercial use and applied as intended. The objective of the study was to select the best material for the detection of capsaicin as a doping substance in horses.

Methods: Nine healthy mature horses were administered 0.1% capsaicin topically in the form of a commercial analgesic gel (15 g of the gel per limb) to the front limbs every 24 hours for five days with a polar fleece bandage. Blood serum and plasma were collected prior to gel application and in the 12th, 18th, 24th, 36th, 42nd, 48th, 60th, 84th, 108th, 132nd, 156th hour after the gel application. Qualitative and quantitative analysis was performed using ultra-high performance liquid chromatography coupled with a triple quadrupole mass spectrometry (UHPLC-QqQ-MS/MS).

Results: The concentration of capsaicin in the serum samples did not exceed the lower limit of quantification. Capsaicin was not detected in the plasma samples during the entire study period. Dihydrocapsaicin was not detected in blood serum or plasma.

Conclusion: The presented results suggest that capsaicin is not detected in horse serum in the 24-hour-periodfollowing its last application according to the dosage regimen used by owners and veterinarians for therapy rather than doping, based on a five day gel application and a polar bandage.

Background

Capsaicin (8-Methyl-N-vanillyl-trans-6-nonenamide) is an alkaloid organic compound from the *Capsicum* family [1]. Large amounts of capsaicin are found in various peppers and give them a sharp taste [2]. Capsaicin is commonly used by humans locally due to its analgesic properties, mainly in the treatment of back pain, neurogenic pain, osteoarthritis and postherpetic neuralgia [2]. It has gastro-protective properties when administered orally and minimalises postoperative nausea and vomiting when used in acupressure [2]. In horses, it is mainly included in analgesic and anti-inflammatory heat rub creams used in lameness, laminitis and arthritis [3]. In a reversible model of equine foot lameness, the degree of lameness and other parameters indicating pain sensation significantly decreased in horses treated topically with capsaicin from 40 to 240 min after its application [3]. Capsaicin is also used as a wood chewing deterrent [4]. Dihydrocapsaicin (N-(4-Hydroxy-3-methoxy-benzyl)-8-methylnonanamide) is a capsaicinoid analog found in spicy peppers. It is a capsaicin derivative that is an irritant but also has analgesic properties. Hence, it is also used in analgesic creams [5].

Capsaicin is included in the 2017 Equine Prohibited Substances List created by the Federation Equestrian International (FEI). Hence, it should be identified easily and rapidly in horses [6]. Ultra-high – performance liquid chromatography – tandem mass spectrometry (UHPLC-

* Correspondence: agnieszka.zak@upwr.edu.pl
[1]Department of Internal Medicine and Clinic of Diseases of Horses, Dogs and Cats, Faculty of Veterinary Medicine, Wroclaw University of Environmental and Life Sciences, pl. Grunwaldzki 47, 50-366 Wroclaw, Poland
Full list of author information is available at the end of the article

MS-MS) is the newest method for the detection of capsaicin in plasma. Due to its selectivity, sensitivity, repeatability and time of detection, it is considered the best method for the rapid analysis of capsaicin [7]. The aim of the study was to determine the time of capsaicin detection in the blood following its topical administration in the form of a gel with the dosage regimen based on the most common guidelines. The second objective was to compare capsaicin detectability in blood serum and plasma and to choose the best diagnostic material.

Methods

All experimental procedures were performed with the approval of the Animal Experimentation Local Ethics Committee in Wroclaw (permit No 97/2016). Nine horses (seven mares and two geldings) with a mean age of 13.7 years (range: 6–18 yo), a mean weight of 436.1 kg (range: 300–517 kg) and of various breeds were included in the study. These horses were nonworking and were kept at the Department of Internal Medicine and Clinic of Diseases of Horses, Dogs and Cats of the Wroclaw University of Environmental and Life Sciences for didactic student training. The horses were maintained in a stable-pasture system with full-time access to hay, water and mineral licks and were fed oats based on their nutrient requirements, body weight and level of physical activity. Prior to inclusion in the study, the horses underwent clinical and orthopaedic examinations, which were unremarkable. Prior to commencing the study, 10 mL of venous blood was collected from each horse at T0 into tubes[1] containing ethylenediaminetetraacetic acid (EDTA) as an anticoagulant and plain tubes[1] using an 18G needle. On the first day of the study, 15 g (weighed using a scale[2]) of a commercial gel containing 0.1% capsaicin[3] was applied to each thoracic limb from the level of the carpal joint to the pastern joint at 8 pm. Latex gloves were worn during the application of the capsaicin gel. The gel was massaged into the limb for five minutes. Then, a polar fleece bandage was placed on the digit for 12 h. The application procedure was repeated every 24 h for five days. Venous blood was collected from the jugular vein prior to the application of the gel (T0), then 12 (T12), 18 (T18), 24 (T24), 36 (T36), 42 (T42), 48 (T48), 60 (T60), 84 (T84), 108 (T108), 132 (T132), 156 (T156) hours following the last gel application. The samples were allowed to settle on ice for 30 min. The plasma (5 min, 2500G) and serum (10 min, 4000G) samples were then centrifuged [7]. The samples were frozen at – 20 °C and then transferred to – 70 °C [7]. After the study the horses were returned to their owner.

Analysis of serum and plasma sample

The analyses were carried out at the laboratory of the Department of Forensic Medicine of the Wroclaw Medical University.

Chemicals and reagents

The following reagents were used: water (Chromasolv® LC–MS)[4], acetonitrile (Chromasolv® LC–MS)[4], methanol (Chromasolv® LC–MS)[4], dichloromethane[4], formic acid[4], ammonium formate[5]; capsaicin[6]; dihydrocapsaicin[7]; phenacetin[8]. Stock solutions of capsaicin, dihydrocapsaicin and phenacetin were prepared in methanol at a concentration of 10 mg/mL. The standard solutions were stored in a freezer at – 20 °C.

Instrumentation

Chromatographic analysis was performed using an ultra-high performance liquid chromatograph UHPLC Shimadzu Nexera X2[9]. The particles were separated using an ACQUITY UPLC® BEH C18[10] 1.7 μm, 2.1 × 50 mm column with a thermostat at 40 °C. A mixture of 10 mL ammonium formate/0.1% formic acid in water (A) and 0.1% formic acid in acetonitrile (B) was used as a mobile phase. The gradient elution was carried out at constant flow 0.5 mL/min. The following gradient was applied: 0 min. - 5% B, 4 min. – 95% B, 4.5 min - 95% B, 5 min. – 5% B and then 7 min 5% B. Return to initial gradient compositions (95% A and 5% B) was performed at 5 min.

A triple-quadrupole mass spectrometer (QqQ, Shimadzu 8050)[11] equipped with an ESI ion source, was used to detect the studied compounds. The MRM mode was employed to establish the presence of the studied substances. The analytes were then quantified by multiple reaction monitoring (MRM). MRM transitions were m/z 306.10 → 137.05; 94.10; 122.05 for capsaicin, m/z 308.40 → 137.10; 94.10; 122.10 for dihydrocapsaicin, and m/z 180.30 → 110.05; 93.10; 65.05 for phenacetin.

The retention times for capsaicin and dihydrocapsaicin were 3.2 and 3.4 min, respectively. Positive ionization was performed. The injected volume was 1 μL. The lower limit of quantification (LLOQ) for capsaicin and dihydrocapsaicin was 0.5 pg/mL and 1 pg/mL, respectively. The LLOQ was considered as the lowest calibration concentration with a relative standard deviation (%RSD) precision below 20%.

Sample preparation

The sample (200 μL) was transferred into a 2-mL plastic tube, and 20 μL of an internal standard (Phenacetin 100 ng/mL) was added. A liquid-liquid extraction using dichloromethane (1 mL) was carried out for five minutes. The samples were then centrifuged at 10000 rpm. The organic phase (0.9 mL) was transferred into a 2-mL Eppendorf tube and evaporated to dryness under a stream of nitrogen (at 40 °C). The extract was dissolved in 25 μL of methanol, transferred to a glass insert and analysed using UHPLC-QqQ-MS/MS. Stock solutions were diluted with methanol to obtain working standard solutions at the following concentrations of capsaicin and dihydrocapsaicin: 5, 10, 100, 500, 1000, 5000 and 10,000 pg/mL. Calibration points were prepared by diluting the appropriate working solution with horse serum. The final concentrations of the calibrators were: 0.5, 1, 10, 50, 100, 500 and 1000 pg/mL horse serum for capsaicin and dihydrocapsaicin. QC samples (capsaicin and dihydrocapsaicin at 1 pg/mL) were run with each batch of samples.

Results

None of the horses showed signs of discomfort, pain or stress associated with the effects of capsaicin. Detailed patient data are presented in Table 1. None of the results exceeded the lower limit of quantification (LLOQ), which amounted to 0.5 pg/mL for capsaicin and 1 pg/mL for dihydrocapsaicin. At T0, capsaicin was not detected in the blood serum or plasma. In two of the nine horses (22.22%), the concentration of capsaicin was quantifiable in the blood serum 12 h (T12) following the last gel administration although the obtained values were lower than the lower limit of quantification and amounted to 0.345 pg/mL and 0.135 pg/mL in horse no.4 and horse no.9, respectively (Fig. 1). Capsaicin was not detected in the blood serum more than 12 h following its last application. Capsaicin was not detected in the blood plasma samples. Dihydrocapsaicin was not detected in any of the samples.

Discussion

Capsaicin is a strong and highly selective exogenous agonist of the transient receptor potential cation channel subfamily V member 1 (TRPV-1), also known as the vanilloid receptor 1 [8–10]. In literature, the local analgesic effect of capsaicin was reported to result from a depletion of substance P in nerve endings in the skin [11]. Currently, this mechanism is thought to play a minor analgesic role [12]. Topical administration of capsaicin decreases skin sensitivity and reduces the sensation of pain in a process known as "defunctionalisation" [13]. Defunctionalisation consists of a temporary loss of the membrane potential, an inability to transport neurotrophic factors and a reversible retraction of the nerve endings from the epidermis and skin [13–15]. The side effects of capsaicin use include a transient, local reaction that appears in the form of a pain, burning or erythema following the first administration [13]. Side effects, such as permanent skin denervation and nociceptive dysfunction, are theoretical and have not been described in clinical scenarios [16]. Higher concentrations of capsaicin (e.g. 10%) cause severe pain in horses [17].

Table 1 Detailed patient data

Number	Sex	Age	Breed	Weight [kg]
1.	gelding	18	german riding horse	500
2.	gelding	14	thoroughbred	517
3.	mare	17	halfbreed	506
4.	mare	14	pony	310
5.	mare	16	halfbreed	453
6.	mare	14	pony	300
7.	mare	15	halfbreed	447
8.	mare	10	arabian horse	440
9.	mare	6	arabian horse	452

In horses, capsaicin is considered a doping substance and is prohibited by the State of Pennsylvania Racing Commissions and FEI due to its strong analgesic effect and the duration of analgesia that extends beyond the time of the race/parkour [6]. Four of the 15 samples collected from show jumping horses at the Beijing Olympics in 2008 during anti-doping tests revealed the presence of capsaicin, causing those horses to be disqualified [7]. Gels containing capsaicin are widely available worldwide in equestrian shops. They are often used by sports and pleasure horses owners for mild lameness, or oedema as a first line treatment prior to a veterinary medical consultation. Such use is intended for curative purposes, including limiting pain and lameness, rather than doping. The described regimen for the use of the gel containing capsaicin, based on a five day gel administration together with polar bandages is the most common treatment scheme used by horse owners. The authors of this study marked capsaicin 12 h after its administration due to the fact that the study focused on detecting capsaicin administered for therapeutic use, not doping. You et al. 2013 studied the topical use of a 0.025% capsaicin gel and found that it reached its peak concentration two hours following its administration (241.9 pg/mL), and that it decreased to baseline values within 24 h (4.7 pg/mL) [7]. Those authors suggested that capsaicin is undetectable in blood plasma 48 h after application [7]. Hence, the purpose of this study was to determine the duration of capsaicin and dihydrocapsaicin detectability during long-term use and to assess whether capsaicin can be detected in blood serum and plasma.

The obtained results suggest a short-term detectability of capsaicin even after its long-term use. The results of this study did not exceed the LLOQ. However, the applied method enabled the authors to detect lower concentrations of the analytes although those concentrations did not meet the LLOQ. Hence, in two cases (horses no. 4 and 9), the chromatography of the studied sample at T12 had a similar analytical signal to the LOQ sample which signifies that capsaicin was present in the tested material at a very low concentration. The obtained results are contrary to those of You et al. who found that the concentration of capsaicin 24 h

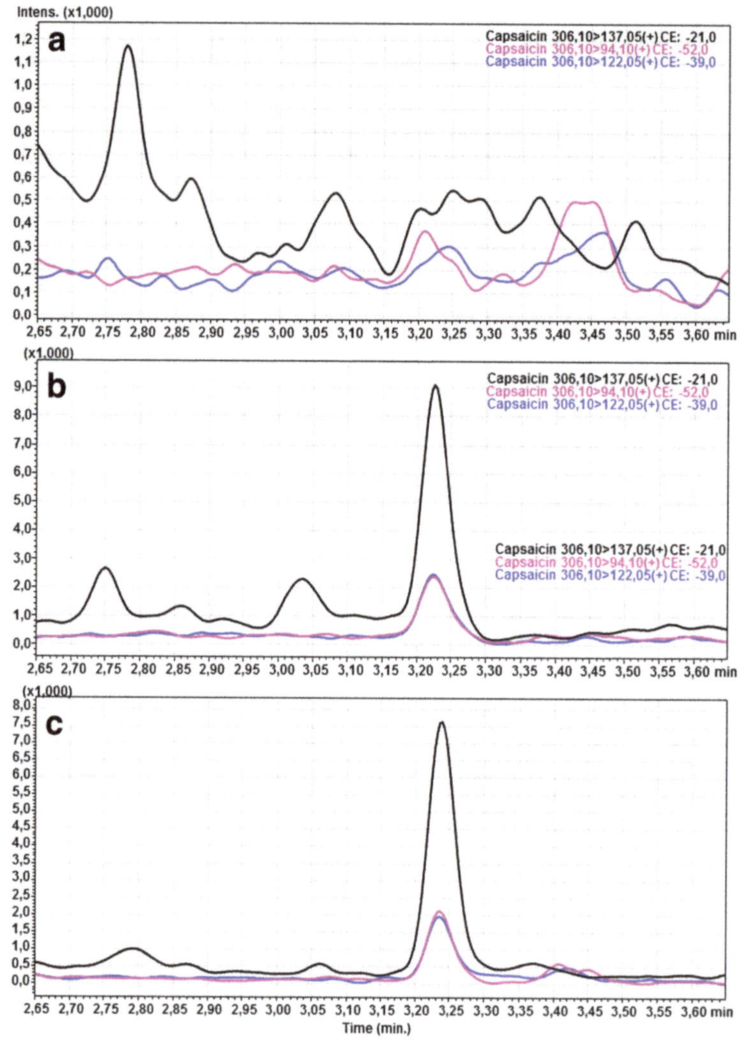

Fig. 1 MRM chromatograms of capsaicin in the blank serum (**a**), spiked serum sample (at a concentration of 0.5 pg/mL) with capsaicin (**b**), case sample –horse number 9 (**c**)

following its administration amounted to 4.7 pg/mL, despite the use a similar methodology and the same detection method in both studies [7]. In addition, dihydrocapsaicin was not detected in any of the samples, which stands in contrast to the findings of You et al., who noted detectable levels of dihydrocapsaicin 24 h after its administration [7]. The obtained results may be caused by breed differences between the studies, the different hair and feathers length as well as differences in the gel ingredients (especially the medium). In this study, isopropanol was the medium of the applied gel, which determined the best gel absorbance in humans [18]. Due to the possibility of local irritation caused by capsaicin, the application area was not shaved despite differing hair length among the studied horses. In addition, a polar bandage was applied to the treated area according to the treatment regimen, potentially causing some of the gel to absorb into the bandage material rather than the skin. Despite the availability of many capsaicin detection methods, such as: colorimetry/spectrophotometry, complexion chromatography, liquid chromatography (LC), gas chromatography and liquid chromatography tandem mass spectrometry (LC-MS) the authors chose to use ultra-high-performance liquid chromatography - tandem mass spectrometry, described in literature as fast, selectively sensitive, reproducible, reliable and fully validated [7, 19–25]. Future studies on the detectability of capsaicin, especially used for anti-doping testing, should consider assessing its metabolites for example vanillyamine and vanillic acid which are present invitro in the human skin [26].

The obtained results may be useful for horse owners and veterinary professionals to determine the period that should lapse from the administration of gels containing capsaicin (long-term therapeutic use) to the participation of the horses in riding events in accordance with the FEI [6]. Based on these results, capsaicin cannot be detected in horse serum 24 h after its administration using the therapeutic five day regimen and a commercially available gel together with a polar bandage. However, the absorbance of capsaicin (as well as its detectability) may be influenced by the type of hair at the application site, the size of the animal, the type of medium and numerous other factors. The second aim of the study was to determine the best study material for the detection of capsaicin and dihydrocapsaicin. According to the current study, serum is the best material for the detection of capsaicin. In the study by You et al. 2013, capsaicin and dihydrocapsaicin were measured just in blood plasma [7]. The studies of capsaicin metabolism have shown that only 0.095% of the unchanged circulating substance is excreted in urine and a small amount is excreted in faeces [27, 28]. Attempts have been made to determine the concentration of capsaicin in horse urine, but the tests have proven to be insensitive. They were found to detect capsaicin at a concentration of 10 ng/mL, indicating this method to be inadequate for routine horse evaluation [29]. The detection of capsaicin in horse hair has not been established so far, which should be the subject of further research. Horse hair from the mane and tail is usually used for the long-term evaluation of doping substances [30–32]. It provides a longer detection time than blood serum, plasma or urine, but it renders the determination of the time of substance intake impossible [30]. Current studies demonstrate the possibility to detect anabolic steroids, clenbuterol and other therapeutic substances not listed in the Prohibited Substances List in horse hair [30–32]. To date, the detectability of capsaicin in horse hair has not been determined. Hence, this should be studied further.

Conclusion

After long-term (five days) use of capsaicin in horses in the form of a topical gel and a polar bandage, the substance is not detected in blood serum 24 h after the last gel application. The absorption and detection time of capsaicin in horses probably depends on the size and breed, hair length on the limbs, the kind of product and type of application.

Endnotes

[1]Sartadt, Germany
[2]Ohaus Explorer, E1RR80, Ohaus, Switzerland
[3]Horse Gel, Eclipse Biofarmab, Sweden
[4]Sigma-Aldrich, Steinheim, Germany
[5]Sigma-Aldrich, India
[6]Sigma-Aldrich, Switzerland
[7]Sigma-Aldrich, China
[8]Aldrich Chemistry, China
[9]Kyoto, Japan
[10]Waters, Ireland
[11]Kyoto, Japan

Abbreviations

FEI: Federation Equestrian International; LC: Liquid chromatography; LC-MS: Liquid chromatography tandem mass spectrometry; LLOQ: Lower limit of quantification; MRM: Multiple reaction monitoring; TRPV-1: Transient receptor potential cation channel subfamily V member 1; UHPLC: Ultra-high performance liquid chromatograph; UHPLC-QqQ-MS/MS: Ultra-high performance liquid chromatography coupled with a triple quadrupole mass spectrometry

Acknowledgements

The results of this study has been presented in preliminary form at the poster report session at 15th World Equine Veterinary Association Congress in Beijing, 21-23 April 2018.

Funding

All necessary reagents were financed by the Department of Internal Medicine and Clinic of Diseases of Horses, Dogs and Cats, Faculty of Veterinary Medicine, Wroclaw University of Environmental and Life Sciences.

Authors' contributions

AZ contributed to study design, study execution, data analysis and interpretation and was a major contributor in writing the manuscript. NS, MS, HB and PS participated in the design of the study, study execution, in the collection of the samples and in analysis and interpretation of data. AN and MZ conceived of the study, and participated in its design and coordination and helped to draft the manuscript. All authors were involved in continuous discussion regarding manuscript. All authors read and approved the final manuscript.

Competing interests

The authors declare that they have no competing interests.

Author details

[1]Department of Internal Medicine and Clinic of Diseases of Horses, Dogs and Cats, Faculty of Veterinary Medicine, Wroclaw University of Environmental and Life Sciences, pl. Grunwaldzki 47, 50-366 Wroclaw, Poland. [2]Department of Forensic Medicine, Wroclaw Medical University, ul. Mikulicza-Radeckiego 4, 50-345 Wroclaw, Poland.

References

1. Cordell GA, Arajou OE. Capsaicin: Identification, nomenclature, and pharmacotherapy. Ann Pharmacother. 1993;27:330–6.
2. Hayman M, Kam PCA. Capsaicin: a review of its pharmacology and clinical applications. Curr Anaesth Crit Care. 2008;19:338–43.
3. Seino KK, Foreman JH, Greene SA, Goetz TE, Benson GJ. Effects of topical Perineural capsaicin in a reversible model of equine foot lameness. J Vet Intern Med. 2003;17:563–6.

4. Aley JP, Adams NJ, Ladyman RJ, Fraser DL. The efficacy of capsaicin as an equine repellent for chewing wood. J Vet Behav. 2015;10:243–7.
5. Govidarajan VS. Capsicum production, technology, chemistry and quality. Part I: history, cultivation and primary processing. CRC Cr Rev Food Sci. 1985;22:109–76.
6. 2017 Equine Prohibited Substances List FEI. https://inside.fei.org/sites/default/files/2018%20Equine%20Prohibited%20Substances%20List.pdf. Accessed 12 June, 2018.
7. You Y, Uboh C, Soma LR, Guan F, Taylor D, Li X, Liu Y, Chen J. Validated UHPLC-MS-MS method for rapid analysis of capsaicin and Dihydrocapsaicin in equine plasma for doping control. J Anal Toxicol. 2013;37:122–32.
8. Alawi K, Keeble J. The paradoxical role of the transient receptor potential vanilloid 1 receptor in inflammation. Pharmacol Ther. 2010; 125:181–95.
9. Caterina MJ, Schumacher MA, Tominaga M, Rosen TA, Levine JD, Julius D. The capsaicin receptor: a heat-activated ion channel in the pain pathway. Nature. 1997;389:816–24.
10. Conway SJ. TRPing the switch on pain: an introduction to the chemistry and biology of capsaicin and TRPV1. Chem Soc Rev. 2008;37: 1530–45.
11. Winters J, Bevan S, Campbell AE. Capsaicin and pain mechanism. Brit J Anaesth. 1995;75:157–68.
12. Hill R. NK1 (substance P) receptor antagonists – why are they not analgesic in humans? Trends Pharmacol Sci. 2000;21:244–6.
13. Anand P, Bley K. Topical capsaicin for pain management: therapeutic potential and mechanisms of action of the new high-concentration capsaicin 8% patch. Brit J Anaesth. 2011;107(4):490–502.
14. Holzer P. The pharmacological challenge to tame the transient receptor potential vanilloid-1 (TRPV1) nocisensor. Br J Pharmacol. 2008; 155:1145–62.
15. Polydefkis M, Hauer P, Sheth S, Sirdofsky M, Griffin JW, McArthur JC. The time course of epidermal nerve fibre regeneration: studies in normal controls and in people with diabetes, with and without neuropathy. Brain. 2004;127:1606–15.
16. Scholzen T, Armstrong CA, Bunnet NW, Luger TA, Olerud JE, Ansel JC. Neuropeptides in the skin: interactions between the neuroendocrine and the skin immune systems. Exp Dermatol. 1998;39:1–6.
17. Gleerup KB, Forkman B, Lindegaard C, Andersen PH. An equine pain face. Vet Anaesth Analg. 2015;42(1):103–14.
18. Pershing LK, Reilly CA, Crolett JL, Crouch DJ. Effects of vehicle on the uptake and elimination kinetics of capsaicinoids in human skin in vivo. Toxicol Appl Pharm. 2004;200:73–81.
19. Gibbs H, O'Garro LO. Capsaicin content of West Indies hot pepper cultivars using colorimetric and chromatographic techniques. HortScience. 2004;39: 132–5.
20. Constant HL, Cordell GA, West DP, Johnson JH. Separation and quantification of capsaicinoids using complexation chromatography. J Nat Prod. 1995;58:1925–8.
21. Cooper TH, Guzinski JA, Fisher C. Improved high performance liquid chromatography method for the determination of major capsaicinoids in capsicum oleoresins. J Agr Food Chem. 1991;39:2253–6.
22. Karnka R, Rayanakorn M, Watanesk S, Vaneesorn Y. Optimization of high-performance liquid chromatographic parameters for the determination of capsaicinoids compounds using the simple method. Anal Sci. 2002;18:661–5.
23. Thomas BV, Schreibe AA, Weisskopf CP. Simple method for quantitation of capsaicinoids in peppers using capillary gas chromatography. J Agr Food Chem. 1998;46:2655–63.
24. Kozukue N, Han JS, Kozukue E, Lee S, Kim LA, Lee KR, Levin CE, Friedman M. Analysis of eight capsaicinoids in peppers and peppercontaining foods by high-performance liquid chromatography and liquid chromatography-mass spectrometry. J Agr Food Chem. 2005;53:9172–81.
25. Reilly CA, Crouch DJ, Yost GS, Fatah AA. Determination of capsaicin, Nonivamide, and Dihydrocapsaicin in blood and tissue by liquid chromatography-tandem mass spectrometry. J Anal Toxicol. 2002;26: 313–9.
26. Chanda S, Bashir M, Babbar S, Koganti A, Bley K. In vitro hepatic and skin metabolism of capsaicin. Drug Metab Dispos. 2008;36(4):670–5.
27. Suresh D, Srinivasan K. Tissue distribution and elimination of capsaicin, piperine and curcumin following oral intake in rats. Indian J Med Res. 2010; 131:682–91.
28. Reyes-Escogido ML, Gonzalez-Mondragon EG, Vazquez-Tzompantzi E. Chemical and pharmacological aspects of capsaicin. Molecules. 2011;16: 1253–70.
29. Stanley SMR, Wee WK, Lim BH, Foo HC. Direct-injection screening for acidic drugs in plasma and neutral drugs in equine urine by differential-gradient LC-LC coupled MS/MS. J Chromatogr B. 2007;848:292–302.
30. Anielski P. Hair analysis of anabolic steroids in connection with doping control—results from horse samples. J Mass Spectrom. 2008;43(7):1001–8.
31. Dunnett M, Lees P. Hair analysis as a novel investigative tool for the detection of historical drug use/misuse in the horse: a pilot study. Equine Vet J. 2004;36(2):113–7.
32. Schlupp A, Anielski P, Thieme D, Mülle RK, Meyer H, Ellendorff F. The β-agonist clenbuterol in mane and tail hair of horses. Equine Vet J. 2004;36(2):118–22.

20

Effect of prolonged submaximal exercise on serum oxidative stress biomarkers (d-ROMs, MDA, BAP) and oxidative stress index in endurance horses

Nika Brkljača Bottegaro[1*], Jelena Gotić[2], Jelena Šuran[3], Diana Brozić[4], Karla Klobučar[5], Krunoslav Bojanić[6] and Zoran Vrbanac[7]

Abstract

Background: Oxidative stress (OS) associated with an intense exercise may have a negative influence on equine health. The aim of this study was to determine the effects of endurance races on oxidative and antioxidative status of horses by evaluating changes in reactive oxygen metabolites (d-ROMs), malondialdehyde (MDA), biological antioxidant potential (BAP) and oxidative stress index (OSI) values. The study was carried out on 53 race starts (28 individual horses) competing at different endurance races according to distance (40 and 80 km) and difficulty (easy and demanding). Blood samples were taken before and after the race.

Results: Compared to levels of OS serum biomarkers before the race, an increase in values of d-ROMs ($P < 0.01$), MDA ($P < 0.01$), and BAP ($P < 0.001$), and a decrease in OSI ($P < 0.001$) have been noted after the race. Contrary to other measured biomarkers, BAP did not show significant individual effects of horses. Horses competing at shorter races have shown a significant change in d-ROMs ($P = 0.002$), BAP ($P < 0.001$) and OSI ($P = 0.004$), whereas those competing at longer races in MDA ($P = 0.002$), BAP ($P < 0.001$) and OSI ($P < 0.001$) post-race values. Endurance racing induced changes in values of d-ROMs, BAP and OSI during both easy and demanding races.

Conclusions: Changes in all measured OS biomarkers indicate that prolonged aerobic exercise during endurance race could contribute to the imbalance between oxidants and antioxidants in horses, mainly characterised by a pronounced antioxidant response. Biological antioxidant potential was found to be the most reliable biomarker of OS in endurance horses in the present study.

Keywords: Horse, Oxidative stress, Reactive oxygen metabolites, Malondialdehyde, Biological antioxidant potential, Oxidative stress index, Endurance race

Background

Endurance riding is an extremely demanding equine discipline since horses compete over distances of up to 160 km in a day. An endurance ride is divided into several phases of 20 to 40 km, each followed by a mandatory resting period. Before the ride, as well as after each phase, horses undergo a mandatory veterinary examination and can be eliminated if their metabolic status or orthopaedic condition are found inadequate to enable them to continue the ride.

Exercise induces increased oxygen consumption which in return affects the oxidant/antioxidant status [5]. During short-term maximal exercise in Thoroughbred racehorses, there is an increase in oxygen consumption that reaches levels 30 times above basal [1]. Even though endurance riding is classified as a prolonged aerobic exercise [2], the high oxygen demands over a longer period of time result in an increased reactive oxygen species (ROS) formed by 1 to 2% of the oxygen that is not completely reduced into carbon dioxide and water [3]. The consequence of

* Correspondence: nikabottegaro@gmail.com
[1]Clinic for Surgery, Orthopaedics and Ophthalmology, Faculty of Veterinary Medicine, University of Zagreb, Heinzelova 55, 10 000 Zagreb, Croatia
Full list of author information is available at the end of the article

increased oxidants is a disarranged equilibrium between oxidants and antioxidants resulting in oxidative stress (OS) [4]. Apart from chronic accumulation of ROS, OS can also be a sequel of insufficient antioxidant defence systems, causing damage to all cell components, especially DNA, lipids, and proteins [5].

OS is characterised by changes in different biomarkers but the determination of reactive oxygen metabolites (d-ROMs), an indirect method evaluating the free radicals in the serum, is considered a 'gold standard' biomarker for measuring total systemic oxidative status [6]. Among other OS biomarkers, the biological antioxidant potential (BAP) test indicates systemic antioxidative properties and reflects the reduction potential in the serum [7], whereas malondialdehyde (MDA) is one of the better-known secondary products of lipid peroxidation [8] used as an indicator of cell membrane injury [9]. The oxidative stress index (OSI) is an arbitrary parameter derived from values of d-ROMs and BAP to represent an individual's antioxidant potential against the condition of OS [10]. In equine medicine, OSI has been investigated in resting Thoroughbred foals [11], Thoroughbred racehorses [12], and Thoroughbred horses after castration [13].

Oxidative stress in horses induced by endurance races has been investigated in a few studies mainly focusing on antioxidant status [14–17]. Oxidative properties have been assessed by measurements of the following biomarkers: lipid hydroperoxides (LPO) [17, 18] and total barbituric acid reactive substances (TBARS) [14, 16]. On the other hand, antioxidant status in endurance horses has been more intensively studied by evaluating several different biomarkers: α-tocopherol (TOC) and glutathione (GSH) [15–18]; glutathione peroxidase (GPx) activity [14, 15, 17, 18], glutathione redox ratio [16], glutathione reductase, superoxide dismutase (SOD), and total antioxidans status (TAS) [14].

However, data on changes of d-ROMs and BAP in relation to endurance exercise in horses, to the best of our knowledge, have not been reported. All the mentioned studies [14–18] have been performed on horses competing at only higher levels of exercise, although a vast majority of horses in this sport compete at lower level national endurance races.

The aim of this study was to determine the effects of endurance races on blood levels of d-ROMs, BAP, and MDA OS biomarkers in horses. It was hypothesised that even endurance races at shorter distances induce OS in horses. The second hypothesis was that endurance exercise causes changes in OSI levels which could emphasise a more pronounced prooxidant or antioxidant response.

Methods

The competition and conditions

The study included horses competing at four national competitions held from April to July 2017 at different locations in Croatia. The horse owners were offered to participate in the study prior to the first veterinary inspection of their horse (s) at the competition sites. The races differed according to terrain condition and configuration, altitude changes, and weather conditions, thus considering these variations, the races were categorised as demanding or easy. Demanding races were those with a considerable part of rocky racetrack, higher altitude differences and/or with higher temperature and humidity. Two races were classified as easy and two as demanding. Also, there were races of both 40 and 80 km distances at each competition.

Horses

All horses competing at the mentioned races were eligible for enrolment in this study. Forty-five horses were enrolled following owner consent. Horses had been transported from varying distances and arrived at the event several hours (at least two) before the first veterinary examination. All the horses participating in the study successfully passed the veterinary inspection prior the start, as well as all the inspections during and after the race. Since the inclusion criterion for horses was to successfully complete the race, only the finisher horses were included in the further study.

The average speed of horses in the study population was also evaluated with a cut-off of 14 km/h which is the minimal average speed required for the Federation Equestre International 4* championship qualification according to the FEI Endurance rules 2017.

Sample collection and analysis

Blood samples (5 ml) were collected by jugular venipuncture into vacuum tubes for serum with gel using a vacutainer (Vacutainer®, Becton Dickenson, USA) 1 h before the start and 30 min after the end of the race. The samples were left to coagulate for 30 min and thereafter centrifuged at 3500 pm for 15 min. The samples were shipped in cooling bags at 4 °C to the laboratory, and within 4 h after sampling stored at – 80 °C until further analyses.

Serum levels of d-ROMs and BAP were measured using Diacron® commercial kits (Grosseto, Italy) on biochemical analyzer Olympus AU640 (Olympus, Japan) according to manufacturer's instructions. The results of the d-ROMs analysis were expressed in Carratelli units (U.Carr.), where 1 U.Carr. corresponds to 0.8 mg/L H_2O_2. The results of the BAP analysis were expressed in μmol/L of the reduced ferric ions. The assays used in this study had been previously validated for use in horses [6, 19].

Malondialdehyde concentration was determined with a slight modification of a previously reported method [8]. Briefly, a volume of 150 μL of serum was added to 50 μL of water and 50 μL of NaOH (3 N). After 30 min of shaking water bath with an incubation at 60 °C, 250 μL 6% phosphoric and 0.8% thiobarbituric acid were added and the mixture was heated at 90 °C for 45 min. The mixture was cooled and extracted by adding 250 µl of pure methanol and 100 µl 10% of sodium dodecyl sulphate, vortex-mixed for 1 min and centrifuged at 3000×g for 10 min. A Shimadzu 2010 LC system equipped with an InertSustain C18 (4.6 mm × 150 mm × 5 µm) column (GL Sciences) and a UV–visible detector was used for analysis with the following conditions: the UV wavelength was set at 532 nm, injection volume of 20 µL, and the column oven temperature at 38 °C. The mobile phase was as follows: pump A, methanol, and pump B, 50 mM potassium dihydrogen phosphate with pH of 6.2 (50:50, *v*/v). The flow rate of the mobile phase was 1 mL/min with constant flow control. The standard curve was prepared using 1,1,3,3-tetraethoxypropane with the linearity of 0.98 and reliability of 97.5% for given standard concentration range (0.5–15.175 µM of standard). MDA concentration was expressed as µmol per L serum. The method was in-house validated for use in horse serum, with the inter- and intra-assay coefficients of variation of less than 5%.

The OSI was calculated from the measured values of d-ROMs and BAP using the ratio of d-ROMs/BAP multiplied by 100 [11].

Statistical methods

All exploratory data analyses and statistical tests were performed using the R programming language version 3.2.21 [20]. The dependent variables of interest were the changes in values after the race compared to before the race for each OS biomarker. The explanatory variables of interest that were collected from horses (gender and average speed) and race events characteristics (race distance and difficulty) were used to evaluate the changes in values of OS biomarkers within each subpopulation category and for comparisons between the respective subpopulation categories of horses. Horses never ran more than once per event and some horses participated at several events. Therefore, the overall average change (post-race value – pre-race value) was analysed via hierarchical linear mixed-effects regression models due to repeated measurements on the same horse subjects and using linear regression models for analysis of data within each race. Modelling was performed using the lme4 software package [21]. Outliers were assessed using leave-one-out method for mixed models and Bonferroni outlier test for linear models. As no significant influence on model coefficient estimates was observed, the data were kept in the model. The alpha level of significance for model fits and regression coefficients was taken at less than 0.05. Data were expressed as an average (±standard error of mean (SEM)).

Results

Study population characteristics

Forty-five horses participated in the study. However, 28 individual horses successfully met the inclusion criteria in 53 race starts, since some horses participated in more races. The characteristics of the study population are presented in the Table 1. The average age of the studied horses was 7.3 years (±0.3 SEM) and their average speed during races was 13.5 (±0.2 SEM) km/h. The average speed was 13.7 (±0.3 SEM) km/h at 40 km races and 13.2 (±0.3 SEM) km/h at 80 km races. There were no significant differences between the average speed at 40 and 80 km races ($P = 0.3$). However, there was a significant association ($P < 0.001$) of the average speed and the race difficulty with the average speed of 14.3 (±0.2 SEM) km/h at easy races and the average speed of 12.3 (±0.4 SEM) km/h at demanding races.

Comparison of pre- and post-race values of OS biomarkers

The values of all oxidative stress markers evaluated in this study have significantly changed after the race compared to the values before the race and their distribution is depicted in Fig. 1. The most significant changes were noted for BAP and OSI ($P < 0.001$). On average, values of BAP, d-ROMs, and MDA increased, while values of OSI decreased after the race. Inspection of random effects of horse subjects showed that the estimates of average change of MDA, d-ROMs, and OSI are extremely variable between and within horses. Ten horses had discordant changes (an increase at one race and a decrease at another) in values of d-ROMs, 11 in MDA and 8 in values of OSI. On the other hand, the individual variation of the subjects was low for BAP.

Values of different OS markers in relation to categories of race distances and difficulty, and of average speed and gender of horses are presented in Tables 2 and 3, respectively. On average, there was a significant change in values of BAP and OSI irrespective of the race distances. However, a significant change in values of d-ROMs was observed only at 40 km and of MDA only at 80 km race distances. Values of d-ROMs, BAP and OSI showed significant differences before and after the race for both easy and demanding races. Changes in concentrations of MDA were significantly different than before the race

Table 1 Characteristics of the study population of horses attending four endurance races

Descriptive variables	Study population
Horses	28 individuals
Breed (Arabian / other breeds)	14 / 14
Gender (females / males)	18 / 10
Races	53 records
Distance (40 km / 80 km)	30 / 23
Race difficulty (easy/demanding)	32 / 21
Average speed (< 14 km/h / ≥ 14 km/h)	35 / 18
Races per horse (one / two / three / four races)	8 / 15 / 5 / 0

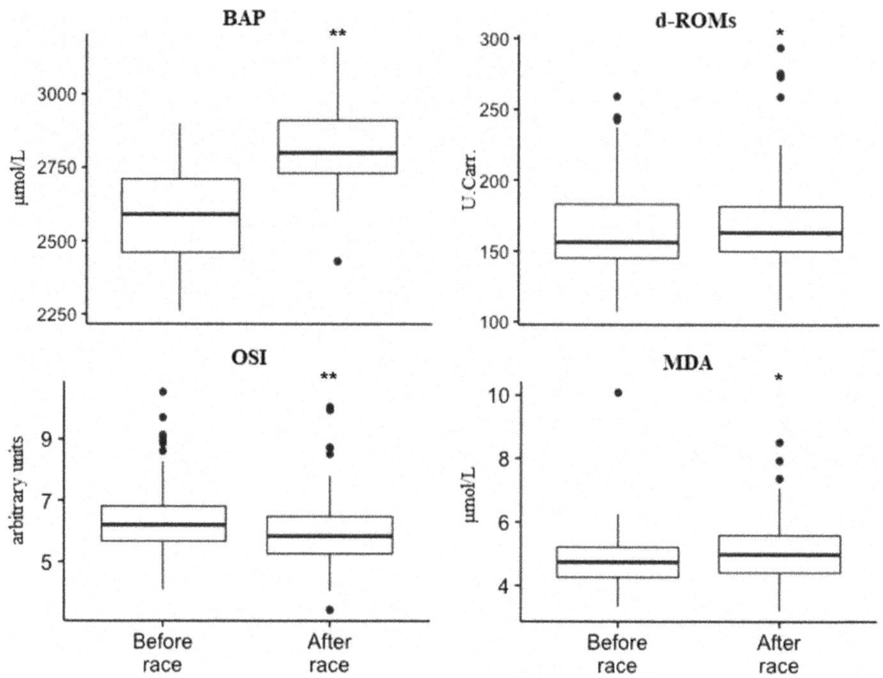

Fig. 1 Boxplots of values of the OS biomarkers in endurance horses pre- and post-races. The data represents measurements from 53 race runs of 28 individual horses over four racing events. The null hypothesis states there is no difference between the two (i.e., the delta change being zero). Significant differences from values at baseline (before race); **P* < 0.01; ***P* < 0.001; d-ROMs, determination of reactive oxidative metabolites; BAP, biological antioxidant potential; MDA, malondialdehyde; OSI, oxidative stress index

Table 2 Values of serum OS biomarkers in horses pre- and post-races by distance and difficulty

Parameter (unit)		Distance		Race difficulty	
		40 km	80 km	Easy	Demanding
d-ROMs (U.Carr.)	Pre-race	164.8 ± 7.6	168.7 ± 9.9	166.2 ± 6.8	166.2 ± 8.2
	Post-race	173.8 ± 8.6	174.2 ± 11.3	171.9 ± 7.8	172.3 ± 8.4
	P	0.002	0.2	0.04	0.03
	P[a]	0.5		0.9	
BAP (μmol/L)	Pre-race	2610.9 ± 33.2	2533.1 ± 30.5	2612.7 ± 32.2	2541.9 ± 32.2
	Post-race	2820.4 ± 28.2	2840.4 ± 31.4	2817.5 ± 28.8	2840.5 ± 22.8
	P	< 0.001	< 0.001	< 0.001	< 0.001
	P[a]	0.01		0.02	
MDA (μmol/L)	Pre-race	5.1 ± 0.2	4.5 ± 0.1	4.7 ± 0.1	5.1 ± 0.3
	Post-race	5.3 ± 0.2	4.7 ± 0.2	5.1 ± 0.2	5.1 ± 0.3
	P	0.1	0.02	0.003	0.7
	P[a]	0.9		0.1	
OSI (arbitrary units)	Pre-race	6.3 ± 0.3	6.7 ± 0.4	6.4 ± 0.3	6.6 ± 0.3
	Post-race	6.1 ± 0.3	6.2 ± 0.4	6.1 ± 0.3	6.1 ± 0.3
	P	0.004	< 0.001	0.005	< 0.001
	P[a]	0.01		0.1	

d-ROMs determination of reactive oxidative metabolites, *BAP* biological antioxidant potential, *MDA* malondialdehyde, *OSI* oxidative stress index
P-significance for the differences within each category division (e.g. change in values at 40 km and at 80 km separately)
P[a]-significance for the difference between the two categories (e.g. change in values at 40 km vs. at 80 km)

Table 3 Values of serum OS biomarkers in horses pre- and post-races by average speed and gender

Parameter (unit)		Average speed		Gender	
		< 14 km/h	≥ 14 km/h	Female	Male
d-ROMs (U.Carr.)	Pre-race	166.5 ± 7.3	160.6 ± 8.2	171.1 ± 8.0	164.1 ± 11.9
	Post-race	171.5 ± 8.2	170.1 ± 8.9	177.2 ± 9.8	170.2 ± 11.8
	P	0.08	0.002	0.03	0.03
	P^a	0.3		0.8	
BAP (μmol/L)	Pre-race	2570.2 ± 28.4	2585.4 ± 39.6	2570.8 ± 29.82	2588.5 ± 46.9
	Post-race	2849.2 ± 23.3	2787.5 ± 41.0	2816.3 ± 21.3	2825.2 ± 51.5
	P	< 0.001	< 0.001	< 0.001	< 0.001
	P^a	0.1		0.9	
MDA (μmol/L)	Pre-race	4.7 ± 0.1	5.2 ± 0.3	4.8 ± 0.1	4.9 ± 0.3
	Post-race	5.0 ± 0.2	5.3 ± 0.3	5.1 ± 0.1	5.1 ± 0.4
	P	0.003	0.3	0.01	0.2
	P^a	0.8		0.9	
OSI (arbitrary units)	Pre-race	6.5 ± 0.3	6.2 ± 0.3	6.7 ± 0.4	6.4 ± 0.5
	Post-race	6.04 ± 0.3	6.1 ± 0.4	6.4 ± 0.5	6.4 ± 0.5
	P	< 0.001	0.116	< 0.001	< 0.001
	P^a	0.02		0.6	

d-ROMs determination of reactive oxidative metabolites, *BAP* biological antioxidant potential, *MDA* malondialdehyde, *OSI* oxidative stress index
P-significance for the differences within each category division (e.g. change in values at female and at male horses separately)
P^a-significance for the difference between the two categories (e.g. change in values at female vs. male horses)

only after easy races. Dividing the horse population in accordance with the average speed obtained, the changes in levels of d-ROMs increased significantly after the race only in horses of the faster speed category, levels of BAP increased significantly in both speed categories, and the concentration of MDA and values of OSI had significant changes only in the slower speed category. Both female and male horses showed significant differences in values of d-ROMs, BAP, and OSI after the race, compared to the values before the race. However, a significant change in concentrations of MDA due to race has been noted just in female horses. Both race distance and difficulty had a significant association with the change of BAP. The changes in values of OSI before and after the race have been influenced by race distance and average speed of horses.

Table 4 presents values of different OS markers in relation to each of the four races. Values of d-ROMs showed significant differences before and after the race during 40 km at the first race. With regard to values of BAP, significant differences before and after the race values have been noted in all but the second race. In addition, the changes in values of BAP were more pronounced after 80 km than after 40 km races at the third and fourth race events. Values of OSI showed significant differences before and after the race during 80 km at the third and after 40 km at the fourth race event.

Discussion

The intense exercise during endurance races in this study induced OS in horses demonstrated by increased values of d-ROMs, MDA, and BAP, and a decreased values of OSI. The increase in values of d-ROMs and MDA in horses undergoing endurance races could be attributed to the enhanced production of free radicals and lipid peroxidation resulting in oxidative damage [6, 22]. On the other hand, the rise in levels of BAP after the race indicates a mobilization of antioxidants as a response to an increased amount of ROS [6]. Oxidative stress depends on the balance between ROS and body's antioxidant system [23], therefore, OSI could more accurately reflect OS status due to a joint measure of oxidants and antioxidants properties. In the present study, a significant decrease in values of OSI shown after the race indicates a pronounced antioxidant response induced by endurance exercise.

Previous studies have evaluated OS in response to endurance exercise. Endurance race of 80 km caused an effect on oxidative status balance characterised by an increase in values of LPO, GSH and a decrease in values of GPx and TOC [17]. A study [18] reported OS after 80 km endurance race by an increase in total plasma LPO. On the other hand, changes in oxidative status markers (lipid hydroperoxides oxygen radical absorbance capacity, vitamin E, and GSH) have not been observed after 80 km endurance race [24]. The discrepancy between the mentioned findings [24] and the present

Table 4 Values of serum OS biomarkers in horses pre- and post-races by separate race events

Parameter (unit)		First race		Second race		Third race		Fourth race	
		40 km ($n = 12$)	80 km ($n = 7$)	40 km ($n = 2$)	80 km ($n = 3$)	40 km ($n = 8$)	80 km ($n = 8$)	40 km ($n = 8$)	80 km ($n = 5$)
d-ROMs (U.Carr.)	Pre-race	160.9 ± 11.8	172.7 ± 16.8	163.8 ± 13.5	184.9 ± 27.4	162.7 ± 9.2	163.3 ± 17.5	174.9 ± 6.2	157.2 ± 12.1
	Post-race	176.0 ± 13.2	175.6 ± 20.2	161.7 ± 0.7	184.2 ± 22.0	169.1 ± 8.9	173.8 ± 19.7	173.0 ± 8.2	156.7 ± 14.7
	P	< 0.001	0.7	0.9	0.9	0.06	0.07	0.7	0.9
	P^a	0.07		0.9		0.5		0.9	
BAP (μmol/L)	Pre-race	2575.8 ± 51.6	2628.6 ± 71.9	2820.0 ± 40.0	2573.3 ± 53.6	2556.3 ± 52.6	2446.3 ± 21.5	2692.5 ± 34.3	2516.0 ± 53.5
	Post-race	2806.7 ± 49.0	2837.1 ± 90.3	2905.0 ± 155.0	2883.3 ± 69.4	2812.5 ± 33.6	2836.3 ± 32.3	2818.8 ± 29.3	2826.0 ± 65.7
	P	< 0.001	0.04	0.7	0.06	< 0.001	< 0.001	< 0.01	< 0.01
	P^a	0.8		0.3		0.03		0.02	
MDA (μmol/L)	Pre-race	4.9 ± 0.2	4.5 ± 0.2	8.2 ± 1.9	3.8 ± 0.2	5.2 ± 0.2	4.6 ± 0.1	4.8 ± 0.1	4.3 ± 0.2
	Post-race	4.9 ± 0.2	4.9 ± 0.3	7.8 ± 0.7	3.5 ± 0.1	5.3 ± 0.2	4.8 ± 0.2	5.7 ± 0.4	5.2 ± 0.3
	P	0.8	0.2	0.8	0.08	0.5	0.5	0.08	0.02
	P^a	0.3		0.9		0.9		0.9	
OSI (arbitraryunits)	Pre-race	6.2 ± 0.4	6.7 ± 0.8	5.8 ± 0.4	7.1 ± 0.9	6.4 ± 0.4	6.7 ± 0.7	6.5 ± 0.3	6.2 ± 0.4
	Post-race	6.2 ± 0.4	6.3 ± 0.9	5.6 ± 0.3	6.4 ± 0.7	6.0 ± 0.3	6.1 ± 0.6	6.1 ± 0.3	5.5 ± 0.5
	P	0.9	0.07	0.2	0.07	0.08	0.02	0.01	0.2
	P^a	0.05		0.1		0.4		0.4	

d-ROMs determination of reactive oxidative metabolites, *BAP* biological antioxidant potential, *MDA* malondialdehyde, *OSI* oxidative stress index
P-significance for the differences within each category division (e.g. change in pre-race and post-race values)
P^a-significance for the difference between the two categories (e.g. change in values at 40 km vs. 80 km)

study could be attributed to OS biomarker selection, a small number of horses (four), and a relatively low average speed of 10.5 km/h in the former compared to the present study which had a larger sample size and horses competing with almost 30% faster average speed. A previous report [2] assessed the levels of SOD, GPx, reduced GSH, catalase, and the plasma MDA in endurance horses, performing at different distances. Interestingly, that study did not record any significant changes in values of OS biomarkers after the race, although higher values of SOD, GPx, GSH, and MDA were recorded after the 80 km race but not at 160 km. These results could also be attributed to the relatively small number (15) of studied horses and potentially individual differences in measured OS biomarkers could have precluded the true association of OS biomarkers and race distance. As the effect of individual horses could vary between OS biomarkers, the selection of the markers should be an important aspect for consistent results. The results of the present study showed only BAP not to have a significant horse effect, hence, making it a more consistent OS biomarker than d-ROMs, MDA, and OSI. In this study d-ROMs, MDA, and OSI showed discordant changes in approximately one-third of the horses. These findings indicate that there are other determinants and explanatory variables also influencing the changes in d-ROMs, MDA, and OSI that have not been followed in this study. Similarly, a report [25] noted fluctuation of values of d-ROMs especially, but also of BAP to a lesser degree, in ten Thoroughbred horses. Studies involving a large number of horses are needed in order to better understand the physiological levels and changes of OS biomarkers.

In the present study, MDA has also shown increased concentrations after the race indicating a potential cumulative cell membrane lipid peroxidation. A rise in erythrocyte MDA concentration was also recorded after short-term heavy exercise in jumping horses [22], as well as by an increased intensity ride in ten Maremmana stallions [26]. A previous study [27] reported significant changes of TBARS, a test that measures mainly but not only MDA, induced at 80 km endurance race, but not in horses competing at 160 km. That study suggested horses competing at higher speed (which occurred at 80 km races) are more prone to rise in TBARS levels after exercise. Contrary, a study on five horses competing at 160 km race, noted an increase in values of TBARS after the race [14]. In the present study, an increased concentration of MDA was noted only in horses at 80 km races with no significant difference in average speed between race distances thus, it is not possible to confirm the suggestion of influence of speed [27]. Lack of increased lipid peroxidation after half race distance and an increase after 80 km endurance race have also been noted by measuring values of plasma LPO [17]. Actually, the present study showed a significant rise in the concentration of MDA after the race only at slower average speed. Furthermore, the present study also adds race difficulty as another factor potentially influencing changes of MDA, as a significant change was shown only at easy races. These unexpected results raise suspicion for a potential individual variation in lipid peroxidation between horses and the possibility that training status or other parameters not followed in this study, could influence concentrations of MDA. On the other hand, lower values of LPO have been recorded in an easier race in comparison to a more difficult race as defined by the riders [17]. However, a clear definition of race difficulty could be challenging, especially considering possible subjectivity of riders opinion.

In previous studies of OS subsequent endurance races, opponent changes in antioxidant status have been noted. After 80 km race an increase in values of GSH [17, 18] and GPx [15, 18] has been noted, contrary, another study showed a decrease in values of GPx [17]. A study on 5 horses competing at 160 km endurance race noted a drop in values of TAS at 80 km, but a rise of both values of SOD and TAS at the race end [14]. The inconsistent findings of previous studies could be attributed to the antioxidant biomarker selection, since some of the vast antioxidant biomarkers show lower antioxidant capacity caused by exercise, but simultaneously other biomarkers enhance their activity. Therefore we believe that BAP could be more effective in delineating general antioxidant status, since it provides an overall measurement of different antioxidants including uric acid, ascorbic acid, proteins, TOC, bilirubin and others [7]. The increase in levels of BAP after the race in the present study indicates an increase in the antioxidant potential as a consequence of exercise. It suggests that during prolonged exercise OS induces a response to the increase in ROS thus, results in an increase in levels of BAP. Similar findings have also been noted in Thoroughbred horses after a short-term maximum treadmill exercise [28]. In that study, the rise in BAP levels was more pronounced immediately after exercise, with a clear drop in levels of BAP 30 min after exercise. After finishing the race endurance horses have a maximum of 30 min rest before presenting to the final inspection. Owners did not allow blood sampling during that rest period in the present study since it could negatively affect the horse's attitude thus compromising the final results of the forthcoming mandatory veterinary inspection. Therefore, our sampling time was 30 min after the end of the race and it could have been the reason of our comparably lower recorded levels of BAP, although still significantly different from the values before the race. Similarly, in the study were samples have been taken after 80 km race as well as after 60 min of recovery period, there was an evident decrease in GPx levels during the last sampling [15]. Though, in the abovementioned study it is not defined if the samples have been taken immediately after finishing the race or after passing the mandatory veterinary inspection. The time preceding the veterinary inspection after each loop is very short (often lasting up to few minutes) and considerable effort is put on minimising any discomfort to horses. In the study on Thoroughbred horses undertaking treadmill exercise [28], an increase in levels of d-ROMs in response to exercise has been reported. Interestingly, in that study, at 30 min after exercise, the levels of d-ROMs returned to baseline. Therefore, the sampling time could have also influenced the results of d-ROMs in the present study. Similarly, in a previous study an effect of 3 h recovery period after 80 km endurance race on prooxidative biomarker values has been noted since values of LPO also decreased to baseline [17]. An overall increase in levels of d-ROMs in the present study has been noted after the race but, taking into account race distances, this increase was significant only in horses competing at shorter distances. Horses competing at 80 km races were running twice longer and furthermore had one more compulsory rest period according to endurance rules that could have influenced our results. In addition, the lack of the elevation of d-ROMs after 80 km races could also indicate that the horses used in this study had an adaptive antioxidant response. Horses competing at 80 km are trained for longer periods than those competing at 40 km what could induce adaptations to OS, supported by a study in Standardbred horses that proved that training increase antioxidant capacity [29].

In the present study, a decrease in values of OSI has been noted after the race similarly to a previous report on Thoroughbred horses [28] using the BAP/d-ROMs ratio, which is an inverse formula of the one used in the present study. Nevertheless, the results of both studies indicate a more pronounced antioxidant status as a result of OS induced by exercise. Reference ranges of d-ROMs, BAP and OSI values have been studied in 372 Thoroughbreds racehorses [12]. That study measured values of OS biomarkers only during resting and the levels of both BAP and d-ROMs OS markers have been very similar to our findings.

The observed changes in levels of BAP after exercise were associated with both race distance and difficulty with higher values recorded on longer and demanding races. These findings indicate that the type and intensity of exercise can also influence OS in endurance horses. In a previous study, comparing results between two 80 km races of unequal difficulties, a more pronounced rise in GSH has also been recorded after a more difficult race [17]. More pronounced changes in antioxidant biomarkers were noted in eventing horses competing at a higher intensity exercise indicating the influence of increased anaerobic and musculoskeletal demands [30]. Expectably, a more pronounced increase in levels of BAP coupled with a more pronounced decrease in values of OSI was recorded on longer compared to the shorter races, as well as on demanding compared to easy races, which could indicate a stronger antioxidant response due to a higher intensity of exercise. However, even when considering only horses participating at 40 km races an increase in values of BAP and OSI was evident after the race, which emphasises that horses competing at lower distances are also susceptible to OS, especially characterised by a pronounced activation of antioxidant defence mechanisms. Accordingly, OS could still have negative consequences on performance and health in shorter races.

Taking into account the speed of horses, more pronounced changes in values of OSI have been noted at slower average speed. Since less experienced horses often compete at a lower speed, this result could indicate that those horses are more susceptible to OS than more experienced horses. The same was proved in mice chronically exposed to exercise, that showed less oxidative damage after exhaustive exercise than the untrained animals. This is largely due to the upregulation of endogenous antioxidant enzymes such as mitochondrial SOD, GPx, and γ-glutamylcysteine synthetase [31]. On the other hand, more pronounced adaptation to OS in horses competing at higher speed could be explained by more intense regular training. An adaptive response to training has also been noted in serum amyloid A concentrations in endurance horses after training session resembling a race [32].

The results did not show gender to influence any studied OS markers, however, when separately evaluating female and male horses, only concentrations of MDA was significantly different from pre-race values in mares only. This could be a spurious finding due to a lesser number of male horses. A previous study on Warmblood horses noted increased SOD activities in response to exercise only in mares [33].

When comparing individually the 4 races results with the whole testing population, an inconsistency is present, especially in the second race. However, there were only 2 horses competing at 40 km and 3 horses competing at 80 km at the second race. Nevertheless, we find noteworthy mentioning that the BAP levels in all races, except the second race, have shown consistent results what additionally accentuates the reliability of this biomarker in representing antioxidative status in endurance horses.

There were some limitations of this study that need to be emphasised. Lack of standardised feeding or detailed nutritional surveys to determine the intake level of antioxidants can be considered a limitation. However, most of the horses were fed balanced formulas for endurance horses. In light of standardisation, as this study was not performed in an experimentally controlled setting, the heterogeneity of horses may have introduced bias or confounding of the results. Since all horses have been transported to the event from different locations, the transport per se could have also influenced the results despite that all the horses have been given a period of rest in proportion to the distance and stress caused by the transportation before the start of the competition. However, it was taken that the transportation effect had ceased since all the animals have successfully passed the veterinary examination and were declared fit to compete prior to the blood sampling. Another limitation was that the enrolment into the study was voluntary and it did not include all the participating horses.

Conclusions

The submaximal endurance exercise-induced simultaneous changes in all measured OS markers in this study. Unlike d-ROMs, MDA, and OSI biomarkers, BAP was observed to have low individual variations between and within horses, making it the most reliable biomarker of OS in endurance horses in the present study. Our results indicate that disequilibrium between oxidants and antioxidants also occurred during shorter races and those performed on less demanding racetracks, both in particular characterised by a pronounced antioxidant defence.

Abbreviations

BAP: Biological antioxidant potential; d-ROMs: Reactive oxygen metabolites; GPx: Glutathione peroxidase; GSH: Total glutathione; MDA: Malondialdehyde; OS: Oxidative stress; OSI: Oxidative stress index; ROS: Reactive oxygen species; TBARS: Total barbituric acid reactive substances

Authors' contributions

NBB and ZV conceived and designed the study. NBB, JG, KK and ZV contributed in an equal manner to the acquisition of data. JŠ and DB: contributed in laboratory analysis. KB conceived the statistical analysis, analysed the data and contributed to the interpretation of the data. NBB was a major contributor in writing the manuscript. ZV supervised the manuscript. All of the authors were involved in drafting the article and revising it critically for important intellectual content. All authors read and approved the final manuscript.

Consent for publication

Not applicable.

Competing interests

The authors declare that they have no competing interests.

Author details

[1]Clinic for Surgery, Orthopaedics and Ophthalmology, Faculty of Veterinary Medicine, University of Zagreb, Heinzelova 55, 10 000 Zagreb, Croatia. [2]Clinic for Internal Diseases, Faculty of Veterinary Medicine, University of Zagreb, Heinzelova 55, 10 000 Zagreb, Croatia. [3]Department of Pharmacology and Toxicology, Faculty of Veterinary Medicine, University of Zagreb, Heinzelova 55, 10 000 Zagreb, Croatia. [4]Department of Animal Nutrition and Dietetics, Faculty of Veterinary Medicine, University of Zagreb, Heinzelova 55, 10 000 Zagreb, Croatia. [5]Faculty of Veterinary Medicine, University of Zagreb, Heinzelova 55, 10 000 Zagreb, Croatia. [6]School of Veterinary Science, Massey University, Palmerston North, New Zealand. [7]Department of Radiology,

Ultrasound Diagnostic and Physical Therapy, Faculty of Veterinary Medicine, University of Zagreb, Heinzelova 55, 10 000 Zagreb, Croatia.

References

1. Butler PJ, Woakes AJ, Smale K, Roberts CA, Hillidge CJ, Snow DH, Marlin DJ. Respiratory and cardiovascular adjustments during exercise of increasing intensity and during recovery in thoroughbred racehorses. J Exp Biol. 1993; 179:159–80.
2. Siqueira RF, Weigel RA, Nunes GR, Mori CS, Fernandes WR. Oxidative profiles of endurance horses racing different distances. Arq Bras Med Vet Zootec. 2014;66:455–61.
3. Sjodin B, Hellsten Westing Y, Apple FS. Biochemical mechanisms for oxygen free radical formation during exercise. Sports Med. 1990;10:236–54.
4. Sies H. Oxidative stress: from basic research to clinical application. Am J Med. 1991;91:31S–8S.
5. Williams CA. The effect of oxidative stress during exercise in the horse. J Anim Sci. 2016;94:4067–75.
6. Celi P, Sullivan M, Evans D. The stability of the reactive oxygen metabolites (d-ROMs) and biological antioxidant potential (BAP) tests on stored horse blood. Vet J. 2010;183:217–8.
7. Benzie IFF, Strain JJ. The ferric reducing ability of plasma (FRAP) as a measure of "antioxidant power": the FRAP assay. Anal Biochem. 1996;239:70–6.
8. Grotto D, Santa Maria LD, Boeira S, Valentini J, Char MF, Moro AM, Nascimento PC, Pomblum VJ, Garcia SC. Rapid quantification of malondialdehyde in plasma by high performance liquid chromatography – visible detection. J Pharm Biomed Anal. 2007;43:619–24.
9. Esterbauer H, Schaur RJ, Zollner H. Chemistry and biochemistry of 4-hydroxynonenal, malonaldehyde and related aldehydes. Free Radic Biol Med. 1991;11:81–128.
10. Lee S, Hashimoto J, Suzuki T, Satoh A. The effects of exercise load during development on oxidative stress levels and antioxidant potential in adulthood. Free Radic Res. 2017;51:179–86.
11. Po E, Williams C, Muscatello G, Celi P. Assessment of oxidative stress biomarkers in exhaled breath condensate and blood of thoroughbred foals. Vet J. 2013;196:269–71.
12. Kusano K, Yamazaki M, Kiuchi M, Kaneko K, Koyama K. Reference range of blood biomarkers for oxidative stress in thoroughbred racehorses (2–5 years old). J Equine Vet Sci. 2016;27:125–9.
13. Tsuzuki N, Sasaki N, Kusano K, Endo Y, Torisu S. Oxidative stress markers in thoroughbred horses after castration surgery under inhalation anesthesia. J Equine Vet Sci. 2016;27:77–9.
14. Frankiewicz-Jozko A, Szarska E. Anti-oxidant level to exercise in the blood of endurance horses. Biol Sport. 2000;17:217–27.
15. Hargreaves BJ, Kronfeld DS, Waldron JN, Lopes MA, Gay LS, Saker KE, Cooper WL, Sklan DJ, Harris PA. Antioxidant status and muscle cell leakage during endurance exercise. Equine Vet J. 2002;34:116–21.
16. Marlin DJ, Fenn K, Smith N, Deaton CD, Roberts CA, Harris PA, Dunster C, Kelly FJ. Changes in circulatory antioxidant status in horses during prolonged exercise. J Nutr. 2002;132(Suppl 2):1622–7.
17. Williams CA, Kronfeld DS, Hess TM, Saker KE, Waldron JE, Crandell KM, Harris PA. Comparison of oxidative stress and antioxidant status in endurance horses in three 80-km races. Equine Comp Exerc Physiol. 2005;2:153–7.
18. Williams CA, Kronfeldt DS, Hess TM, Saker KE, Waldron JN, Crandell KM, Hoffman RM, Harris PA. Antioxidant supplementation and subsequent oxidative stress of horses during an 80-km endurance race. J Anim Sci. 2004; 82:588–94.
19. Bergero D, Miraglia N, Schiavone A, Polidori M, Prola L. Effect of dietary polyunsaturated fatty acids and vitamin E on serum oxidative status in horses performing very light exercise. Ital J Anim Sci. 2004;3:141–5.
20. Team RDC. R: a language and environment for statistical computing. Vienna: R Foundation for Statistical Computing; 2015.
21. Bates D, Maechler M, Bolker B, Walker S. Fitting linear mixed-effects models using lme4. J Stat Softw. 2015;67:1–48.
22. Soares JCM, Zanella R, Bondan C, Alves LP, de Lima MR, da Motta AC, Zanella EL. Biochemical and antioxidant changes in plasma, serum, and erythrocytes of horses before and after a jumping competition. J Equine Vet Sci. 2011;31:357–60.
23. McBride JM, Kraemer WJ. Free radicals, exercise, and antioxidants. J Str Cond Res. 1999;13:175–83.
24. Kinnunen S, Atalay M, Hyyppä S, Lehmuskero A, Hänninen O, Oksala N. Effects of prolonged exercise on oxidative stress and antioxidant defence in endurance horse. J Sports Sci Med. 2005;4:415–21.
25. Tsuzuki N, Endo Y, Kikkawa L, Korosue K, Kaneko Y, Kitauchi A, Katamoto H, Hidaka Y, Hagio M, Torisu S. Effects of ozonated autohemotherapy on the antioxidant capacity of thoroughbred horses. J Vet Med Sci. 2015;77:1647–50.
26. Chiaradia E, Avellini L, Rueca F, Spaterna A, Porciello F, Antonioni MT, Gaiti A. Physical exercise, oxidative stress and muscle damage in racehorses. Comp Biochem Physiol B Biochem Mol Biol. 1998;119:833–6.
27. Holbrook TC, McFarlane D, Schott HC. Neuroendocrine and non-neuroendocrine markers of inflammation associated with performance in endurance horses. Equine Vet J. 2010;42:123–8.
28. Tsubone H, Hanafusa M, Endo M, Manabe N, Hiraga A, Ohmura H, Aida H. Effect of treadmill exercise and hydrogen-rich water intake on serum oxidative and anti-oxidative metabolites in serum of thoroughbred horses. J Equine Vet Sci. 2013;24:1–8.
29. Janiak M, Suska M, Dudzińska W, Skotnicka E. Blood glutathione status and activity of glutathione-metabolizing antioxidant enzymes in erythrocytes of young trotters in basic training. J Anim Physiol Anim Nutr. 2010;94:137–45.
30. Williams CA, Burk AO. Antioxidant status in elite three-day event horses during competition. Oxid Med Cell Long. 2012;2012: Article ID 572090, 8 pages. https://doi.org/10.1155/2012/572090.
31. Salminen A, Vihko V. Lipid peroxidation in exercise myopathy. Exp Mol Pathol. 1983;38:380–8.
32. Cywinska A, Witkowski L, Szarska E, Schollenberger A, Winnicka A. Serum amyloid a (SAA) concentration after training sessions in Arabian race and endurance horses. BMC Vet Res. 2013;9:91.
33. Andriichuk A, Tkachenko H, Kurhaluk N. Gender differences of oxidative stress biomarkers and erythrocyte damage in well-trained horses during exercise. J Equine Vet Sci. 2014;34:978–85.

Deletion of the ORF2 gene of the neuropathogenic equine herpesvirus type 1 strain Ab4 reduces virulence while maintaining strong immunogenicity

Christiane L. Schnabel[1], Christine L. Wimer[1], Gillian Perkins[2], Susanna Babasyan[1], Heather Freer[1], Christina Watts[1], Alicia Rollins[1], Nikolaus Osterrieder[3] and Bettina Wagner[1*]

Abstract

Background: Equine herpesvirus type 1 (EHV-1) induces respiratory infection, abortion, and neurologic disease with significant impact. Virulence factors contributing to infection and immune evasion are of particular interest. A potential virulence factor of the neuropathogenic EHV-1 strain Ab4 is ORF2.

This study on 24 Icelandic horses, 2 to 4 years of age, describes the infection with EHV-1 Ab4, or its deletion mutant devoid of ORF2 (Ab4ΔORF2) compared to non-infected controls (each group $n = 8$). The horses' clinical presentation, virus shedding, viremia, antibody and cellular immune responses were monitored over 260 days after experimental infection.

Results: Infection with Ab4ΔORF2 reduced fever and minimized nasal virus shedding after infection compared to the parent virus strain Ab4, while Ab4ΔORF2 established viremia similar to Ab4. Concurrently with virus shedding, intranasal cytokine and interferon α (IFN-α) production increased in the Ab4 group, while horses infected with Ab4ΔORF2 expressed less IFN-α.

The antibody response to EHV-1 was evaluated by a bead-based multiplex assay and was similar in both infected groups, Ab4 and Ab4ΔORF2. EHV-1 specific immunoglobulin (Ig) G1 was induced 8 days after infection (d8 pi) with a peak on d10–12 pi. EHV-1 specific IgG4/7 increased starting on d10 pi, and remained elevated in serum until the end of the study. The intranasal antibody response to EHV-1 was dominated by the same IgG isotypes and remained elevated in both infected groups until d130 pi. In contrast to the distinct antibody response, no induction of EHV-1 specific T-cells was detectable by flow cytometry after ex vivo re-stimulation of peripheral blood mononuclear cells (PBMC) with EHV-1 in any group. The cellular immune response was characterized by increased secretion of IFN-γ and interleukin10 in response to ex vivo re-stimulation of PBMC with EHV-1. This response was present during the time of viremia (d5–10 pi) and was similar in both infected groups, Ab4 and Ab4ΔORF2.

Conclusions: ORF2 is a virulence factor of EHV-1 Ab4 with impact on pyrexia and virus shedding from the nasal mucosa. In contrast, ORF2 does not influence viremia. The immunogenicity of the Ab4ΔORF2 and parent Ab4 viruses are identical.

Keywords: Horse, EHV-1, ORF2, Virulence factor, Immunoglobulin, Mucosal immune response

* Correspondence: bw73@cornell.edu
[1]Department of Population Medicine and Diagnostic Sciences, College of Veterinary Medicine, Cornell University, Ithaca, NY 14853, USA
Full list of author information is available at the end of the article

Background

Equine herpesvirus type 1 (EHV-1), an alphaherpesvirus, frequently induces clinical disease in horses and other equids. Clinical manifestations include upper respiratory tract disease with pyrexia as well as late-term abortions, or neurologic disease (equine herpesvirus myeloencephalopathy, EHM) [1–4]. EHV-1 is transmitted from horse to horse during close contact via respiratory secretions. The virus initially infects the epithelium of the upper respiratory tract. Subsequently, it spreads through the underlying connective tissue to the regional lymphoid tissues, and next establishes cell-associated viremia in leukocytes [5–8]. Viremia is a prerequisite for the infection of endothelial cells by EHV-1 and induction of vasculitis in the pregnant uterus or central nervous system. These events can lead to the severe manifestations of EHV-1, abortions or EHM, respectively [1, 3, 5, 9–11]. Furthermore, EHV-1 establishes latency. Latent EHV-1 is believed to reside in the trigeminal ganglia and in lymphoid tissues of the respiratory tract [12–16].

Latency and immune evasion are two important characteristics of EHV-1 preventing EHV-1 elimination by the immune system of the horse. Latently infected horses serve as a reservoir of the virus. EHV-1 can be reactivated with or without signs of clinical disease in the reservoir horse [17]. Lack of vaccination and incomplete protection after vaccination provide sufficient susceptible hosts within the population for EHV-1 to spread periodically after reactivation from latency or from newly infected horses [4, 17]. The prevalence of EHV-1 is high with detection in 54–88% of horses post mortem [14, 18, 19]. Outbreaks of EHV-1 associated clinical disease occur in variable frequencies but the severity of EHV-1 manifestations seems to be increasing [4, 14, 20]. Consequently, EHV-1 has a significant impact on animal health and the equine economy worldwide and EHM has been classified as an emerging disease in the USA [20, 21]. Management practices and biosecurity measures are applied to contain EHM outbreaks. Many horses are regularly vaccinated in order to prevent infection and disease. However, the use of inactivated or attenuated live virus vaccines has not been sufficient to completely protect from EHV-1 infection and outbreaks [22, 23]. Protective immune responses after vaccination with available vaccines and after natural infection in horses are transient and re-infection is possible between 4 and 8 weeks post vaccination or 6–18 months post infection of adult horses [22, 24–27].

Recent research has focused on the identification of EHV-1 virulence factors such as the mutation of the DNA polymerase gene (ORF30) at nucleotide position 2554, which has been associated with the manifestation of EHM and the establishment of latency after EHV-1 infection. Strains of the genotype G2554 are thus classified as neuropathogenic [2, 28–31]. However, this classification rather points to an increased risk of the G2554 strains causing EHM. During neurological outbreaks between 3 and 45% of the horses developed EHM [32–34] while non-neuropathogenic strains can also cause neurologic disease [2]. The identification and better characterization of virulence factors underlying the pathogenicity, immune modulation, and establishment of latency by EHV-1 can enable targeted approaches to develop a vaccine providing advanced protection from the virus.

In vitro infection of equine peripheral blood mononuclear cells (PBMC) with the non-neuropathogenic EHV-1 strain RacL11 reduced the down-regulation of MHC class I expression [35], and increased interleukin10 (IL-10) secretion [36] and chemokine expression [37] compared to infection with the neuropathogenic strain Ab4. These findings point to differences of EHV-1 strains in modulating the host immune response, which may also be relevant in vivo. Potential virulence factors of the Ab4 strain are the ORF1 and ORF2 gene products. ORF1 and ORF2 genes are absent in RacL11, the RacL11-derived modified-live vaccine strain RacH [38], as well as in the apathogenic strain KyA [39, 40]. Hence, the ORF1 and ORF2 genes could encode for immunomodulatory proteins leading to differences in virulence or pathogenicity. The ORF1 gene product pUL56 of Ab4 is responsible for MHC class I down-regulation and seems to be involved in the suppression of cytokine expression and chemokine secretion in vitro [35, 41]. The ORF2 gene is related to ORF1, but is less well characterized. The ORF2 gene product was classified as a membrane protein due to its predicted hydrophobic profile [40]. The deletion of ORF1 and ORF2 genes from EHV-1 Ab4 significantly reduced pyrexia duration and viral shedding compared to the parent Ab4 virus in experimentally infected ponies [42]. This effect could be attributed to either gene or the combination of ORF1 and ORF2.

Here, we performed an experimental infection in horses using a single ORF2 gene deletion mutant of the EHV-1 strain Ab4, Ab4ΔORF2 [35], in comparison to the parent Ab4 virus. The goals of the approach were to identify ORF2 as a potential virulence factor, to characterize the effects of ORF2 on EHV-1 infection and immune responses in horses with known EHV-1 infection and immunity background.

Methods

Horses

Twenty-four horses from the Cornell University herd of Icelandic horses with known EHV-1 status were enrolled in this study [43, 44]. All horses were offspring of one stallion, were born at Cornell University, and kept at an isolated facility as a group from birth. They had no contact with other horses in the USA prior to and for the duration of this study. All horses were previously experimentally infected with the EHV-1 strain NY03 after weaning at 7 months of age [44, and unpublished data]. By the time of the present study, the horses were 2–4 years of age and their EHV-1 specific immunity had waned to values typically observed in EHV-1 susceptible horses. EHV-1 specific protective immunity is short-lasting and believed to be maintained for 6–9 months after a single experimental infection [27]. The horses were vaccinated annually against rabies, tetanus, West Nile virus, Eastern and Western Encephalitis virus. They were also dewormed twice a year. Prior to the EHV-1 infection described here, the horses were kept on pasture and were clinically healthy. Grass hay was fed to the horses *ad libitum*. All horses survived and were kept for research purposes at Cornell University after the study.

Virus preparation

Viruses for experimental infection were EHV-1 Ab4, a neuropathogenic strain isolated from a quadriplegic mare [9, 28, 45] and its ORF2 gene deletion mutant, Ab4Δ ORF2, produced by two-step Red-mediated recombination of a recombinant BAC Ab4 as previously described in detail [35]. The viruses were propagated in rabbit kidney cells (RK13) in MEM medium supplemented with 0.292 g/l L-glutamine, 1 mM sodium pyruvate, 100 U/ml Penicillin, 100 μg/ml Streptomycin, 50 μg/ml gentamycin, (all Thermo Fisher Scientific Waltham, MA, USA), and 10% FCS (Atlanta biological, Flowery Branch, GA, USA). Virus titers were determined in RK13 cells as previously described [22, 37].

The identity of each virus preparation was confirmed by conventional PCR on 2 μg of extracted viral DNA, using primers flanking the ORF1 and ORF2 gene region of the EHV-1 strain Ab4: forward primer – 5'-AACA ACCCTGGGCTCTTTA -3′ and reverse primer – 5'-G ATTCGCACCTCATCTCCAC -3′. The resulting PCR products had a distinct size of 2043 bp for Ab4 and of 1446 bp for Ab4ΔORF2. The deletion of ORF2 in Ab4Δ ORF2 was furthermore confirmed by sequencing of the PCR product at the Biotechnology Core Facility of Cornell University. After infection, the titer of each virus preparation was confirmed by titration on a plaque assay as described below.

Experimental EHV-1 infection

For experimental EHV-1 infection horses were assigned to three groups of eight horses each with four mares and four geldings per group. One group of horses served as non-infected controls (control group) and had a median age of 2 years (range 2–3 years). Horses in a second group were infected with the neuropathogenic EHV-1 strain Ab4 (Ab4 group) and had a median age of 2.5 years (range 2–3 years). The third group was infected with the Ab4ΔORF2 deletion mutant virus (Ab4ΔORF2 group) and had a median age of 2.5 years (range 2–4 years).

One day prior to EHV-1 infection the horses were moved into the isolation barns with individual box stalls and allowed to acclimate. The stalls did not allow the horses to have direct nose-to-nose contact. The non-infected control group was housed in a separate barn and always handled prior to the other groups. The two infected groups were housed in a barn with two separate sections, one for each group, divided by a barrier. Airflow between sections was not possible. Each section had a separate entry with an area for donning personal protective equipment including disposable coveralls, boots, hats, and gloves. Biosecurity precautions were taken and people changed personal protective equipment upon entering and exiting each of the sections to prevent the spread of viruses between the three groups. Within each barn section, the barn had a center hallway with the same airspace and horses were handled in each section as one group. No other specific care was taken to prevent spread of virus from horse-to-horse within each group by animal handlers except for changing gloves after taking nasal swab samples from each horse.

Baseline serum and nasopharyngeal secretion (NS) samples were taken the day before EHV-1 infection (d-1). Baseline physical examination measurements were taken on d-1 and again immediately before EHV-1 infection on d0. Then, Ab4 group horses were infected intranasally with 1×10^7 plaque-forming units (PFU) of EHV-1 strain Ab4 in 4 ml of saline with a mucosal atomizer device (Wolfe Tory Medical, Salt Lake City, UT). Horses in the Ab4ΔORF2 group received $1x10^7$ PFU of the Ab4ΔORF2 virus in 4 ml saline intranasally. Horses in the control group received 4 ml saline without any virus using the mucosal atomizer device.

Clinical examination

Rectal temperatures were recorded on d-1, twice daily from d0 until d10 post infection (pi) and then daily on days 11 to 14 pi. Fever was defined as body temperatures exceeding 101.5 °F (38.6 °C). Clinical scores were recorded daily between d-1 and d14 pi by veterinarians according to the grade of depression (0–3), reduced appetite (0–2), nasal discharge (0–6), eye discharge (0–3), mandibular lymph node swelling (0–3) and ataxia (0–5). To determine ataxia, neurologic examinations were performed as previously described [46]. The gait was evaluated by walking the horses individually in the aisle, including turns to both sides. A summarized clinical score for each horse and day was calculated as the sum of the scores (possible range 0–22). Due to the housing arrangement examiners were aware of the identity of the control group, but blinded to the identification of the Ab4 or Ab4ΔORF2 groups. The control group was always worked with first.

Sample collection and processing

The horses were assigned randomized ID numbers, which did not reveal the group affiliation to the laboratory personnel. Samples were processed based on ID numbers and the group affiliation remained blinded until data collection and assay evaluation was completed.

NS were acquired on d-1, and d1-d9 pi using two polyester tipped applicators (Puritan Medical Products Company, Gullford, Maine, USA) swabbed on the nasal mucosa of one nostril for 5 s. Swabs were immediately put into 1 ml of sterile phosphate buffered saline (PBS) in a polystyrene tube. The swabs (in PBS) were transported to the laboratory, maintained at 4 °C and processed within 2 hours after sampling. For virus isolation, NS samples were not further processed. For the analysis of intranasal cytokines and antibodies, NS samples were spun at 1000×g at 4 °C for 5 min leaving the swabs in the collection tube. The diluted NS surrounding the swabs was collected into a 1.5 ml centrifuge tube and the cleared fluid supernatant was stored at − 20 °C until analysis.

Blood samples were collected on d-8, d-1, d1 to d14, d24, d60, d100, d130, d170, d220, and d260 pi by jugular venipuncture with a vacutainer system (20G Monoject Covidien, Medtronic Minneapolis, MN, USA) into glass tubes without anticoagulant or with sodium heparin (both BD, Franklin Lakes, NJ, USA). Serum and heparinized blood samples were processed as described previously [43, 44].

Within 3 h after blood collection PBMC were isolated by density gradient centrifugation (Ficoll-Paque Plus, GE Healthcare, Piscataway, NJ, USA) and re-suspended in cell culture media (DMEM supplemented with 1% (*v*/*v*) non-essential amino acids, 2 mM L-glutamine, 50 μM 2-mercaptoethanol, 50 μg/ml Gentamicin, 100 U/ml penicillin, 100 μg/ml streptomycin, all from all Thermo Fisher Scientific Waltham, MA, USA and 10% FCS, Atlanta biological, Flowery Branch, GA, USA) at 6×10^6 cells/ml for the following experiments. For PCR detection of viremia, 1×10^7 PBMC were washed twice in PBS, dry cell pellets were snap frozen in liquid nitrogen, and stored at − 80 °C until PCR was performed.

Serum samples were stored in aliquots at − 20 °C until analysis.

Virus isolation

To determine nasal virus shedding virus isolation was performed as plaque assays as previously described [22]. Briefly, vortexed NS were titrated on confluent RK13 cells. Virus penetration into the cells was allowed during an incubation for 2-4 h at 37 °C before supernatants were aspirated and an overlay with 0.5% Methylcellulose (Sigma Aldrich, St. Louis, MO, USA) in cell culture medium (Gibco MEM medium, supplemented with 0.292 g/l L-glutamine, 1 mM sodium pyruvate, 50 μg/ml Gentamicin, 300 U/ml Penicillin, 300 μg/ml Streptomycin, 0.75 μg/ml Amphotericin B, all from Thermo Fisher Scientific and 10% inactivated FCS, Atlanta biological) was performed. After 5 days of incubation, plaques were counted after fixation and staining in crystal violet staining solution (PBS containing 4% *v*/v Paraformaldehyde, 1% *v*/*w* Methanol, 0.05% *w*/*v* Crystal Violet, all Sigma Aldrich). Viral titers were expressed as PFU per ml NS.

EHV-1 PCR

Cell associated viremia was evaluated by real-time PCR targeting the gB gene of EHV-1 performed exactly as previously described [44, 47]. A total of 1×10^7 PBMC was used for DNA extraction per horse and time point. The PCR was performed at the Animal Health Diagnostic Center at Cornell University.

EHV-1 multiplex assay

EHV-1 specific antibodies in serum and NS samples were determined using a fluorescent bead-based EHV-1 multiplex assay. The assay was previously described in detail [43] and has been slightly modified for this approach: Instead of coupling viral glycoproteins directly to the beads a monoclonal anti-equine IL-4 antibody (clone 25, [48]) was coupled to all three beads numbered 33, 35 and 36 (Luminex Corp., Austin, TX, USA). Bead 33 was then incubated with IL-4/EHV-1 glycoprotein B (gB), bead 35 with IL-4/EHV-1 gC, and bead 36 with IL-4/EHV-1 gD. All three IL-4/EHV-1 glycoproteins were expressed as described previously [43]. The EHV-1 gB, gC and gD coated beads were then multiplexed, and incubated with diluted serum (1:400) or undiluted NS samples, followed by the respective detection antibodies as described [43]. Total EHV-1 specific immunoglobulin (Ig) was detected by biotinylated goat anti-horse IgG (H + L) antibody (Jackson Immunoresearch Laboratories, WestGrove, PA, USA). Ig isotype detection was performed by biotinylated monoclonal antibodies specific for IgG1 (CVS45) and IgG4/7 (CVS39) [both [49]], IgG1/3 (clone 159–4), IgG3/5 (clone 586), IgG6 (clone 267) [all three [50]], and IgA (BVS2) [51]. Data were reported as median fluorescent intensities (MFI).

Ex vivo re-stimulation of PBMC with EHV-1

EHV-1 re-stimulation of PBMC was performed as previously described in detail [36, 44, 52]. In brief, PBMC were re-stimulated ex vivo with EHV-1 strain Ab4 at a multiplicity of infection of 1 for 48 h. As controls culture in medium alone or stimulation with Phorbol12-myristate13-acetate (PMA; 25 ng/ml) and ionomycin (1 μM; both Sigma Aldrich) were included. Supernatants were collected for the analysis of secreted cytokines. In assays for intracellular staining of cytokines and flow cytometric analysis, protein secretion was blocked by Brefeldin

A (10 µg/ml; Sigma Aldrich) added for the last 24 h of incubation, before the cells were fixed.

Multiplex assay detection of cytokine secretion

Equine cytokines were quantified in PBMC supernatants, serum and NS samples using a fluorescent bead-based multiplex cytokine assay as previously described [53]. Interferon (IFN)-α, IL-4, and IL-10 were reported in pg/ml and IFN-γ, and IL-17 were reported as U/ml. Furthermore, soluble cluster of differentiation 14 (sCD14), an inflammatory marker, was determined in serum and NS samples by a separate bead-based assay as previously described in detail [54] and was reported in ng/ml. All samples were measured undiluted except for the sCD14 analysis in serum. The latter samples were diluted at 1:200.

EHV-1-specific T-cell detection by flow cytometry

Flow cytometric analysis of EHV-1 specific cells was performed as previously described [43, 44, 52]. In brief, after EHV-1 re-stimulation PBMC were fixed, permeabilized, and triple stained for CD4, and CD8 on the cell surface and intracellular IFN-γ production or intracellular IL-10, IL-17A, and IL-4, respectively. For each sample 100,000 events were analyzed in a FACS Canto II flow cytometer (BD Biosciences, San Diego, CA). The evaluation of the data was performed using FlowJo version 10.2 (FlowJO LLC, Ashland, OR, USA). A small lymphocyte gate was set by morphology and the proportions of IFN-γ positive lymphocytes after re-stimulation, compared to medium controls were assessed as previously described [52]. The average intracellular cytokine detection in PMA/ionomycin stimulated PBMC (positive controls) was 14.7% IFN-γ positive, 2.2% IL-4 positive, 2.2% IL-10 positive, and 0.1% IL-17 positive lymphocytes.

Statistical analysis

The data sets of this study were not normally distributed according to D'Agostino and Pearson normality tests. All data sets were analyzed by repeated measures ANOVA with Tukey's Post-hoc test to identify group differences for each time point analyzed. P-values < 0.05 were considered significant. Correlation of peak virus shedding and peak nasal cytokine concentrations were analyzed by two-tailed Spearman rank correlation. The statistical analysis was performed using GraphPad Prism software version 6 or higher (GraphPad Software Inc., La Jolla, CA, USA).

Results

Clinical signs after EHV-1 infection

Horses in both infected groups, Ab4 and Ab4ΔORF2, displayed clinical signs of mild upper respiratory tract disease as well as fever, while horses in the control group did not (Fig. 1a and b). Horses in the Ab4 group developed an initial fever peak from d1.5 to d2.5 pi, which was significantly higher than the control group ($p <$ 0.0001). The initial fever peak was less pronounced in the Ab4ΔORF2 group. Although Ab4ΔORF2 horses had increased body temperatures compared to the control group on d1.5 ($p <$ 0.001), d2 and d2.5 pi (both $p <$ 0.0001), the fever in the Ab4ΔORF2 group was lower than in the Ab4 group on d2 ($p <$ 0.001) and d2.5 pi (p < 0.01). Afterwards, body temperatures in both infected groups were significantly increased compared to the control group at several time points until d5 pi (Fig. 1a).

Horses in the control group had low clinical scores between 0 and 5 (median 3) resulting from mild nasal or ocular discharge while stabled. In both infected groups, nasal discharge and enlarged submandibular lymph nodes were the most prevalent clinical signs of disease. Clinical scores in the Ab4 and Ab4ΔORF2 groups were similar over the course of the study and higher than in the controls on days 2 to 7 pi (Fig. 1b). Mild signs of ataxia (score 1) were seen in one horse each in the infected groups. These were noted in the Ab4 group on days 9 and 10 pi and in the Ab4ΔORF2 group on days 5, 6 and 9 pi, respectively. These symptoms resolved completely in both horses by d10 pi.

Viral shedding and viremia after EHV-1 infection

Virus was isolated from the NS samples of horses in the infected groups between d1to d6 pi and peaked on d2 pi (Fig. 1c). Viral shedding on d2 pi was reduced in Ab4ΔORF2 infected horses compared to those in the Ab4 group ($p < 0.0001$). Although viral shedding was detected in horses of the Ab4ΔORF2 group, their NS virus amounts were low and not different from the control group throughout the study. No virus was detected in NS of the control group.

Cell-associated viremia was detected by PCR. Viral DNA was detectable in PBMC of both infected groups and significantly elevated compared to the control group between days 4 to 8 pi ($p < 0.01$, Fig. 1d). On d8 pi, higher amounts of viral DNA were detected in PBMC of the Ab4ΔORF2 group than in those of the Ab4 group (p < 0.01) and on d10 pi viral DNA was still elevated in this group compared to the control group ($p < 0.01$). Viremia was not detected in the control group.

Intranasal cytokine production in response to EHV-1 infection

Within the first days after intranasal infection with Ab4 or Ab4ΔORF2 cytokines were detected in the NS. As the first cytokine, IFN-α was detected and peaked in both infected groups at d2 pi, (Fig. 2a) which corresponds with the time of maximum nasal virus shedding (Fig. 1c). The intranasal IFN-α concentration of the Ab4 group exceeded that of the Ab4ΔORF2 group on d2 pi ($p < 0.001$) and it remained significantly higher than in the control group on d3 pi (p

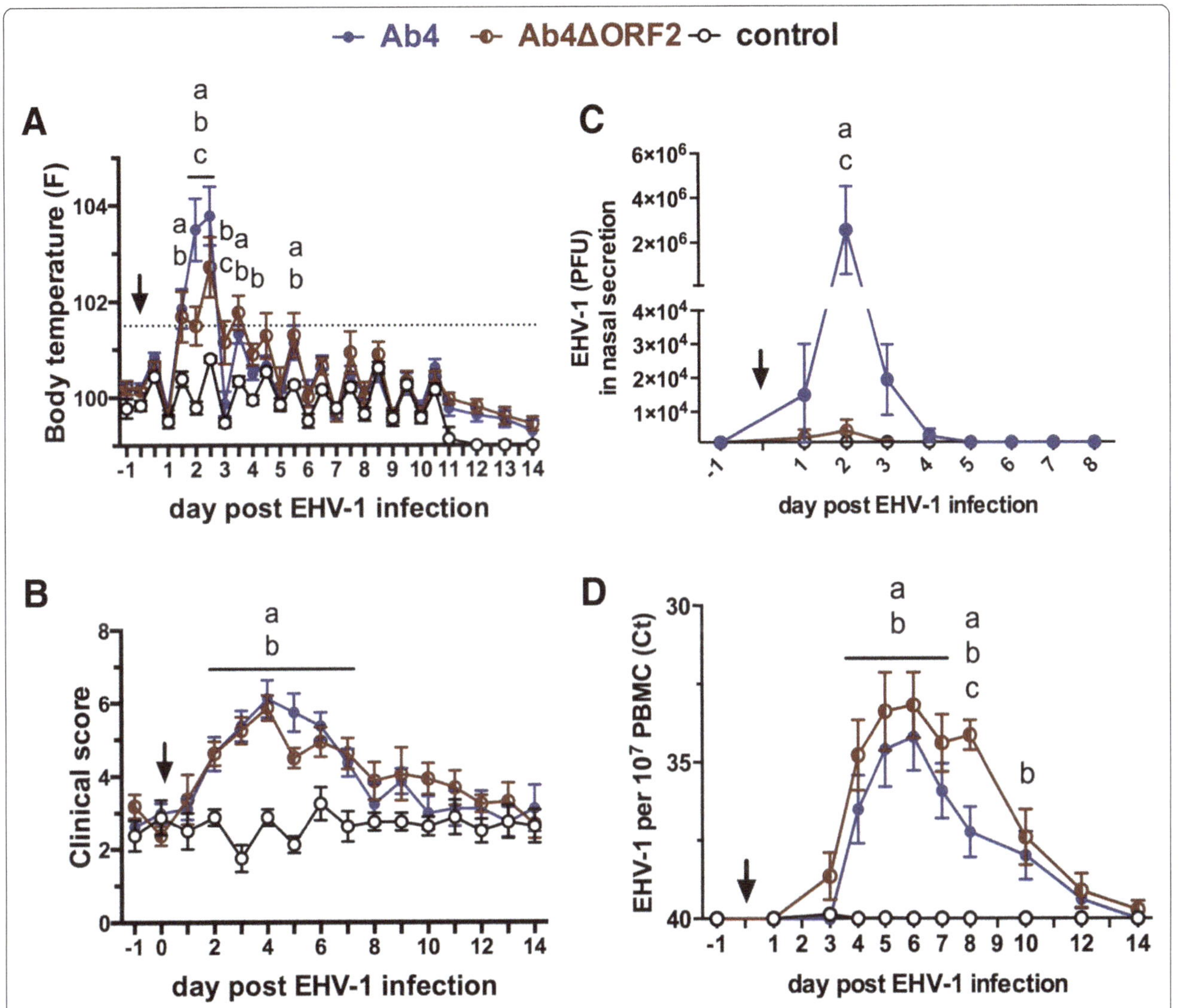

Fig. 1 Clinical findings and virus detection after experimental infection with the EHV-1 strain Ab4 or its deletion mutant Ab4ΔORF2. Three groups of horses ($n = 8$/group) were either not infected (control), infected with the neuropathogenic EHV-1 strain Ab4 (Ab4) or infected with its ORF2 deletion mutant (Ab4ΔORF2) on day 0 (arrow). Clinical signs were monitored and virus detection was performed before (day − 1) and after infection (day 1 to 14). Body temperature (**a**); clinical score calculated as the sum of numerical scores for nasal discharge, ocular discharge, lymph node enlargement, ataxia, depression and reduced appetite ranging from 0 to 22 (**b**); viral shedding expressed as plaque forming units (PFU/ml) detected by virus isolation in nasopharyngeal secretions (**c**); viremia determined in 1×10^7 PBMC by real time PCR and shown as Ct values for the gB gene (**d**) were analyzed; Mean and standard errors are plotted for each group. Dotted horizontal lines represent a cutoff of fever at 101.5 °F (**a**) and EHV-1 detection at a Ct 38 (D). Significant differences between groups are marked: **a** = Ab4 vs. control, **b** = Ab4ΔORF2 vs. control, **c** = Ab4 vs. Ab4ΔORF2

< 0.001). IFN-α concentrations declined to undetectable values by d4 pi. Intranasal IFN-α was not detected in the control group.

The inflammatory protein sCD14 increased in NS of both infected groups starting on d2 pi to peak on d3 pi, and remained elevated compared to the control group until d5 pi (Fig. 2b). In Ab4ΔORF2 infected horses sCD14 was increased on d2 pi in comparison to the control and Ab4 groups (both $p < 0.001$), while the peak sCD14 concentration on d3 pi was higher in Ab4 infected than in Ab4ΔORF2 infected horses ($p < 0.05$).

Nasal IFN-γ increased as a sharp peak on d3 pi in the Ab4 and Ab4ΔORF2 groups while it remained undetectable in the control group. Differences in IFN-γ values between the Ab4 and Ab4ΔORF2 infected groups were not observed (Fig. 2c). At d4 pi IL-10 concentrations peaked in NS of both infected groups compared to the control group ($p > 0.0001$) and declined gradually to baseline values by d6 pi in the Ab4 or d7 pi in the Ab4ΔORF2 group. Differences in IL-10 values between the Ab4 and Ab4ΔORF2 infected groups were not observed (Fig. 2d). Intranasal IL-4 and IL-17 were overall

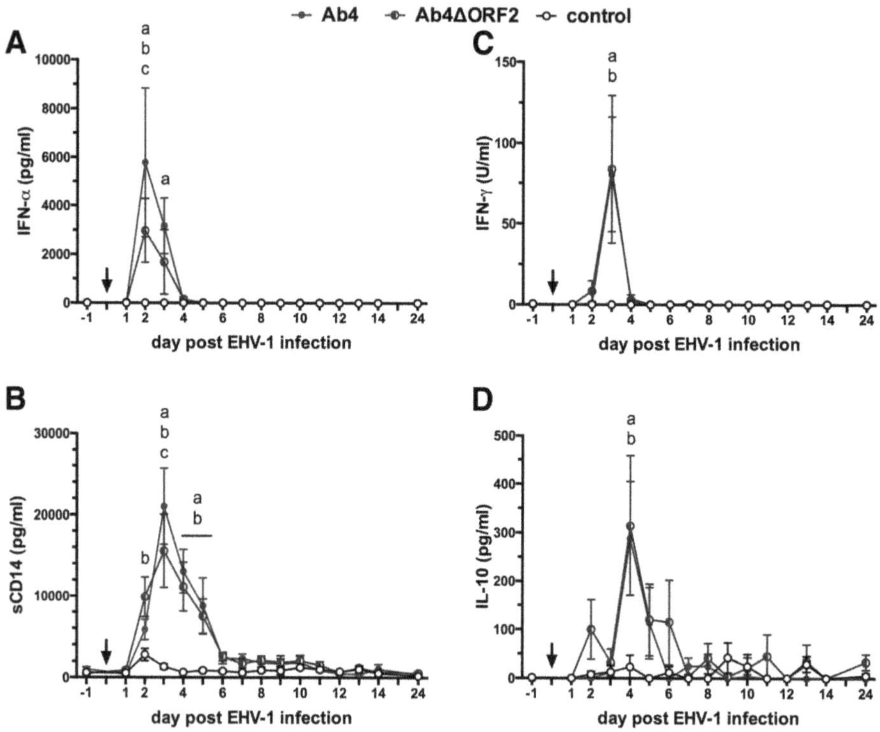

Fig. 2 Intranasal cytokine secretion after infection with EHV-1 strain Ab4 or its deletion mutant Ab4ΔORF2. Three groups of horses ($n = 8$/group) were either not infected (control), infected with the neuropathogenic EHV-1 strain Ab4 (Ab4) or infected with its ORF2 deletion mutant (Ab4ΔORF2) on day 0 (arrow). Before (day − 1) and after (day 1–24) experimental infection IFN-α (**a**), sCD14 (**b**), IFN-γ (**c**), and IL-10 (**d**) were quantified in nasopharyngeal secretions by fluorescent bead-based assays. Mean and standard errors are plotted for each group. Significant differences between groups are marked: **a** = Ab4 vs. control, **b** = Ab4ΔORF2 vs. control, **c** = Ab4 vs. Ab4ΔORF2

low or not detectable and similar between all three groups (data not shown).

The maximum values of virus shedding on d2 pi positively correlated with the peak values of nasal IFN- α and sCD14 (Table 1).

Antibody response to infection

Antibodies specific for EHV-1 gB, gC, and gD were determined before and following infection in NS and serum. The antibody responses to the three EHV-1 antigens were slightly different in magnitude (gB = gD < < gC) but followed a similar overall pattern. Thus, representative results of EHV-1 gC-specific antibodies are presented here. Total EHV-1 gC-specific Ig in NS and serum were low

Table 1 Spearman rank correlations of peak nasal cytokines and sCD14 with maximum virus shedding on day 2 pi

day pi	IFN-α d2 pi	sCD14 d3 pi	IFN-γ d3 pi	IL-10 d4 pi
r	0.8066	0.7866	0.6442	0.4860
95% CI[a]	0.5894 to 0.9150	0.5526 to 0.9056	0.3140 to 0.8354	0.09026 to 0.7492
P value	< 0.0001	< 0.0001	0.0007	0.0160

[a]*CI* confidence interval

before infection (d-8, d-1) in all horses and EHV-1 specific antibody induction was not detected in the control group throughout the study (Fig. 3).

Intranasal antibody response to infection

Infection with either virus, Ab4 or Ab4ΔORF2, induced EHV-1 gC-specific intranasal antibodies following a wave-like pattern, characterized by several peaks on days 3–4, 10, 24–60, and 130 pi (Fig. 3a). The highest magnitude of the anti-gC antibody response was observed at d24 pi (Ab4ΔORF2 group) or d60 pi (Ab4 group). The increase of total anti-gC antibodies was significantly higher in both infected groups compared to the control group from days 8 to 130 pi, and in the Ab4 group also on days 4 and 7 pi. After d170 pi, nasal EHV-1 specific antibodies in both infected groups returned to low values until the end of the study period on d260 pi, which were not significantly different from the control group. Intranasal gC-specific antibodies were higher in the Ab4ΔORF2 group than in the Ab4 group on d13 pi ($p < 0.01$) and d24 pi ($p < 0.05$), while the Ab4 group exceeded the Ab4ΔORF2 group on d60 pi ($p < 0.05$) (Fig. 3a).

The initial intranasal antibody increase on d3-4pi was composed of low amounts of different gC-specific IgG

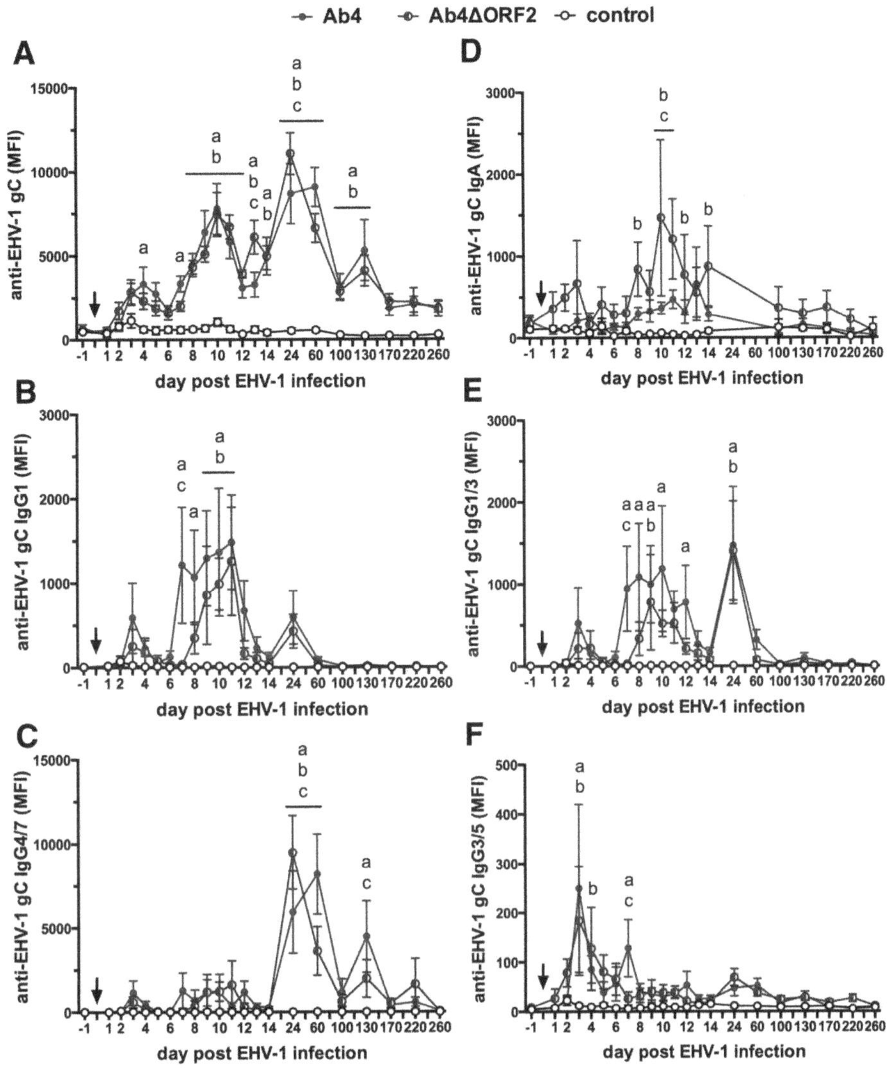

Fig. 3 Anti-EHV-1 gC antibodies in nasal secretions after infection with EHV-1 strain Ab4 or its deletion mutant Ab4ΔORF2. Three groups of horses ($n = 8$/group) were either not infected (control), infected with the neuropathogenic EHV-1 strain Ab4 (Ab4) or infected with its ORF2 deletion mutant (Ab4ΔORF2) on day 0 (arrow). Before (day − 1) and after (day 1–260) experimental infection EHV-1 gC specific total Ig (**a**), IgG1 (**b**), IgG4/7 (**c**), IgA (**d**), IgG1/3 (**e**), and IgG3/5 (**f**) were quantified in nasopharyngeal secretions by fluorescent bead-based multiplex assays and expressed as median fluorescent intensities (MFI). Note that Y-axes scales differ depending on isotype. Mean and standard errors are plotted for each group. Significant differences between groups are marked: **a** = Ab4 vs. control, **b** = Ab4ΔORF2 vs. control, **c** = Ab4 vs. Ab4ΔORF2

isotypes, and in the Ab4ΔORF2 group also some IgA (Fig. 3b-f). Afterwards, the intranasal anti-gC total antibody peaks (Fig. 3a) were each dominated by different isotypes. The second anti-gC peak on day 10 pi corresponded to an increase of primarily IgG1 in both infected groups (Fig. 3b). IgG1 was induced slightly earlier (d7, $p < 0.001$) in the Ab4 than in the Ab4 ΔORF2 group (d8). Similarly, the third anti-gC antibody peak on days 24 (Ab4ΔORF2) and 60 pi (Ab4), respectively, and the fourth peak on d130 pi were mainly composed of IgG4/7 (Fig. 3c). Anti-gC IgG4/7 antibodies almost mimicked the total anti-gC response shown in Fig. 3a from d24 pi on. This included the higher IgG4/7 in the AbΔORF2 compared to the Ab4 group on d24 pi ($p < 0.01$), the lower intranasal IgG4/7 in the AbΔORF2 group on d60 pi ($p < 0.0001$), and on d130 pi ($p < 0.05$) compared to the Ab4 group.

Notably, intranasal gC-specific IgA after infecting horses with the Ab4 virus was not different from the control horses at any time after infection. In contrast, gC-specific IgA was increased in the Ab4ΔORF2 group compared to the control group on days 8, 10 to 12 and 14 pi. It was also higher than in the Ab4 group on d10 and d11 pi (Fig. 3d). Overall the intranasal EHV-1 specific IgA response in the Ab4ΔORF2 group appeared to be of low magnitude and short duration.

EHV-1 gC-specific IgG1/3 induction mimicked IgG1 antibodies until d14 pi (Fig. 3e). IgG3/5 antibodies were overall low including the early peak on d3pi (Fig. 3f). This supports that the IgG1/3 antibody response until d14 pi was mainly composed of IgG1. On d24 pi, an increase in IgG1/3 antibodies was observed in both infected groups ($p < 0.0001$). This increase was higher than the corresponding IgG1 response in Fig. 3b and likely results from a mixture of anti-gC IgG1 and IgG3 on d24 pi.

EHV-1 specific IgG6 in nasal secretion was low for all groups throughout the study and no group differences were observed (data not shown).

Serum antibody response to EHV-1 infection

Infection with either Ab4 or Ab4ΔORF2 resulted in the induction of EHV-1 gC-specific Ig in serum which was first increased on d8 pi compared to the control group ($p < 0.01$). Anti-gC antibodies peaked on d24 pi, slowly declined until d170 pi, and then remained constant in both infected groups for the remainder of the study (Fig. 4a) with p-values< 0.0001 between days 10 to

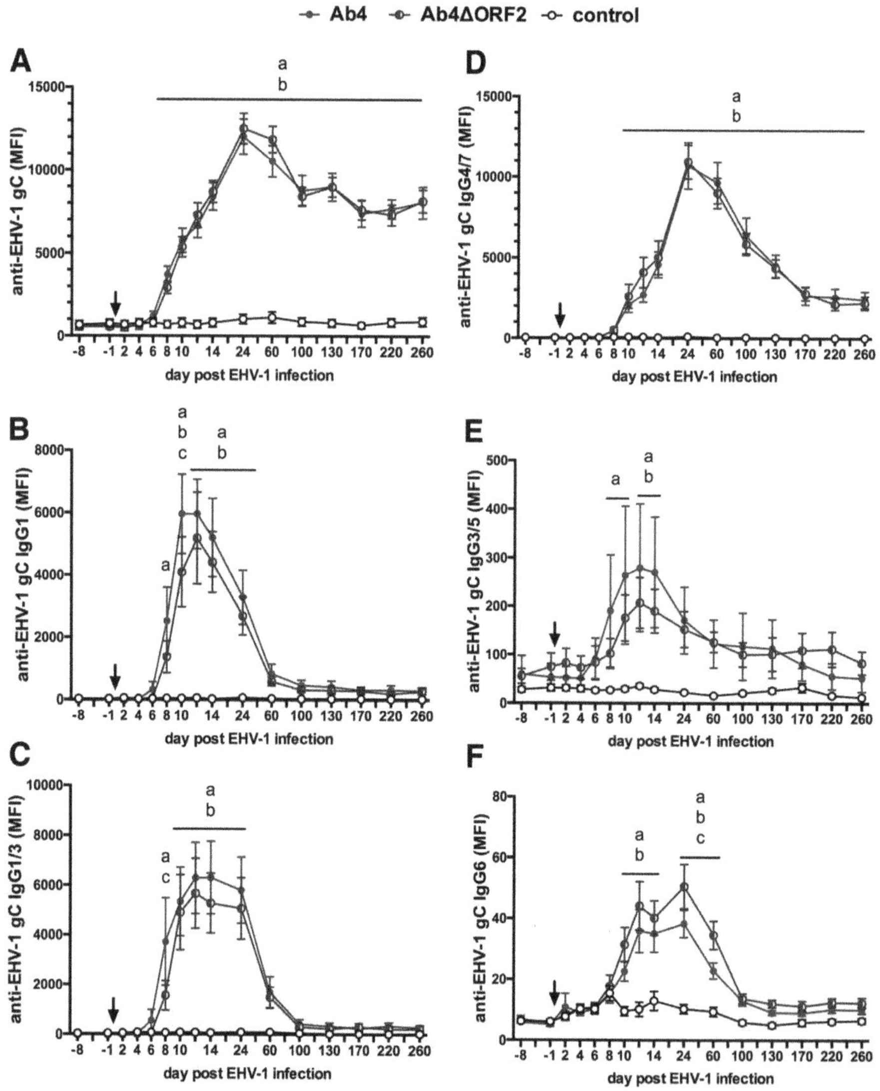

Fig. 4 Anti-EHV-1 gC antibodies in serum after EHV-1 infection. Three groups of horses ($n = 8$/group) were either not infected (control), infected with the neuropathogenic EHV-1 strain Ab4 (Ab4) or infected with its ORF2 deletion mutant (Ab4ΔORF2) on day 0 (arrow). Before (day − 1) and after (day 1–260). EHV-1 gC specific total Ig (**a**), IgG1 (**b**), IgG1/3 (**c**), IgG4/7 (**d**), IgG3/5 (**e**) and IgG6 (**f**) were quantified in serum by fluorescent bead-based multiplex assays and expressed as median fluorescent intensities (MFI). Note that Y-axes scales differ depending on isotype. Mean and standard errors are plotted for each group. Significant differences between groups are marked: **a** = Ab4 vs. control, **b** = Ab4ΔORF2 vs. control, **c** = Ab4 vs. Ab4ΔORF2

260 pi for both infected groups in comparison to the control group. Total gC-specific Ig was similar between the Ab4 or Ab4ΔORF2 groups for the duration of the study and until d260 pi.

In both infected groups, Ab4 and Ab4ΔORF2, IgG1 (Fig. 4b) and IgG1/3 isotypes (Fig. 4c) were detectable early after infection. They peaked on d10 or d12 pi and declined by d60 pi to baseline values. In the Ab4 group, IgG1 and IgG1/3 induction was slightly earlier than in the Ab4ΔORF2 group (d8 vs. d10) and reached higher IgG1 values on d10 pi ($p < 0.05$).

Anti-gC IgG4/7 antibodies were first detectable on d10 pi in both infected groups and peaked on d24 pi. IgG4/7 primarily contributed to the long-term EHV-1 specific antibody response with both infected groups exceeding the antibody values of the control group from d10 until d260 pi (p-values between $p < 0.0001$ and $p < 0.05$) (Fig. 2d). Differences in the IgG4/7 response of the Ab4 and Ab4ΔORF2 groups were not observed.

EHV-1 specific IgG3/5 was weakly induced and elevated in comparison to the control group in the Ab4 group on days 8 to 14 pi ($p < 0.05$ to $p < 0.001$), and in the Ab4ΔORF2 group on d12 and d14 pi (both $p < 0.05$) (Fig. 4e). IgG6 was present in low quantities, which were increased compared to the control group in the Ab4 and Ab4ΔORF2 group on days 10–60 pi ($p < 0.05$). On days 24 and 60 pi the IgG6 antibodies in the Ab4ΔORF2 group exceeded those in the Ab4 group (both $p < 0.05$) (Fig. 4f). EHV-1-specific IgA was not detected in serum at any time point (data not shown).

Cellular immune response to EHV-1

Cellular immune responses of all groups were analyzed by ex vivo re-stimulation of PBMC with the EHV-1 strain Ab4. Between 5 and 10 days pi, IFN-γ secretion was elevated in horses of both the Ab4 group (days 5, 6, 8 pi, $p < 0.05$) and the Ab4ΔORF2group (days 6, 10 pi, $p < 0.05$) (Fig. 5a) compared to the control group. The IL-10 secretion displayed a similar biphasic response with the IL-10 peak on d5 pi slightly preceding the maximal IFN-γ response on d6pi. Both infected groups had greater IL-10 secretion from re-stimulated PBMC on days 5 and 6 pi ($p < 0.05$), and IL-10 was again elevated at d8 pi in the Ab4 group ($p < 0.01$) while the Ab4ΔORF2 group had a non-significant trend at d8 pi (Fig. 5b) compared to the control group. At later time points (d60-d260 pi) induction of cytokine secretion from PBMC in response to ex vivo re-stimulation was not observed for any group. Significant IL-4 and IL-17 secretion was not induced at any time (data not shown). Secretion of IFN-α was induced by the in vitro infection of PBMC at all time points, but there was no difference between the three groups of horses during the study (data not shown).

In addition, EHV-1 specific T-cells were analyzed by flow cytometry for intracellular IFN-γ, IL-4, IL-10 and IL-17 production. IFN-γ production by lymphocytes in PBMC were overall low and not different between the three groups for the duration of the study (Fig. 5c). IL-4, IL-10 or IL-17 were not induced in lymphocytes after ex vivo re-stimulation with EHV-1 (data not shown).

Cytokine responses in serum

Cytokine measurements in serum were overall similar between the control and the infected groups with a few exceptions. IFN-α was increased in serum early after infection on days 2 and 4 pi in the Ab4ΔORF2 group compared to the control group ($p < 0.05$, Additional file 1: Figure S1A). In the Ab4 group there was a non-significant trend of increased serum IFN-α on d2 pi. It remained at a similar level to uninfected controls for the remainder of the measurements in both infected groups. Serum sCD14 was elevated in the Ab4 group compared to the control group at d3 pi ($p < 0.01$). In addition, sCD14 was higher in the control group compared to the Ab4ΔORF2 group at d13 pi ($p < 0.05$) but was increased in the Ab4ΔORF2 group compared to both, Ab4 and the controls at d24 pi, while it was still significantly higher in Ab4 than in controls at that day ($p < 0.05$, Additional file 1: Figure S1B). The levels of serum IFN-γ did not differ between the three groups at any point during the experiment (Additional file 1: Figure S1C). IL-4 and IL-17 concentrations were overall low in serum and without any differences between the three groups (data not shown).

Discussion

An experimental infection with Ab4ΔORF2 deletion mutant virus in comparison to the parental neuropathogenic Ab4 virus was performed in horses to characterize the effects of the ORF2 gene product on EHV-1 infection and host immunity. Clinical signs, nasal shedding, and viremia induced by both viruses were evaluated to identify differences in virulence. Host immunity was analyzed locally and systemically summarizing the influence of both viruses on innate and adaptive cellular immunity and antibody induction. Overall, the infection of the horses with the parent Ab4 strain in the present study induced similar clinical signs, virus shedding, and viremia as previously described for EHV-1 infections of susceptible hosts [14, 42, 55]. Most importantly, the deletion of ORF2 reduced the virulence of the neuropathogenic EHV-1 Ab4 virus. This was evident by clearly reduced magnitude of the initial fever and significantly lower viral shedding in horses infected with the Ab4 ΔORF2 deletion mutant virus. The reduced virulence observed here for Ab4ΔORF2 was comparable to the

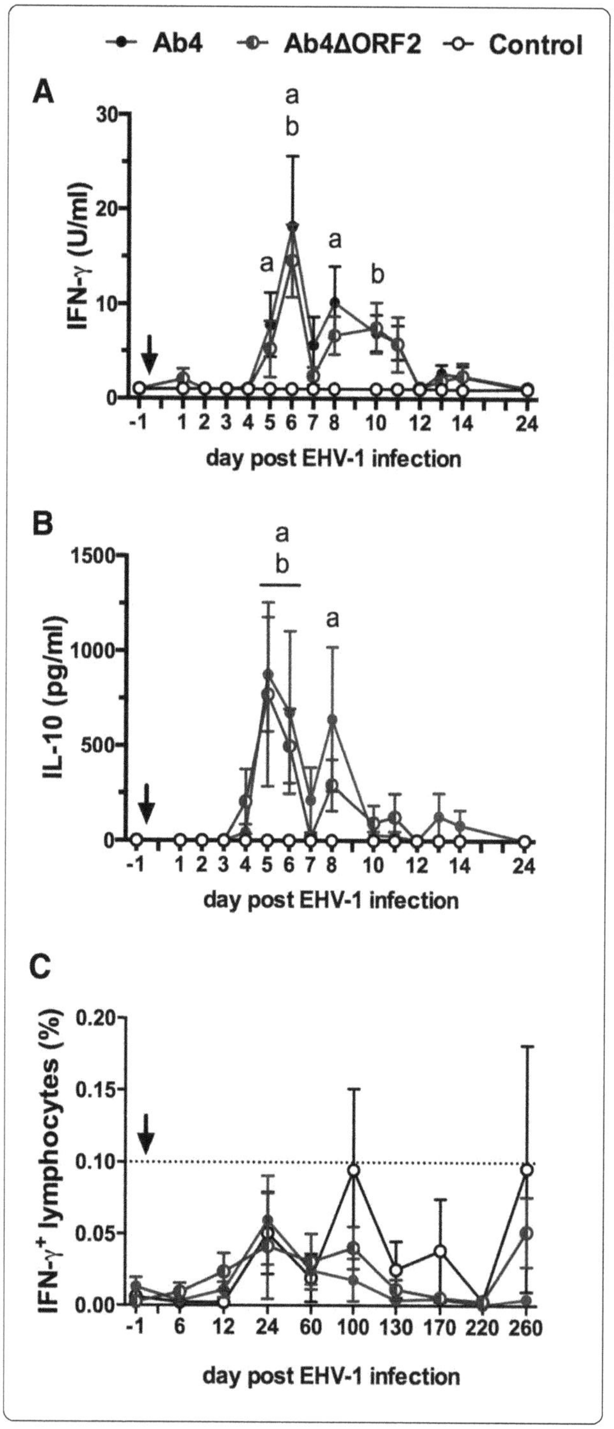

Fig. 5 Cellular immune responses to infection in PBMC. Three groups of horses ($n = 8$/group) were either not infected (control), infected with the neuropathogenic EHV-1 strain Ab4 (Ab4) or infected with its ORF2 deletion mutant (Ab4ΔORF2) on day 0 (arrow). PBMC were isolated before (day – 1) and after infection (day 1–24) and re-stimulated ex vivo for 48 h with EHV-1 (strain Ab4). Cytokine production (**a** and **b**) was analyzed in supernatants by a fluorescent bead-based multiplex assay. All values are cell culture medium control corrected. Mean and standard errors of secreted IFN-γ (**a**) and IL-10 (**b**) are shown. EHV-1 stimulated PBMC were also fixed, stained for intracellular IFN-γ and analyzed by flow cytometry. Total EHV-1 specific IFN-γ producing lymphocytes are displayed (**c**). Values have been corrected by the respective medium control of each horse. The dotted horizontal line (**c**) represents a cutoff of 0.1% IFN-γ + lymphocytes. PBMC controls kept in medium alone typically result in values below this cutoff value. Significant differences between groups are marked: **a** = Ab4 vs. control, **b** = Ab4ΔORF2 vs. control, **c** = Ab4 vs. Ab4ΔORF2

shorter duration of pyrexia, decreased magnitude and duration of virus shedding but unaffected viremia after infection of ponies with Ab4ΔORF1/2 compared to Ab4 in an earlier approach [42]. Differences in the magnitude of the fever induced by Ab4 infection may be based on individual differences between the ponies used by Soboll et al. [42] and the Icelandic horses used in the present study. Importantly, the reduction of virus shedding after deleting the ORF2 gene from the Ab4 virus versus using an Ab mutant deprived of ORF1 and ORF2 in the previous study [42] should be directly compared: Soboll et al. quantified EHV-1 DNA by PCR, while we used a plaque assay for virus isolation from nasal secretion, which measures infectious virus and thus correlates more directly with viral shedding and the ability of the infected horse to infect other horses.

The virulence factor ORF2 impacted the infection of the upper respiratory tract and local innate immune response to EHV-1 infection. In this study, we provided a comprehensive evaluation of intranasal cytokine and antibody secretion to longitudinally describe local innate and adaptive host immune responses during EHV-1 infection and potential differences induced by the ORF2 gene product. The onset of the local innate immune response against Ab4 was observed during virus shedding as determined by isolation of infectious virus from nasal discharge on days 2–4 pi. Intranasal innate immunity was characterized by sequential increases of the cytokines IFN-α (peak on d2pi), IFN-γ (peak on d3pi), and IL-10 (peak on d4pi) followed by a quick decline of all three cytokines. The nasal secretion of bioactive IFN was correlated with the time of EHV-1 shedding in a prior experimental infection study by Edington [56]. Here, we identified that high IFN-α in nasal secretion positively correlates with EHV-1 shedding in the Ab4 infected horses. In comparison, Ab4ΔORF2 infection induced low viral shedding combined with decreased IFN-α secretion. It has been demonstrated earlier, that equine IFN-α detected by the mAb pair used for quantification here, is also bioactive [57]. This suggests that intranasal IFN-α serves as an early danger signal during EHV-1 infection and, as previously pointed out by Randall [58], likely supports the induction of an antiviral state in the infected mucosal tissues. Notably, local IFN-γ and IL-10 induction was similar in both infected groups, Ab4ΔORF2 and Ab4, indicating that the ORF2 gene product does not influence these two immune regulatory cytokines.

Besides the short-term and pointed IFN-α, IFN-γ, and IL-10 secretion, elevated sCD14 concentrations were locally induced by EHV-1 between days 2–5 pi, while the horses were shedding virus. Peak intranasal sCD14 concentrations were slightly reduced in Ab4ΔORF2 infected horses supporting a possible direct or indirect influence of the ORF2 gene on intranasal sCD14 secretion. Lipopolysaccharide (LPS) binding sCD14 is produced by exocytosis and shedding of CD14 from monocyte or macrophage cell surfaces [59–61] and has a neutralizing, anti-inflammatory role for protection against LPS-induced cell death [62, 63]. The exact mechanisms of sCD14 induction after EHV-1 infection still needs to be unraveled in future approaches. However, EHV-1 was described to infect monocytes in vitro and in vivo [7, 64, 65] and the nasal sCD14 increase could thus result from a stress response of EHV-1 infected macrophages of the upper respiratory tract or reflect responses related to phagocytosis of virus and infected cells.

Overall, these findings point towards a role of ORF2 in the lytic virus replication cycle in the nasopharyngeal epithelia, which supported the release of infectious virions and increased the local viral load and nasal shedding. Nevertheless, the transmission of EHV-1 to mononuclear cells resulting in cell-associated viremia [7, 66] appears to be independent of ORF2. Here, viremia in the Ab4ΔORF2 and Ab4 groups was similar. Additionally, in vitro replication of Ab4ΔORF2 or Ab4ΔORF1/2 in RK13 cells and equine dermal cells (NBL-6) was not impaired compared to Ab4 [42]. Thus, the impact of ORF2 on nasal virus replication can only be shown after in vivo infection of the mucosal epithelial tissue including an intact local innate immune response.

In contrast to the clear reduction of virulence and intranasal IFN-α and sCD14 secretion, the immunogenicity of EHV-1 was not influenced by the deletion of the ORF2 gene from Ab4, with the only exception of a slightly earlier onset of IgG1 antibody responses in serum and during the second intranasal IgG1 peak in Ab4 infected horses. Local EHV-1 specific antibodies were detectable intranasally within 3–4 days after infection at low values. At this time, antibody responses were not yet detectable in serum. This very early nasal Ig secretion was similarly composed of different IgG isotypes and seems to be derived from local cells, such as *lamina propria* memory B-cells or those located in the regional lymph nodes. It should again be pointed out that all horses in this study were infected with EHV-1 about 1.5 to 2.5 years prior to the EHV-1 infection performed here. This confirmed the transient nature of protective immunity to EHV-1 described earlier [3, 27]. Although our horses did not mount a rapid and sufficiently high immune response to be protected from the EHV-1 infection performed here, they likely still had EHV-1 specific memory B-cells in their lymphatic tissues. Binding of viral antigens may have stimulated these memory B-cells to quickly secrete EHV-1 specific antibodies in low, but detectable quantities 3–4 days after infection.

Interestingly, this initial local IgG response waned quickly and was followed by peaks of higher local antibody responses that were dominated by IgG1 between d8–12 pi and IgG4/7 after d14 pi. Our findings clearly indicate that mucosal immune responses against the EHV-1strain Ab4 are dominated by IgG antibodies. Intranasal EHV-1 specific IgG antibodies are secreted in waves for several months post infection. The reason for the wave-like local antibody pattern is not yet clear. It requires further evaluation to identify if they could reflect local subclinical EHV-1 re-activation stages or other mechanisms. In contrast, the nasal EHV-1 specific IgA response after Ab4 infection was of overall low magnitude with slightly higher values for some horses that were infected with Ab4ΔORF2 during d8–14 pi. EHV-1 specific IgA was not detectable at later time points supporting that IgA is neither a major nor a long-lasting antibody during the mucosal EHV-1 specific immune response. This finding is contradictory to Breathnach [67], who found significant induction of nasal EHV-1 specific IgA together with IgG isotypes 2 weeks after primary infection with the neuropathogenic EHV-1 strain Army 183. Both, Breathnach [67] and our analysis here were based on the same monoclonal IgA antibody, BVS2, but used different assays for antibody detection. Future research could analyze if the observed difference in the EHV-1 specific IgA response are related to viral strain differences or if they have other reasons.

Robust systemic EHV-1 specific antibody responses were found in serum by d8 pi and were still detectable at high values by 9 months after infection with Ab4 or Ab4ΔORF2. Similar to nasal secretions, IgG1 was the dominating isotype of the early systemic response to EHV-1 infection, while IgG4/7 maintained the long-term response for more than 9 months. Similar serum antibody kinetics and IgG isotypes were previously reported after experimental EHV-1 infection with different EHV-1 strains including Ab4 and NY03 [42, 44]. This report exceeds the longitudinal observation period of previous articles of monitoring the immune response for weeks [67, 68] to 2–3 months [42, 69] after EHV-1 infection.

The glycoproteins gC and gD of EHV-1 are entry receptors of EHV-1 and highly immunogenic [70, 71]. In horses, gC and gD specific serum Ig and IgG4/7 highly correlate with EHV-1 serum neutralization titers [43]. Here, EHV-1 gC and gD specific antibodies were measured as representative markers for local and systemic EHV-1 specific immune responses with anti-gD responses closely resembling those of anti-gC. Equine IgG1 and IgG4/7 antibodies expressed effector functions such as strong complement activation and a high affinity to bind to IgG Fc-receptors in vitro [72]. EHV-1 specific IgG1 and IgG4/7 were thus assumed to have neutralizing activity and mediate protection against EHV-1 infection [22, 52]. Furthermore, IgG1 and IgG4/7 have been associated with a Th1 dominated B-cell response against EHV-1 [43].

In contrast to the robust antibody induction, an EHV-1 specific T-cell response could not be detected after infection with Ab4 or Ab4ΔORF2. Similarly, only weak T-cell responses were observed in former studies after experimental infection with EHV-1 Ab4 or NY03 [42, 44, 69]. However, we and others demonstrated increased EHV-1 specific cytotoxic T-cells frequencies or IFN-γ producing T-cells during viremia in older horses [13, 73], a few weeks after infection [74–76], or in horses that experienced an EHM outbreak and several vaccinations afterwards [52]. This suggests that EHV-1 specific T-cell activation is not consistently observed after infection, can be a slow process, and may require several immune recognition and activation events with EHV-1 or an EHV vaccine to develop substantial EHV-1 specific T-cell immunity. In the present study, a detectable cellular response was limited to EHV-1 re-stimulation ex vivo, occurred during viremia, and was represented by increased secretion of IL-10 and IFN-γ from PBMC. Likely, the transient stimulations of cytokine secretion from PBMC resulted from an innate response to the viral infection of circulating immune cells. It is believed that EHV-1 down-regulates cellular immunity and thereby allows the establishment of cell-associated viremia, which is a prerequisite for infection of endothelial cells prior to developing EHM or abortions [27, 66]. Current modified-live vaccines containing deletions of ORF2 and other genes [38, 40], similar to Ab4ΔORF2, fail to induce T-cell immunity and thus do not prevent viremia [22, 23, 52]. The EHV-1 genes responsible for the down-regulation of T-cell immunity still need to be identified to achieve robust cellular host immunity against EHV-1.

Conclusions

The ORF2 gene of the EHV-1 strain Ab4 is a virulence factor which contributes to fever and viral shedding but does not affect viremia or adaptive host immunity against the virus (Graphical abstract). Infection with EHV-1 Ab4 and its ORF2 deletion mutant induces high local and systemic antibody responses dominated by early IgG1 and long-lasting IgG4/7 isotypes. Local EHV-1 specific IgA antibodies are only a minor component of the mucosal immune response. Antibody responses dominate EHV-1 immunity while T-cell immunity is overall low.

Additional file

Additional file 1: Figure S1. Cytokines in serum after infection with EHV-1 strain Ab4 or its deletion mutant Ab4ΔORF2. Horses (*n* = 8/group) were infected with one of two EHV-1 strains (Ab4) or Ab4ΔORF2 at day 0 or kept as uninfected controls. Serum was sampled at several times before (days − 8, − 1) and after (day 1- day 24) infection, cytokines and the inflammatory marker sCD14 were evaluated with fluorescent bead-based assays. Mean and standard errors for IFN-α (A), sCD14 (B), and IFN-γ (C) in the serum are displayed. Significant differences between groups are marked: a = Ab4 vs. control, b = Ab4ΔORF2 vs. control, c = Ab4 vs. Ab4ΔORF2. (JPG 332 kb)

Abbreviations

Ab4ΔORF2: Deletion mutant of Ab4 devoid of open reading frame 2; CI: Confidence interval; d pi: Days post infection; EHM: Herpesvirus myeloencephalopathy; EHV-1: Equine herpesvirus type 1; gB/C/D: Envelope glycoprotein B/C/D; IFN: Interferon; Ig: Immunoglobulin; IL: Interleukin; LPS: Lipopolysaccharide; MFI: Median fluorescent intensities; NS: Nasopharyngeal secretion; ORF: Open reading frame; PBMC: Peripheral blood mononuclear cells; PBS: Phosphate buffered saline; PFU: Plaque-forming units; PMA: Phorbol12-myristate13-acetate; RK13: Rabbit kidney 13 cells; sCD14: Soluble cluster of differentiation 14

Acknowledgements

The authors would like to thank Fahad Raza, Franziska Kaiser, Karynn Kilts and Adam Alejandro for their help with sample collection and processing; Laura Goodman and Amy Glaser at the Animal Health Diagnostic Center at Cornell for performing the EHV-1 PCR on PBMC; Kevin Yager, Larry Carlisle, and Brooke Burke for excellent horse care and barn management; Rob Felt, Christine Bellezza, Debra Dwyer, and the IACUC committee at Cornell University for their input and recommendation; and USDA/APHIS for their support and help to establish an EHV-1-free herd at Cornell University and supervision during import quarantine.

Funding

The experimental EHV-1 infection study described in this manuscript was performed with funding from the Agriculture and Food Research Initiative Competitive Grant no. 2015–67015-23091 'The effect of ORF1 and ORF2 gene expression on Innate and Adaptive Immunity and Protection Against Equine Herpes Virus Type 1 (EHV-1)' supported by the USDA National Institute of Food and Agriculture. Reagent development for horse Ig isotype and cytokine monoclonal antibodies was supported by Agriculture and Food Research Initiative Competitive Grants no. 2005–01812 'The US Veterinary Immune Reagent Network' and no. 2015–67015-23072 'Equine Immune Reagents: Development of monoclonal antibodies to improve the analysis of immunity in horses' also supported by the USDA National Institute of Food and Agriculture. The horses used for this project were obtained through funding provided by the Harry M. Zweig Memorial Fund for Equine Research at Cornell University.

Authors' contributions

BW and GP designed the study. NO constructed and provided the ORF2 deletion mutant virus. CLS, GP, and BW performed the experimental infection study and sample acquisition. Cytokine, antibody and cellular assays were performed by CLS, SB, HF, CW, CLW and AR. CLS and BW performed the data analyses. CLS and BW drafted the manuscript. All authors read and approved the final version.

work also followed the Guide for Care and Use of Animals in Agricultural Research and Teaching. All efforts were made to minimize suffering of the animals, for example by sedation of nervous or excited horses before sampling. At the end of this experimental study, all horses were kept at the facility at Cornell University as research horses.

Consent for publication
Not applicable.

Competing interests
The authors declare that they have no competing interests with the content of this article.

Author details
[1]Department of Population Medicine and Diagnostic Sciences, College of Veterinary Medicine, Cornell University, Ithaca, NY 14853, USA. [2]Department of Clinical Sciences, College of Veterinary Medicine, Cornell University, Ithaca, NY 14853, USA. [3]Institut für Virologie, Freie Universität Berlin, Philippstrasse 13, 10115 Berlin, Germany.

References
1. Borchers K, Thein R, Sterner-Kock A. Pathogenesis of equine herpesvirus-associated neurological disease: a revised explanation. Equine Vet J. 2006;38: 283–7.
2. Perkins GA, Goodman LB, Tsujimura K, Van de Walle GR, Kim SG, Dubovi EJ, et al. Investigation of the prevalence of neurologic equine herpes virus type 1 (EHV-1) in a 23-year retrospective analysis (1984-2007). Vet Microbiol. 2009; 139:375–8.
3. Pusterla N, David Wilson W, Madigan JE, Ferraro GL. Equine herpesvirus-1 myeloencephalopathy: a review of recent developments. Vet J. 2009;180: 279–89.
4. Lunn DP, Davis-Poynter N, Flaminio MJBF, Horohov DW, Osterrieder K, Pusterla N, et al. Equine Herpesvirus-1 Consensus Statement. J Vet Intern Med. 2009;23:450–61.
5. Patel JR, Edington N, Mumford JA. Variation in cellular tropism between isolates of equine herpesvirus-1 in foals. Arch Virol. 1982;74:41–51.
6. Kydd JH, Smith KC, Hannant D, Livesay GJ, Mumford JA. Distribution of equid herpesvirus-1 (EHV-1) in the respiratory tract of ponies: implications for vaccination strategies. Equine Vet J. 1994;26:466–9.
7. Kydd JH, Smith KC, Hannant D, Livesay GJ, Mumford JA. Distribution of equid herpesvirus-1 (EHV-1) in respiratory tract associated lymphoid tissue: implications for cellular immunity. Equine Vet J. 1994;26:470–3.
8. Vandekerckhove AP, Glorieux S, Gryspeerdt AC, Steukers L, Van Doorsselaere J, Osterrieder N, et al. Equine alphaherpesviruses (EHV-1 and EHV-4) differ in their efficiency to infect mononuclear cells during early steps of infection in nasal mucosal explants. Vet Microbiol. 2011;152:21–8.
9. Edington N, Bridges CG, Patel JR. Endothelial cell infection and thrombosis in paralysis caused by equid herpesvirus-1: equine stroke. Arch Virol. 1986; 90:111–24.
10. Edington N, Smyth B, Griffiths L. The role of endothelial cell infection in the endometrium, placenta and foetus of equid herpesvirus 1 (EHV-1) abortions. J Comp Pathol. 1991;104:379–87.
11. Smith KC, Mumford JA, Lakhani K. A comparison of equid herpesvirus-1 (EHV-1) vascular lesions in the early versus late pregnant equine uterus. J Comp Pathol. 1996;114:231–47.
12. Slater JD, Borchers K, Thackray AM, Field HJ. The trigeminal ganglion is a location for equine herpesvirus 1 latency and reactivation in the horse. J Gen Virol. 1994;75:2007–16.
13. Allen GP. Risk factors for development of neurologic disease after experimental exposure to equine herpesvirus-1 in horses. Am J Vet Res. 2008;69:1595–600.
14. Allen GP. Antemortem detection of latent infection with neuropathogenic strains of equine herpesvirus-1 in horses. Am J Vet Res. 2006;67:1401–5.
15. Pusterla N, Mapes S, Wilson WD. Prevalence of equine herpesvirus type 1 in trigeminal ganglia and submandibular lymph nodes of equids examined postmortem. Vet Rec. 2010;167:376–9.
16. Pusterla N, Mapes S, David WW. Prevalence of latent alpha-herpesviruses in thoroughbred racing horses. Vet J. 2012;193:579–82.
17. Edington N, Bridges CG, Huckle A. Experimental reactivation of equid herpesvirus 1 (EHV 1) following the administration of corticosteroids. Equine Vet J. 1985;17:369–72.
18. Edington N, Welch HM, Griffiths L. The prevalence of latent equid herpesviruses in the tissues of 40 abattoir horses. Equine Vet J. 1994;26:140–2.
19. Taouji S, Collobert C, Gicquel B, Sailleau C, Brisseau N, Moussu C, et al. Detection and isolation of equine herpesviruses 1 and 4 from horses in Normandy: an autopsy study of tissue distribution in relation to vaccination status. J Vet Med B Infect Dis Vet Public Health. 2002;49:394–9.
20. USDA APHIS. Equine herpesvirus Myeloencephalopathy: a potentially emerging disease. Vet. Serv. Info Sheet. 2007. https://www.aphis.usda.gov/animal_health/emergingissues/downloads/ehv1final.pdf. Accessed 11 Aug 2018.
21. Patel JR, Heldens J. Equine herpesviruses 1 (EHV-1) and 4 (EHV-4) – epidemiology, disease and immunoprophylaxis: a brief review. Vet J. 2005; 170:14–23.
22. Goodman L, Wagner B, Flaminio M, Sussman K, Metzger S, Holland R, et al. Comparison of the efficacy of inactivated combination and modified-live virus vaccines against challenge infection with neuropathogenic equine herpesvirus type 1 (EHV-1). Vaccine. 2006;24:3636–45.
23. Goehring LS, Wagner B, Bigbie R, Hussey SB, Rao S, Morley PS, et al. Control of EHV-1 viremia and nasal shedding by commercial vaccines. Vaccine. 2010;28:5203–11.
24. Kydd JH, Hannant D, Mumford JA. Residence and recruitment of leucocytes to the equine lung after EHV-1 infection. Vet Immunol Immunopathol. 1996; 52:15–26.
25. Heldens JG, Hannant D, Cullinane AA, Prendergast MJ, Mumford JA, Nelly M, et al. Clinical and virological evaluation of the efficacy of an inactivated EHV1 and EHV4 whole virus vaccine (Duvaxyn EHV1,4). Vaccination/challenge experiments in foals and pregnant mares. Vaccine. 2001;19:4307–17.
26. Breathnach CC, Yeargan MR, Timoney JF, Allen GP. Detection of equine herpesvirus-specific effector and memory cytotoxic immunity in the equine upper respiratory tract. Vet Immunol Immunopathol. 2006;111:117–25.
27. Kydd JH, Townsend HGG, Hannant D. The equine immune response to equine herpesvirus-1: the virus and its vaccines. Vet Immunol Immunopathol. 2006;111:15–30.
28. Nugent J, Birch-Machin I, Smith KC, Mumford JA, Swann Z, Newton JR, et al. Analysis of equid herpesvirus 1 strain variation reveals a point mutation of the DNA polymerase strongly associated with Neuropathogenic versus Nonneuropathogenic disease outbreaks. J Virol. 2006;80:4047–60.
29. Goodman LB, Loregian A, Perkins GA, Nugent J, Buckles EL, Mercorelli B, et al. A point mutation in a herpesvirus polymerase determines Neuropathogenicity. PLoS Pathog. 2007;3:e160.
30. Van de Walle G, Goupil R, Wishon C, Damiani A, Perkins G, Osterrieder N. A single nucleotide polymorphism in a herpesvirus DNA polymerase is sufficient to cause lethal neurological disease. J Infect Dis. 2009;200:20–5.
31. Allen GP, Breathnach CC. Quantification by real-time PCR of the magnitude and duration of leucocyte-associated viraemia in horses infected with neuropathogenic vs. non-neuropathogenic strains of EHV- 1. Equine Vet J. 2006;38:252–7.
32. Henninger RW, Reed SM, Saville WJ, Allen GP, Hass GF, Kohn CW, et al. Outbreak of neurologic disease caused by equine herpesvirus-1 at a university equestrian center. J Vet Intern Med. 2007;21:157–65.
33. Gryspeerdt A, Vandekerckhove A, Van Doorsselaere J, Van de Walle G, Nauwynck H. Description of an unusually large outbreak of nervous system disorders caused by equine herpesvirus 1 (EHV1) in 2009 in Belgium. Vlaams Diergeneeskd Tijdschr. 2011;80:147–53.
34. Barbić L, Lojkić I, Stevanović V, Bedeković T, Starešina V, Lemo N, et al. Two outbreaks of neuropathogenic equine herpesvirus type 1 with breed-dependent clinical signs. Vet Rec. 2012;170:227.
35. Ma G, Feineis S, Osterrieder N, Van de Walle GR. Identification and characterization of equine herpesvirus type 1 pUL56 and its role in virus-induced downregulation of major histocompatibility complex class I. J Virol. 2012;86:3554–63.
36. Wagner B, Wimer C, Freer H, Osterrieder N, Erb HN. Infection of peripheral blood mononuclear cells with neuropathogenic equine herpesvirus type-1

strain Ab4 reveals intact interferon-α induction and induces suppression of anti-inflammatory interleukin-10 responses in comparison to other viral strains. Vet Immunol Immunopathol. 2011;143:116–24.
37. Wimer CL, Damiani A, Osterrieder N, Wagner B. Equine herpesvirus type-1 modulates CCL2, CCL3, CCL5, CXCL9, and CXCL10 chemokine expression. Vet Immunol Immunopathol. 2011;140:266–74.
38. Hübert PH, Birkenmaier S, Rziha HJ, Osterrieder N. Alterations in the equine herpesvirus type-1 (EHV-1) strain RacH during attenuation. Zentralblatt vet. Reihe B J Vet Med Ser B. 1996;43:1–14.
39. Yalamanchili RR, Raengsakulrach B, O'Callaghan DJ. Equine herpesvirus 1 sequence near the left terminus codes for two open reading frames. Virus Res. 1991;18:109–16.
40. Telford EA, Watson MS, McBride K, Davison AJ. The DNA sequence of equine herpesvirus-1. Virology. 1992;189:304–16.
41. Soboll Hussey G, Ashton LV, Quintana AM, Van de Walle GR, Osterrieder N, Lunn DP. Equine herpesvirus type 1 pUL56 modulates innate responses of airway epithelial cells. Virology. 2014;464–465:76–86.
42. Soboll Hussey G, Hussey SB, Wagner B, Horohov DW, Van de Walle GR, Osterrieder N, et al. Evaluation of immune responses following infection of ponies with an EHV-1 ORF1/2 deletion mutant. Vet Res. 2011;42:23.
43. Wagner B, Goodman LB, Babasyan S, Freer H, Torsteinsdóttir S, Svansson V, et al. Antibody and cellular immune responses of naïve mares to repeated vaccination with an inactivated equine herpesvirus vaccine. Vaccine. 2015; 33:5588–97.
44. Wagner B, Perkins G, Babasyan S, Freer H, Keggan A, Goodman LB, et al. Neonatal Immunization with a Single IL-4/Antigen Dose Induces Increased Antibody Responses after Challenge Infection with Equine Herpesvirus Type 1 (EHV-1) at Weanling Age. Murthy AK, editor. PLOS ONE. 2017;12:e0169072.
45. Crowhurst FA, Dickinson G, Burrows R. An outbreak of paresis in mares and geldings associated with equid herpesvirus 1. Vet Rec. 1981;109:527–8.
46. Furr M, Reed S. Equine Neurology: Wiley; 2008. Available from: https://books.google.de/books?id=atRmlbfxu5UC
47. Elia G, Decaro N, Martella V, Campolo M, Desario C, Lorusso E, et al. Detection of equine herpesvirus type 1 by real time PCR. J Virol Methods. 2006;133:70–5.
48. Wagner B, Hillegas JM, Babasyan S. Monoclonal antibodies to equine CD23 identify the low-affinity receptor for IgE on subpopulations of IgM+ and IgG1+ B-cells in horses. Vet Immunol Immunopathol. 2012;146:125–34.
49. Lunn DP, Holmes MA, Antczak DF, Agerwal N, Baker J, Bendali-Ahcene S, et al. Report of the second equine leucocyte antigen workshop, squaw valley, California, July 1995. Vet Immunol Immunopathol. 1998;62:101–43.
50. Keggan A, Freer H, Rollins A, Wagner B. Production of seven monoclonal equine immunoglobulins isotyped by multiplex analysis. Vet Immunol Immunopathol. 2013;153:187–93.
51. Lunn DP, Holmes MA, Schram B, Duffus WP. Monoclonal antibodies specific for equine IgG sub-isotypes including an antibody which recognizes B lymphocytes. Vet Immunol Immunopathol. 1995;47:239–51.
52. Goodman LB, Wimer C, Dubovi EJ, Gold C, Wagner B. Immunological correlates of vaccination and infection for equine herpesvirus 1. Clin Vaccine Immunol. 2012;19:235–41.
53. Wagner B, Freer H. Development of a bead-based multiplex assay for simultaneous quantification of cytokines in horses. Vet Immunol Immunopathol. 2009;127:242–8.
54. Wagner B, Ainsworth DM, Freer H. Analysis of soluble CD14 and its use as a biomarker in neonatal foals with septicemia and horses with recurrent airway obstruction. Vet Immunol Immunopathol. 2013;155:124–8.
55. Brosnahan MM, Damiani A, van de Walle G, Erb H, Perkins GA, Osterrieder N. The effect of siRNA treatment on experimental equine herpesvirus type 1 (EHV-1) infection in horses. Virus Res. 2010;147:176–81.
56. Edington N, Bridges CG, Griffiths L. Equine interferons following exposure to equid herpesvirus-1 or –4. J Interf Res. 1989;9:389–92.
57. Wagner B, Hillegas JM, Flaminio MJ, Wattrang E. Monoclonal antibodies to equine interferon-alpha (IFN-alpha): new tools to neutralize IFN-activity and to detect secreted IFN-alpha. Vet Immunol Immunopathol. 2008;125(3–4): 315–25.
58. Randall RE, Goodbourn S. Interferons and viruses: an interplay between induction, signalling, antiviral responses and virus countermeasures. J Gen Virol. 2008;89:1–47.
59. Bazil V, Horejsí V, Baudys M, Kristofová H, Strominger JL, Kostka W, et al. Biochemical characterization of a soluble form of the 53-kDa monocyte surface antigen. Eur J Immunol. 1986;16:1583–9.
60. Landmann R, Fisscher AE, Obrecht JP. Interferon-gamma and interleukin-4 down-regulate soluble CD14 release in human monocytes and macrophages. J Leukoc Biol. 1992;52:323–30.
61. Labeta MO, Durieux J-J, Fernandez N, Herrmann R, Ferrara P. Release from a human monocyte-like cell line of two different soluble forms of the lipopolysaccharide receptor, CD14. Eur J Immunol. 1993;23: 2144–51.
62. Schütt C, Schilling T, Grunwald U, Schönfeld W, Krüger C. Endotoxin-neutralizing capacity of soluble CD14. Res Immunol. 1992;143:71–8.
63. Haziot A, Rong GW, Lin XY, Silver J, Goyert SM. Recombinant soluble CD14 prevents mortality in mice treated with endotoxin (lipopolysaccharide). J Immunol. 1995;154:6529.
64. Wilsterman S, Soboll-Hussey G, Lunn DP, Ashton LV, Callan RJ, Hussey SB, et al. Equine herpesvirus-1 infected peripheral blood mononuclear cell subpopulations during viremia. Vet Microbiol. 2011;149:40–7.
65. Baghi HB, Nauwynck HJ. Impact of equine herpesvirus type 1 (EHV-1) infection on the migration of monocytic cells through equine nasal mucosa. Comp Immunol Microbiol Infect Dis. 2014;37:321–9.
66. Osterrieder N, Van de Walle GR. Pathogenic potential of equine alphaherpesviruses: the importance of the mononuclear cell compartment in disease outcome. Vet Microbiol. 2010;143:21–8.
67. Breathnach CC, Yeargan MR, Sheoran AS, Allen GP. The mucosal humoral immune response of the horse to infective challenge and vaccination with equine herpesvirus-1 antigens. Equine Vet J. 2001;33:651–7.
68. Coombs DK, Patton T, Kohler AK, Soboll G, Breathnach C, Townsend HGG, et al. Cytokine responses to EHV-1 infection in immune and non-immune ponies. Vet Immunol Immunopathol. 2006;111:109–16.
69. Paillot R, Daly JM, Juillard V, Minke JM, Hannant D, Kydd JH. Equine interferon gamma synthesis in lymphocytes after in vivo infection and in vitro stimulation with EHV-1. Vaccine. 2005;23:4541–51.
70. Packiarajah P, Walker C, Gilkerson JR, Whalley JM, Love DN. Immune responses and protective efficacy of recombinant baculovirus expressed glycoproteins of equine herpesvirus 1 (EHV-1) gB gC and gD alone or in combinations in BALB/c mice. Vet Microbiol. 1998;61:261–78.
71. Tewari D, Nair SV, De Ungria MC, Lawrence CL, Hayden M, Love DN, Field HJ, Whalley JM. Immunization with glycoprotein C of equine herpesvirus-1 is associated with accelerated virus clearance in a murine model. Arch Virol. 1995;140:789–97.
72. Lewis MJ, Wagner B, Woof JM. The different effector function capabilities of the seven equine IgG subclasses have implications for vaccine strategies. Mol Immunol. 2008;45(3):818–27.
73. Breathnach CC, Soboll G, Suresh M, Lunn DP. Equine herpesvirus-1 infection induces IFN-γ production by equine T lymphocyte subsets. Vet Immunol Immunopathol. 2005;103:207–15.
74. Allen G, Yeargan M, Costa LR, Cross R. Major histocompatibility complex class I-restricted cytotoxic T-lymphocyte responses in horses infected with equine herpesvirus 1. J Virol. 1995;69:606–12.
75. Kydd J, Wattrang E, Hannant D. Pre-infection frequencies of equine herpesvirus-1 specific, cytotoxic T lymphocytes correlate with protection against abortion following experimental infection of pregnant mares. Vet Immunol Immunopathol. 2003;96:207–17.
76. O'Neill T, Kydd JH, Allen GP, Wattrang E, Mumford JA, Hannant D. Determination of equid herpesvirus 1-specific, CD8+, cytotoxic T lymphocyte precursor frequencies in ponies. Vet Immunol Immunopathol. 1999;70:43–54.

Lower plasma trans-4-hydroxyproline and methionine sulfoxide levels are associated with insulin dysregulation in horses

Ákos Kenéz[1,3†], Tobias Warnken[2*†], Karsten Feige[2] and Korinna Huber[1]

Abstract

Background: Insulin dysregulation in horses is a metabolic condition defined by high insulin concentrations in the blood and peripheral insulin resistance. This hyperinsulinemia is often associated with severe damage in the hooves, resulting in laminitis. However, we currently lack detailed information regarding the potential involvement of particular metabolic pathways in pathophysiological causes and consequences of equine insulin dysregulation. This study aimed to assess the dynamic metabolic responses given to an oral glucose test (OGT) in insulin-sensitive and insulin-dysregulated horses by a targeted metabolomics approach to identify novel metabolites associated with insulin dysregulation.

Results: Oral glucose testing triggered alterations in serum insulin (26.28 ± 4.20 vs. 422.84 ± 88.86 µlU/mL, $p < 0.001$) and plasma glucose concentrations (5.00 ± 0.08 vs. 9.43 ± 0.44 mmol/L, $p < 0.001$) comparing basal and stimulated conditions after 180 min. Metabolome analyses indicated OGT-induced changes in short-chain acylcarnitines (6.00 ± 0.53 vs. 3.99 ± 0.23 µmol/L, $p < 0.001$), long-chain acylcarnitines (0.13 ± 0.004 vs. 0.11 ± 0.002 µmol/L, $p < 0.001$) and amino acids (2.18 ± 0.11 vs. 1.87 ± 0.08 µmol/L, $p < 0.05$). Kynurenine concentrations increased (2.88 ± 0.18 vs. 3.50 ± 0.19 µmol/L, $p < 0.01$), whereas spermidine concentrations decreased during OGT (0.09 ± 0.004 vs. 0.08 ± 0.002 µmol/L, $p < 0.01$), indicating proinflammatory conditions after oral glucose load. Insulin dysregulation was associated with lower concentrations of trans-4-hydroxyproline (4.41 ± 0.29 vs. 6.37 ± 0.71 µmol/L, $p < 0.05$) and methionine sulfoxide (0.40 ± 0.06 vs. 0.87 ± 0.13 µmol/L, $p < 0.01$; mean ± SEM in insulin-dysregulated vs. insulin-sensitive basal samples, respectively), two metabolites which are related to antioxidant defense mechanisms.

Conclusion: Oral glucose application during OGT resulted in profound metabolic and proinflammatory changes in horses. Furthermore, insulin dysregulation was predicted in basal samples (without OGT) by pathways associated with trans-4-hydroxyproline and methionine sulfoxide, suggesting that oxidative stress and oxidant–antioxidant disequilibrium are contributing factors to insulin dysregulation. The present findings provide new hypotheses for future research to better understand the underlying pathophysiology of insulin dysregulation in horses.

Keywords: Metabolome, Metabolomics, Insulin sensitivity, Insulin dysregulation, Oral glucose test, Horses

Background

The equine metabolic syndrome (EMS) in horses is associated with severe disturbances in glucose and lipid homeostasis and adiposity. Equine insulin dysregulation reflected by hyperinsulinemia is a key symptom [1, 2]. However, the underlying causal pathomechanisms of EMS and metabolic consequences of insulin dysregulation are currently poorly understood. Laminitis, a chronic painful disease of the hooves, develops as a severe consequence and is often a cause for euthanasia [3]. Insulin and glucose concentrations in plasma as indicators of metabolic dysregulation are frequently measured parameters in the clinical routine of equine medicine. Oral glucose tests and other challenges were used to assess insulin dysregulation exactly [1]; however, these tests assay only the disturbed insulin response, but do not provide information about potential causal pathways leading to insulin dysregulation

* Correspondence: tobias.warnken@tiho-hannover.de
[†]Equal contributors
[2]Clinic for Horses, University of Veterinary Medicine Hannover, Foundation, Bünteweg 9, 30559 Hannover, Germany
Full list of author information is available at the end of the article

in horses. Furthermore, data on the metabolic consequences of equine insulin dysregulation, other than hyperinsulinemia, are still limited.

Metabolomics can be used as a powerful tool to depict the phenotypic image of metabolism under the current conditions [4]. This is provided by a snapshot-like quantification of an extensive set of metabolites which are products or substrates of various metabolic pathways. Due to the known links between metabolites and pathways, hypotheses about underlying (patho)physiological processes can be proposed. Furthermore, new hypotheses concerning affected pathways can be established as a basis for new experimental designs by introducing novel associations between certain metabolites and phenotypic observations. These new scientific studies will help to identify pathophysiological causes or consequences of equine insulin dysregulation. The identification of potential biomarkers capturing the acute health status is another benefit of metabolomics approaches.

Therefore, the aims of the present study were to (1) elucidate the dynamic metabolic response to a defined oral glucose challenge in horses previously classified as insulin-sensitive (IS) or -dysregulated (ID), and (2) to identify novel metabolites associated with the ID state. To accomplish these aims, the blood plasma metabolome of IS and ID horses was analyzed using a targeted metabolomics approach, important metabolites were selected by multivariate statistical methods, and pathways associated with the selected metabolites were described based on previous knowledge in the context of insulin homeostasis and metabolic disorders.

Methods

Animals and sample collection

Twenty horses of various breeds, ages and body weights (BW) were included in the study. Body condition scoring (BCS) (according to Henneke et al. [5]), (BW) and ages were recorded for each horse. The ages of the horses ranged from 6 to 23 years, and their BW ranged from 147 to 695 kg. Horses had unknown insulinemic status prior to testing and blood samples were collected during routine diagnostic procedures for the assessment of insulin dysregulation in the Clinic for Horses (University of Veterinary Medicine, Hannover). Informed consent was obtained from the owners for scientific use and publication. Horses were fasted overnight before sampling and, in connection with the sample collection, an oral glucose testing (OGT) procedure [6] was conducted. The OGT was carried out by administering 1.0 g/kg BW glucose powder dissolved in 2 L of water by a nasogastric tube. Blood samples were collected from each horse three times: immediately before starting the OGT (BASAL), 120 min after starting the OGT (OGT-120) and 180 min after starting the OGT (OGT-180). Samples were collected via a jugular vein catheter, transferred into plain tubes (Vacuette® Greiner Bio One, Frickenhausen, Germany) for serum and EDTA tubes (Vacuette® Greiner Bio One, Frickenhausen, Germany) for plasma preparation. Blood for serum was incubated at room temperature for 1 h before centrifugation at 3000 g for 6 min at room temperature and blood for plasma was immediately centrifuged at 3000 g for 6 min. Serum and plasma samples were shock-frozen in liquid nitrogen and subsequently continuously stored at − 80 °C until analysis.

Blood serum and plasma analysis

Plasma glucose concentrations were analyzed with a colorimetric assay (GLUC3, Cobas, Roche Diagnostics GmbH, Mannheim, Germany) on an automated discrete analyzer (Cobas Mira, Roche Diagnostics GmbH, Mannheim, Germany). Serum insulin concentrations were analyzed by an equine-optimized ELISA (Equine Insulin ELISA, Mercodia, Uppsala, Sweden) previously validated for use in horses [7].

Plasma metabolome analysis was performed on all samples by the AbsoluteIDQ p180 Kit (Biocrates Life Science AG, Innsbruck, Austria) in the laboratory of Biocrates Life Science AG, according to the manufacturer's standard protocol. This kit format-targeted metabolomics measurement was used to identify and quantify 188 metabolites belonging to five compound classes: acylcarnitines (40), proteinogenic and modified amino acids (19), glycerophospho- and sphingolipids (76 phosphatidylcholines, 14 lyso-phosphatidylcholines, 15 sphingomyelins), biogenic amines (19) and hexoses (1). A detailed list of the compounds is shown in the supplementary material (Additional file 1: Online Resource 1). The fully automated assay was based on phenylisothiocyanate derivatization in the presence of internal standards followed by FIA-MS/MS [acylcarnitines, (lyso-) phosphatidylcholines, sphingomyelins, hexoses] and LC-MS/MS (amino acids, biogenic amines) using a SCIEX 4000 QTRAP® (SCIEX, Darmstadt, Germany) or a Xevo TQ-S Micro (Waters, Vienna, Austria) instrument with electrospray ionization. The experimental metabolomics measurement technique is described in detail by patent US 2007/0004044 [8]. All pre-analytical and analytical procedures were performed, documented and reviewed according to the ISO 9001:2008 certified in-house quality management rules and guidelines of Biocrates Life Sciences AG.

Data analysis and visualization

Plasma glucose and serum insulin concentrations of BASAL, OGT-120 and OGT-180 samples were used as a marker of insulin sensitivity status to assign horses to either of the two experimental groups: IS or ID. This was conducted by a hierarchical clustering algorithm using the Euclidean distance measure and Ward's clustering method in MetaboAnalyst 3.0 [9].

Plasma metabolome data (absolute concentrations of compounds) were analyzed in MetaboAnalyst after normalization by Pareto scaling. All data were tested and confirmed to be normally distributed by use of the Shapiro-Wilk test in GraphPad Prism (GraphPad Prism, Version 6.07 for Windows, Graph-Pad Inc. La Jolla, CA). Principal component analysis (PCA), repeated measures two-way ANOVA (rmTWA) and heatmap generation were conducted in MetaboAnalyst to visualize and evaluate differences in the metabolic profiles due to classification according to insulin sensitivity status and during OGT. The heatmap was created for the top 22 significant metabolites identified by rmTWA; clustering of the metabolites by 'average' clustering algorithm and plotting the tree by Euclidean distance measure were also included. The RmTWA was performed for the factor 'OGT' (BASAL vs. OGT-120 vs. OGT-180) and 'insulin sensitivity status' (IS vs. ID), also considering interactions and applying false discovery rate correction. $P \leq 0.05$ was considered as statistically significant. A volcano plot was created separately for the non-challenged condition (BASAL) to identify metabolites of interest, searching for metabolites having significantly different concentrations between IS and ID horses, even without any OGT influence. Metabolites with a fold change of at least 1.5 and a t-test p value lower than 0.05 were considered to have high importance. Concentrations of important metabolites were plotted in GraphPad Prism. These were tested statistically for the effect of 'insulin sensitivity status' at different levels of the OGT by rmTWA combined with a pairwise Bonferroni post hoc test in GraphPad Prism.

Results

All horses tolerated the OGT procedure without complications. Plasma glucose and serum insulin increased significantly as a response to oral glucose administration observed after 120 and 180 min of the OGT (Fig. 1a+b; both $p < 0.001$). The clustering algorithm applied to these data divided the horses clearly into two groups, which was the basis to assign them to either the IS (n = 10) or ID ($n = 10$) group. The IS group consisted of horses that had low to medium insulin concentrations in response to OGT, while the ID group consisted of horses that had high insulin concentrations in response to OGT. After creating the groups by clustering, the insulin threshold between the two groups was empirically found to be 260 μIU/mL with mean OGT-stimulated insulin concentrations of 131.5 ± 24.8 μIU/mL for IS (mean ± SEM; min. 34.5, max. 257.0 μIU/mL; n = 10) and 675.2 ± 101.2 μIU/mL for ID (mean ± SEM; min. 398.9, max. 1403.0 μIU/mL; $n = 10$), respectively (Mann-Whitney test $p < 0.0001$). The two groups had comparable BCS values as an indicator of obesity (6.8 ± 0.4 vs. 7.1 ± 0.2; mean ± SEM; $n = 10$/group), comparable BW (in kg: 481. 1 ± 29.0 vs. 408.8 ± 53.72; mean ± SEM, $n = 10$/group) and similar ages (in years: 14.2 ± 1.6 vs. 15.7 ± 1.7; mean ± SEM; $n = 10$/group). The OGT-triggered increase in insulin concentration was significantly greater in ID horses compared to IS horses (Fig. 1a; $p < 0.001$), but the increase in glucose concentration was only slightly higher in ID horses (Fig. 1b; $p = 0.98$).

An overview of the changes in the total metabolomic profiles of IS and ID horses during the OGT is given in Fig. 2. The PCA was used to downscale the dimensions of the data matrix resulting in PC1 and PC2 scores. The PCA scores plot showed a clear separation between the BASAL and the OGT-120/OGT-180 metabolic profiles

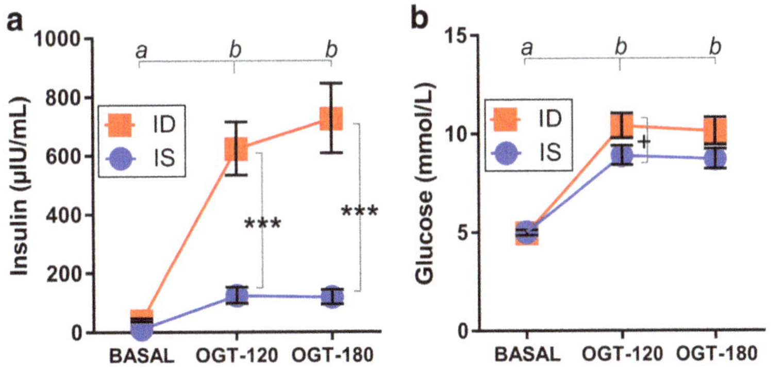

Fig. 1 Increase of (**a**) serum insulin (repeated measure two-way ANOVA (rmTWA) factor oral glucose testing (OGT) $p < 0.001$) and of (**b**) glucose (rmTWA factor OGT $p < 0.001$) concentration in insulin-sensitive (IS) and insulin-dysregulated (ID) horses during OGT. Measurements were carried out at three levels of OGT: immediately before glucose administration (BASAL), and 120 min (OGT-120) and 180 min (OGT-180) after administration. Means ± SEM are shown, $n = 10$, significant differences between IS and ID are indicated as ***($p < 0.001$), and a trend between IS and ID is indicated as $^{+}$(0.05 < p < 0.1). a, b = different superscripts indicate significant differences between basal and challenged conditions as the main OGT effect in both ID and IS horses (Bonferroni post hoc test)

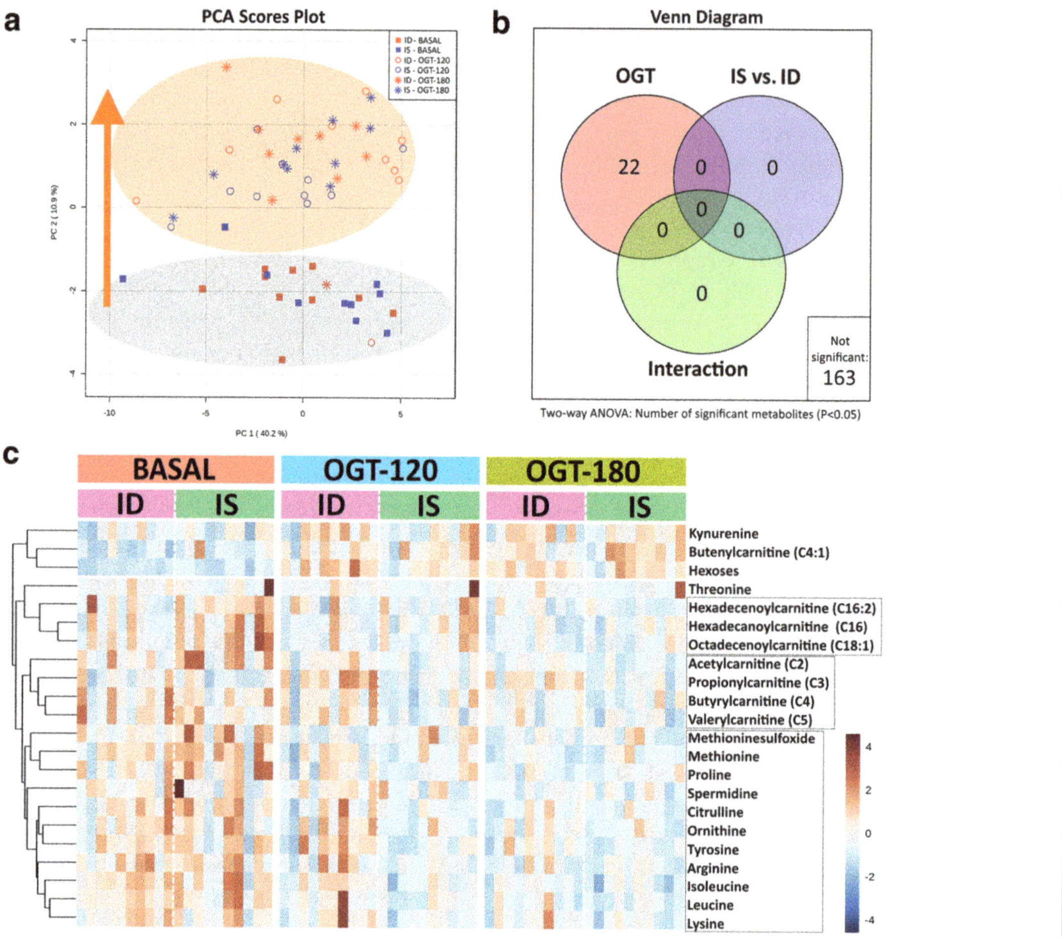

Fig. 2 a Principal component analysis scores plot showing metabolic profiles of insulin-sensitive (IS) and insulin-dysregulated (ID) horses sampled before glucose administration (BASAL), and 120 min (OGT-120) and 180 min (OGT-180) after administration. **b** Number of significant metabolites detected by repeated measure two-way ANOVA with false discovery rate correction, analyzing the effect of OGT, insulin sensitivity status (IS vs. ID) and respective interactions. **c** Heatmap illustrating relative concentrations of top 22 significant metabolites during OGT: BASAL, OGT-120 and OGT-180 for IS and ID horses

with a shift of all OGT-treated horses along the y-axis (PC2; Fig. 2a). Twenty-two metabolites were significantly affected by the OGT (Fig. 2b), which was the underlying cause of the shift observed in the PCA. The insulin sensitivity status did not affect the metabolite profile, and interaction between the two factors also remained non-significant (Fig. 2b). Accordingly, the remaining 163 metabolites studied did not show any significant differences by ANOVA (Fig. 2b). A heatmap was created by using the 22 metabolites identified by the ANOVA to visualize the metabolites responsible for the OGT effect (Fig. 2c). The first level of branches in the clustering tree separated between 3 metabolites having increasing concentrations and 19 metabolites having decreasing concentrations during the OGT. The former group consisted of kynurenine, butenylcarnitine and hexoses, while the latter group consisted of long-chain fatty acylcarnitines (C16-18), short-chain fatty acylcarnitines (C2-5), (modified) amino acids and biogenic amines. Hexose (Fig. 3a; $p < 0.001$) and kynurenine (Fig. 3b; $p = 0.002$) concentrations were increased at OGT-120 and OGT-180, whereas spermidine (Fig. 3c; $p = 0.006$) concentrations decreased during OGT in both IS and ID horses. Proteinogenic amino acids (Fig. 3d; $p = 0.024$), short-chain acylcarnitines (Fig. 3e; $p < 0.001$) and long-chain acylcarnitines (Fig. 3f; $p < 0.001$) also decreased during OGT.

A volcano plot comprised of fold change and t-test statistics was analyzed using the results of BASAL samples to identify the metabolites with the greatest discriminating potential between IS and ID horses (Fig. 4a). This plot shows in its upper left corner the metabolites of interest (MOI), those that have the highest fold change and the lowest p value of the t-test comparing IS vs. ID. Two metabolites were clearly identified: trans-4-hydroxyproline (FC = 1.59, $p = 0.003$) and methionine sulfoxide (FC = 2.09, $p = 0.013$), with higher concentrations in

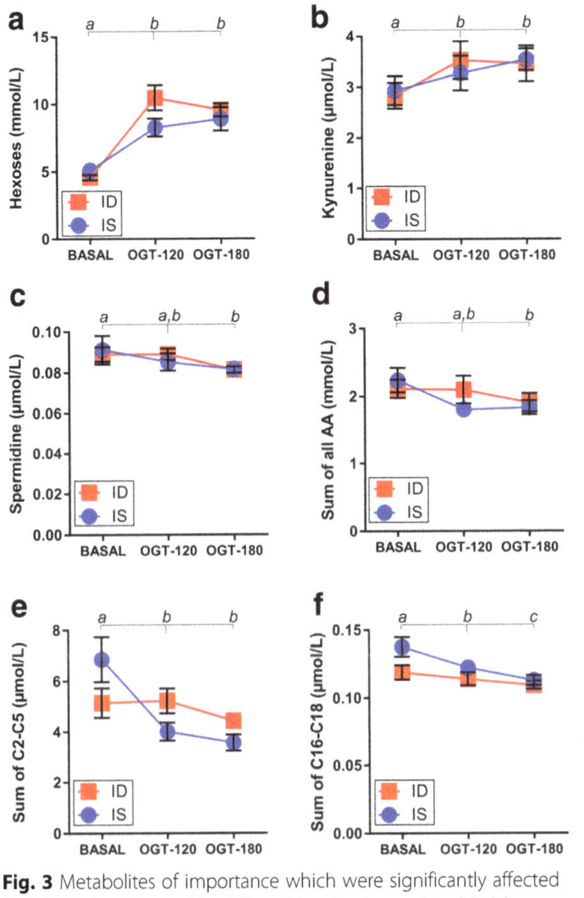

Fig. 3 Metabolites of importance which were significantly affected by OGT in insulin-sensitive (IS) and insulin-dysregulated (ID) horses. Concentration of (**a**) hexoses (repeated measure two-way ANOVA (rmTWA), factor OGT $p < 0.001$) and (**b**) kynurenine (rmTWA, factor OGT $p < 0.01$) increased, while (**c**) spermidine (rmTWA, factor OGT $p < 0.01$), (**d**) sum of all proteinogenic amino acids (rmTWA, factor OGT $p < 0.05$) (**e**) sum of short-chain (C2, C3, C4 and C5) acylcarnitines (rmTWA, factor OGT $p < 0.001$) and (**f**) sum of long-chain (C16, C16:1, C16:2, C:18, C18:1, C:18:2) acylcarnitines (rmTWA, factor OGT $p < 0.001$) decreased. Insulin status had no significant effect. Means ± SEM are shown, $n = 10$; a, b, c = different superscripts indicate significant differences between basal and challenged conditions as the main OGT effect in both ID and IS horses (Bonferroni post hoc test)

IS horses compared to ID horses. Further assessment of the dynamic changes of these MOI during the OGT revealed that both trans-4-hydroxyproline (Fig. 4b) and methionine sulfoxide (Fig. 4c) concentrations decreased significantly in the IS group, but not in the ID group. Analyzing the effect of insulin sensitivity status by Bonferroni post hoc test revealed that these metabolites had significant differences only under BASAL conditions, with lower concentrations in the ID group (Fig. 4b: trans-4-hydroxyproline $p = 0.040$; Fig. 4c: methionine sulfoxide $p = 0.010$).

Discussion

This study used a targeted and quantitative metabolomics analysis to describe novel metabolic profiles associated with equine insulin dysregulation in two approaches. (1) Basal concentrations before OGT challenge are suitable for identifying metabolites which can predict the metabolic characteristics of horses associated with insulin sensitivity or dysregulation. The dynamic responses after oral glucose load seen at OGT-120 and OGT-180 enables the identification of pathways which are causally affected by the oral glucose load as a single challenge in both IS and ID horses. The ID individuals have a higher pancreatic insulin secretion in response to a certain glucose load than IS individuals [1, 2, 6]. These dynamic changes of glucose and insulin were – as expected – clearly visible at 120 (OGT-120) and 180 (OGT-180) min after oral glucose load in the present study, confirming the classification into two groups (IS and ID) by the clustering algorithm. The threshold used here is considered independent of the cut-offs used clinically based on insulin concentrations during OGT and was declared as a differentiation criterion applicable to the animal cohort used in the current study to establish two experimental groups with a physiological and a pathophysiological response, respectively. Therefore, it should be emphasized that the metabolites identified here cannot be extrapolated for clinical diagnostic use; our findings provide rather a basis for hypotheses generation concerning insulin dysregulation mechanisms and potential candidates for future biomarker development. Moreover, the OGT induced a severe insulin increase with a wide dynamic range, reflecting that insulin sensitivity is not a dichotomous state of being either IS or ID, but rather that ID exists in different intensities. The typical condition of hyperinsulinemia was intensive in several ID horses, but this was not accompanied by a correspondingly strong hyperglycemia. This equine-specific pathophysiological phenomenon requires future clarification regarding its underlying mechanisms.

The corresponding metabolic profiles were studied by using the AbsoluteIDQ p180 Kit (Biocrates Life Science AG). This targeted metabolomics assay of kit format was designed by the developer to quantify a defined set of metabolites reflecting specific metabolic disorders, such as obesity, insulin resistance, proinflammation, dysregulation of glucose and lipid metabolism, and mitochondrial dysfunction. Although this assay was developed for human purposes, it is a species-independent analytical tool to accurately detect small molecules in various biological samples and, accordingly, we were able to detect a wide range of circulating acylcarnitines, amino acids, biogenic amines, glycerophospho- and sphingolipids, and hexoses in horses. The assay kit applied was considered suitable for acquiring phenotypic patterns of equine metabolism affected by insulin dysregulation because the metabolic syndrome of horses has been demonstrated to exhibit some common symptoms with

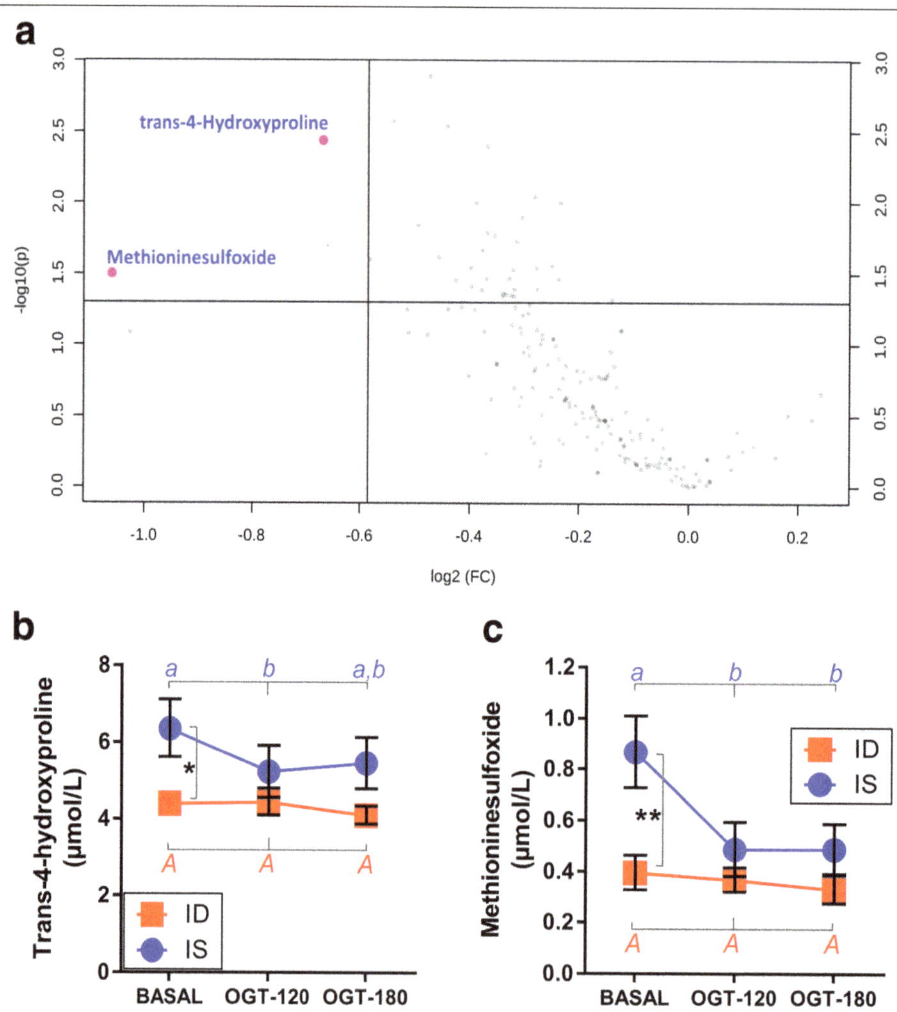

Fig. 4 a Volcano plot highlighting metabolites of importance (trans-4-hydroxyproline $p = 0.0036$; methinonine sulfoxide $p = 0.0134$) which were different between insulin-sensitive (IS) and insulin-dysregulated (ID) horses under unchallenged conditions (only BASAL) (**b**) Trans-4-hydroxyproline (repeated measure two-way ANOVA (rmTWA), factor insulin sensitivity status $p < 0.05$, factor OGT $p = 0.06$) and (**c**) methionine sulfoxide (rmTWA, factor insulin sensitivity status $p < 0.05$, factor OGT $p = 0.06$) concentrations shown in serum of IS and ID horses at three levels of OGT: before glucose administration (BASAL), and 120 min (OGT-120) and 180 min (OGT-180) after administration (means ± SEM, $n = 10$). Significant differences between IS and ID are indicated as ** ($p < 0.01$), * ($p < 0.05$). a, b = different superscripts indicate significant differences ($p < 0.05$) between basal and challenged conditions in IS horses, while no OGT effect was observed (as indicated by A) in ID horses (Bonferroni post hoc test)

the human metabolic syndrome: adiposity, chronic inflammation and mitochondrial dysfunction [1, 2].

Metabolic response to oral glucose challenge

Metabolomic profiles changed overall after the OGT in a similar manner, irrespective of insulin sensitivity status and the time after challenge. Twenty-two metabolites were significantly affected by the OGT. The most obvious pattern seen was the rise of the hexose concentration at OGT-120 and OGT-180 in both IS and ID horses. Since circulating hexoses in adult horses consist mostly of glucose, this increase was expected after the OGT and was in accordance with the result of the serum glucose measurement.

Kynurenine concentrations were also increased due to the OGT in both IS and ID horses. Kynurenine is a known proinflammatory marker originating from tryptophan metabolism, which is increased in human patients with diabetic retinopathy [10]. Kynurenine is generated by indoleamine 2,3-deoxygenase, a rate-limiting enzyme, which is activated by inflammatory signals and reactive oxygen species (ROS) related to obesity and insulin resistance [10, 11]. In general, most of the ID horses are reported to be obese [1, 2]; however, the IS and ID horses in this study had comparable BCS values (as an indicator of obesity) and BW. Adipose tissue in humans was identified as a major source of indoleamine 2,3-deoxygenase activity and inflammatory cytokines [10]. Furthermore, acute short-term oral glucose load led to an inflammatory profile even in nondiabetic humans [12]. Although inflammatory cytokine levels were not determined in the horses of this study, the elevated kynurenine concentrations probably indicate a proinflammatory effect of short-term high glucose load by OGT. Therefore, the acute oral glucose load by OGT appeared to affect the molecular health of the horses and was also reflected by decreased spermidine concentrations in both IS and ID horses. Spermidine is a polycationic biogenic amine which is produced by various tissues. One of its predominant molecular functions is to inhibit nonenzymatic glycation of proteins and nucleic acids [13]. The strong reduction in spermidine concentrations indicated its utilization in the prevention of glycation-derived damage of molecules. To summarize, OGT with 1.0 g glucose/kg BW in horses, even as a short-term challenge, can potentially trigger a shift towards a proinflammatory metabolic phenotype.

Administering glucose in high amounts by OGT necessarily induced a shift in substrate availability for energy-gaining pathways. Horses are known to have a well-developed short-chain fatty acid (SCFA) metabolism, with large amounts of butyrate absorbed from the colon [14]. Thus, their energy metabolism depends largely on SCFA, while glucose must be produced by gluconeogenesis from gluconeogenic amino acids, at least when fed on a diet rich in crude fiber and low in grain. During OGT, gluconeogenic needs decreased and the excess of glucose could be used for ATP generation, while utilization of long-chain fatty acids was diminished, a physiological effect called the Randle cycle [15]. Increasing insulin levels due to the OGT probably accounted for short-term protein anabolism. This was reflected in the metabolomics approach in IS and ID horses by a decrease of all proteinogenic amino acids. A reduced utilization of SCFA was reflected by a decrease of short-chain acylcarnitines, mitochondrial degradation products derived mainly from SCFA metabolism. Moreover, enhanced insulin-mediated anti-lipolysis was reflected by reduced concentrations of long-chain acylcarnitines, mitochondrial degradation products of long-chain fatty acids.

Novel metabolites and pathways associated with insulin resistance

No clear differences between IS and ID horses (if present) could be detected by ANOVA, PCA or the heatmap because of the dominance of the OGT effect on the total metabolome. Physiologically, this means that the response to glucose was equally intensive in both IS and ID horses, superposing any potential effects of insulin dysregulation. However, the metabolomics approach was also performed to identify the basal differences of metabolic profiles as a function of insulin sensitivity status.

Therefore, a volcano plot comprising fold change and t-test statistics was analyzed using the results of BASAL samples to identify the metabolites with the greatest discriminating potential between IS and ID horses. Trans-4-hydroxyproline and methionine sulfoxide attracted attention, with higher concentrations in IS horses compared to ID horses. During the OGT, both MOI, trans-4-hydroxyproline and methionine sulfoxide, decreased significantly in the IS group, but not in the ID group. However, the concentration of these metabolites was only different between IS and ID under BASAL conditions, with lower concentrations in the ID group. It is suggested that both MOI belong to the oxidant–antioxidant system of an organism to cope with oxidative stress directly (methionine sulfoxide) or indirectly (trans-4-hydroxyproline; synthesis needs vitamin C, an antioxidant) [11, 16, 17]. Oxidative stress is defined as oxidant–antioxidant disequilibrium due to either increased ROS production, a decrease in the capacity of antioxidant system to defend the organisms against ROS or both. The ROS promote cell damage and can modify endocrine signaling pathways, such as the insulin response cascade leading to insulin resistance [18], one of the potential underlying processes for insulin dysregulation. Disturbances of the oxidant–antioxidant equilibrium have been well-studied in horses suffering from various diseases, but decent markers are still not sufficiently characterized [19, 20]. It is especially not clear yet whether a disequilibrium of the oxidant–antioxidant system is causally relevant for the development of insulin dysregulation in horses. Equine insulin dysregulation and disturbances in the oxidant–antioxidant system are currently understood poorly. Banse et al. [21] suggested that obesity-associated hyperinsulinemia was connected with oxidative stress in skeletal muscle, but they did not find any evidence for oxidative damage to skeletal muscle in obese hyperinsulinemic horses. However, protein carbonyls – products of protein oxidation by ROS – were significantly reduced in the skeletal muscle of obese hyperinsulinemic horses [21]. Consistent with these findings, both MOI found in this study decreased with increasing glucose and insulin concentrations due to the OGT in serum, with a more pronounced decrease in IS horses. Accepting the former suggestion that acute oral glucose load causally promoted a shift into a more inflammatory status even in healthy horses, a decrease in plasma markers of oxidative stress might reflect an increasing disequilibrium in tissue oxidant–antioxidant balance in IS horses; this condition is intensified in ID horses.

Conclusion

It can be concluded as a working hypothesis that higher concentrations of trans-4-hydroxyproline and methionine sulfoxide under basal conditions are a positive sign for health and insulin sensitivity, indicating that healthy horses can remove oxidatively modified amino acids from cells more effectively to avoid cell damage. The lower basal values of these MOI and their decrease during OGT, as well as the concurrent increase of the proinflammatory marker kynurenine are assessed as negative signs and indicate that oxidative stress is involved in the pathway to insulin dysregulation of horses. However, the underlying mechanisms by which oxidized amino acids were produced and modulated in equine tissues is widely unknown so far. Furthermore, because of the small sample size ($n = 10$ per group) and the lack of a clinical diagnosis, the potential significance of trans-4-hydroxyproline and methionine sulfoxide as biomarkers for equine insulin dysregulation has to be tested in larger populations under clinical settings. However, the high diversity of the experimental horse population presented regarding breed, age and BW suggests a universal pathophysiological potential for equine insulin dysregulation. Further research based on these implications should focus on defining criteria for the early diagnosis of insulin dysregulation in horses.

Abbreviations

ANOVA: Analysis of variance; BCS: Body condition score; BW: Body weight; FIA-MS/MS: Flow-injection analysis and tandem mass spectrometry; ID: Insulin-dysregulated; IS: Insulin-sensitive; LC-MS/MS: Liquid chromatography and tandem mass spectrometry; MOI: Metabolite of interest; OGT: Oral glucose test; rmTWA: Repeated measures two way analysis of variance; ROS: Reactive oxygen species; SCFA: Short-chain fatty acids; SEM: Standard error of the mean

Acknowledgements

The authors thank the patients' owners for their support of this work and language editor Philip Saunders for the validation of the English language used in this article.

Authors' contributions

TW, KF and KH designed the study; ÁK and KH analyzed and interpreted the data; ÁK, TW, KF and KH drafted and revised the data; ÁK, TW, KF and KH approved the final version of the manuscript and agreed to be accountable for all aspects of the study.

Consent for publication

Owners gave written informed consent for publication by means of signing our official client acceptance form.

Competing interests

The authors declare that they have no competing interests.

Author details

[1]Institute of Animal Science, Faculty of Agricultural Sciences, University of Hohenheim, Fruwirthstraße 35, 70593 Stuttgart, Germany. [2]Clinic for Horses, University of Veterinary Medicine Hannover, Foundation, Bünteweg 9, 30559 Hannover, Germany. [3]Present address: College of Veterinary Medicine and Life Sciences, City University of Hong Kong, Kowloon Tong, Hong Kong SAR.

References

1. Frank N, Geor RJ, Bailey SR, Durham AE, Johnson PJ. Equine metabolic syndrome. J Vet Intern Med. 2010;24(3):467–75.
2. Frank N, Tadros EM. Insulin dysregulation. Equine Vet J. 2014;46(1):103–12.
3. Johnson PJ, Messer NT, Slight SH, Wiedmeyer C, Buff P, Ganjam VK. Endocrinopathic laminitis in the horse. Clin Tech Equine Prac. 2004; 3(1):45–56.
4. Patti GJ, Yanes O, Siuzdak G. Metabolomics: the apogee of the omics trilogy. Nat Rev Mol Cell Biol. 2012;13(4):263–9.
5. Henneke DR, Potter GD, Kreider JL, Yeates BF. Relationship between conditions score, physical measurements and body fat percentage in mares. Equine Vet J. 1983;15:371–2.
6. Insulin RSL. glucose regulation. Vet Clin North Am Equine Pract. 2002; 18(2):295–304.
7. Warnken T, Huber K, Feige K. Comparison of three different methods for the quantification of equine insulin. BMC Vet Res. 2016;12(1):196.
8. Ramsay SL, Stoeggl WM, Weinberger KM, Graber A, Guggenbichler W. Apparatus and method for analyzing a metabolite profile. 2007; Patent No. EP 1875401 A2, US 2007/0004044.
9. Xia J, Sinelnikov IV, Han B, Wishart DS. MetaboAnalyst 3.0 – making metabolomics more meaningful. Nucleic Acids Res. 2015;43(W1):251–7.
10. Oxenkrug G. Insulin resistance and dysregulation of tryptophan – kynurenine and kynurenine – nicotinamide adenine dinucleotide metabolic pathways. Mol Neurobiol. 2013;48:294–301.
11. Stadtman ER, Levine RL. Free radical-mediated oxidation of free amino acids and amino acid residues in proteins. Amino Acids. 2003;25:207–18.
12. Sesti G, Fiorentino TV, Succurro E, Perticone M, Arturi F, Sciacqua A, et al. Elevated 1-h post-load plasma glucose levels in subjects with normal glucose tolerance are associated with unfavorable inflammatory profile. Acta Diabetol. 2014;51:927–32.
13. Gugliucci A, Menini T. The polyamines spermine and spermidine protect proteins from structural and functional damage by AGE precursors: a new role for old molecules. Life Sci. 2003;72:2603–16.
14. Argenzio RA, Southworth M, Stevens CE. Sites of organic acid production and absorption in the equine gastrointestinal tract. Am J Phys. 1974;226(5):1043–50.
15. Randle PJ, Garland PB, Newsholme EA, Hales CN. The glucose fatty acid cycle in obesity and maturity onset diabetes mellitus. Ann N Y Acad Sci. 1965;131:324–33.
16. Levine RL, Mosoni L, Berlett BS, Stadtman ER. Methionine residues as endogenous antioxidants in proteins. Proc Nat Acad Sci U S A. 1996; 93(26):15036–40.
17. Padh H, Vitamin C. Newer insights into its biochemical functions. Nutr Rev. 1991;49(3):65–70.
18. Vinayagamoorthi R, Bobby Z, Sridhar MG. Antioxidants preserve redox balance and inhibit c-JUN-N-terminal kinase pathway while improving insulin signaling in fat-fed rats: evidence for the role of oxidative stress on IRS-1 serine phosphorylation and insulin resistance. J Endocrinol. 2008;197:287–96.
19. Kusano K, Yamazaki M, Kiuchi M, Kanelo K, Koyama K. Reference range of blood biomarkers for oxidative stress in thoroughbred racehorses (2–5 years old). J Equine Sci. 2016;27:125–9.
20. El-Bahr SM, El-Deeb WM. Acute-phase proteins, oxidative stress biomarkers, proinflammatory cytokines, and cardiac troponin in Arabian mares affected with pyometra. Theriogenology. 2016;86:1132–6.
21. Banse HE, Frank N, Kwong GPS, McFarlane D. Relationship of oxidative stress in skeletal muscle with obesity and obesity-related hyperinsulinemia in horses. Can J Vet Res. 2015;79:329–38.

PERMISSIONS

All chapters in this book were first published in VR, by BioMed Central; hereby published with permission under the Creative Commons Attribution License or equivalent. Every chapter published in this book has been scrutinized by our experts. Their significance has been extensively debated. The topics covered herein carry significant findings which will fuel the growth of the discipline. They may even be implemented as practical applications or may be referred to as a beginning point for another development.

The contributors of this book come from diverse backgrounds, making this book a truly international effort. This book will bring forth new frontiers with its revolutionizing research information and detailed analysis of the nascent developments around the world.

We would like to thank all the contributing authors for lending their expertise to make the book truly unique. They have played a crucial role in the development of this book. Without their invaluable contributions this book wouldn't have been possible. They have made vital efforts to compile up to date information on the varied aspects of this subject to make this book a valuable addition to the collection of many professionals and students.

This book was conceptualized with the vision of imparting up-to-date information and advanced data in this field. To ensure the same, a matchless editorial board was set up. Every individual on the board went through rigorous rounds of assessment to prove their worth. After which they invested a large part of their time researching and compiling the most relevant data for our readers.

The editorial board has been involved in producing this book since its inception. They have spent rigorous hours researching and exploring the diverse topics which have resulted in the successful publishing of this book. They have passed on their knowledge of decades through this book. To expedite this challenging task, the publisher supported the team at every step. A small team of assistant editors was also appointed to further simplify the editing procedure and attain best results for the readers.

Apart from the editorial board, the designing team has also invested a significant amount of their time in understanding the subject and creating the most relevant covers. They scrutinized every image to scout for the most suitable representation of the subject and create an appropriate cover for the book.

The publishing team has been an ardent support to the editorial, designing and production team. Their endless efforts to recruit the best for this project, has resulted in the accomplishment of this book. They are a veteran in the field of academics and their pool of knowledge is as vast as their experience in printing. Their expertise and guidance has proved useful at every step. Their uncompromising quality standards have made this book an exceptional effort. Their encouragement from time to time has been an inspiration for everyone.

The publisher and the editorial board hope that this book will prove to be a valuable piece of knowledge for researchers, students, practitioners and scholars across the globe.

LIST OF CONTRIBUTORS

Katarzyna Paschalis-Trela, Michał Czopowicz, Jerzy Kita and Lucjan Witkowski
Laboratory of Veterinary Epidemiology and Economics, Faculty of Veterinary Medicine, Warsaw University of Life Sciences-SGGW, Warsaw, Nowoursynowska 159c, 02-786 Warsaw, Poland
TRELA VETs Referrals, Warsaw, Poland

Anna Cywińska
Division of Pathophysiology, Department of Pathology and Veterinary Diagnostics, Faculty of Veterinary Medicine, Warsaw University of Life Sciences-SGGW, Nowoursynowska 159c, 02-786 Warsaw, Poland

Jan Trela
TRELA VETs Referrals, Warsaw, Poland

Ronita Adams and Agricola Odoi
Department of Biomedical and Diagnostic Sciences, University of Tennessee, College of Veterinary Medicine, 2407 River Dr., Knoxville, TN 37996, USA

Jackie Smith, Stephen Locke, Erica Phillips, Erdal Erol and Craig Carter
University of Kentucky, Veterinary Diagnostic Laboratory, 1490 Bull Lea Rd., Lexington, KY 40511, USA

T. Warnken
Department of Physiology, University of Veterinary Medicine Hannover, Foundation, Bischofsholer Damm 15, 30173 Hannover, Germany
Clinic for Horses, University of Veterinary Medicine Hannover, Foundation, Bünteweg 9, 30559 Hannover, Germany

K. Hube
Institute of Animal Science, Faculty of Agricultural Sciences, University of Hohenheim, Fruwirthstraße 35, 70593 Stuttgart, Germany

K. Feige
Clinic for Horses, University of Veterinary Medicine Hannover, Foundation, Bünteweg 9, 30559 Hannover, Germany

Line Nymann Thomsen and Lise Charlotte Berg
Department of Veterinary Clinical Sciences, University of Copenhagen, Hoejbakkegaards alle 5, 2630 Taastrup, Denmark

Preben Dybdahl Thomsen
Department of Veterinary and Animal Sciences, University of Copenhagen, Groennegaardsvej 7, 1870 Frederiksberg C, Denmark

Alison Downing and Richard Talbot
Edinburgh Genomics, Ashworth Laboratories, University of Edinburgh, Edinburgh EH9 3FL, UK

Sanjie Caol
Department of Population Medicine and Diagnostic Sciences, College of Veterinary Medicine, Cornell University, Ithaca, NY 14853, USA
Currently Research Center of Swine Disease, College of Veterinary Medicine, Sichuan Agricultural University, Chengdu 611130, China

Thomas Divers
Department of Clinical Sciences College of Veterinary Medicine, Cornell University, Ithaca, NY 14853, USA

Mark Crisman
Virginia-Maryland College of Veterinary Medicine, Blacksburg, VA 24061, USA

Yung-Fu Chang
Department of Population Medicine and Diagnostic Sciences, College of Veterinary Medicine, Cornell University, Ithaca, NY 14853, USA

Ellen J. van der Zaag
Veterinary Clinic De Delta, Foppenpolder 1, NL-3155 EA Maasland, The Netherlands

Erik A. W. S. Weerts
Department of Pathobiology, Division Pathology, Faculty of Veterinary Medicine, Utrecht University, Yalelaan 1, NL-3584 CL Utrecht, The Netherlands

Antoon J. M. van den Belt
Department of Clinical Sciences of Companion Animals, Faculty of Veterinary Medicine, Utrecht University, Yalelaan 108, NL-3584 CM Utrecht, The Netherlands

Willem Back
Department of Equine Sciences, Faculty of Veterinary Medicine, Utrecht University, Yalelaan 112-114, NL-3584 CM Utrecht, The Netherlands
Department of Surgery and Anaesthesia of Domestic Animals, Faculty of Veterinary Medicine, Ghent University, Salisburylaan 133, B-9820 Merelbeke, Belgium

Dominique De Clercq, Barbara Broux, Lisse Vera, Annelies Decloedt and Gunther van Loon
Department of Large Animal Internal Medicine, Ghent University, Salisburylaan 133, B-9820 Merelbeke, Belgium

A. Zak, N. Siwinska, M. Slowikowska, H. Borowicz, K. Kubiak and A. Niedzwiedz
Department of Internal Medicine and Clinic of Diseases of Horses, Dogs and Cats, Faculty of Veterinary Medicine, Wroclaw University of Environmental and Life Sciences, pl. Grunwaldzki 47, 50-366 Wroclaw, Poland

J. Hildebrand and M. Popiolek
Department of Parasitology, Institute of Genetics and Microbiology, University of Wroclaw, ul. S. Przybyszewskiego 63/77, 51-148 Wroclaw, Poland

Rebeka R. Zsoldos
Department of Sustainable Agricultural Systems, Section Livestock Sciences, University of Natural Resources and Life Sciences Vienna, Vienna, Austria

Anna Voegele and Andreas Weber
Multimedia, Simulation and Virtual Reality Group, Institute of Computer Science II, University of Bonn, Bonn, Germany

Bjoern Krueger
Gokhale Method Institute, Stanford, CA, USA

Ulrike Schroeder
Department for Companion Animals and Horses, University of Veterinary Medicine Vienna, Vienna, Austria

Theresia F. Licka
Department for Companion Animals and Horses, University of Veterinary Medicine Vienna, Vienna, Austria
Royal (Dick) School of Veterinary Studies, The University of Edinburgh, Edinburgh, Scotland, UK

A. Kaspar
Comparative Tropical Medicine and Parasitology, Veterinary Faculty, Ludwig-Maximilians-University, Munich, Germany

K. Pfister
Comparative Tropical Medicine and Parasitology, Veterinary Faculty, Ludwig-Maximilians-University, Munich, Germany
Parasite Consulting GmbH, Wendschatzstrasse 8, CH-3006 Berne, Switzerland

M. K. Nielsen
Department of Veterinary Science, M.H. Gluck Equine Research Center, University of Kentucky, Lexington, KY, USA

C. Silaghi
Comparative Tropical Medicine and Parasitology, Veterinary Faculty, Ludwig-Maximilians-University, Munich, Germany
National Centre of Vector Entomology, Institute for Parasitology, Vetsuisse Faculty, University of Zurich, CH-8006 Zurich, Switzerland

H. Fink
Department of Statistics, Ludwig-Maximilians-University, Munich, Germany

M. C. Scheuerle
Comparative Tropical Medicine and Parasitology, Veterinary Faculty, Ludwig-Maximilians-University, Munich, Germany
ParaDocs Laboratory, Ismaning, Germany

Joachim Kaminsky, Astrid Bienert-Zeit, Maren Hellige and Bernhard Ohnesorge
Clinic for Horses, University of Veterinary Medicine Hannover, Foundation, Bünteweg 9, Hannover D-30559, Germany

Karl Rohn
Institute for Biometry, Epidemiology and Information Processing, University of Veterinary Medicine Hannover, Foundation, Bünteweg 2, Hannover D-30559, Germany

Ulrike Meister, Bernhard Ohnesorge and Michael H. Boevé
Stiftung Tierärztliche Hochschule Hannover, Klinik für Pferde, Bünteweg 9, 30559 Hannover, Germany

Christiane Görig
De Graafschap dierenartsen, Schimmeldijk 1, 7251 MX Vorden, The Netherlands

Christopher J. Murphy
UC Davis Health, 2315 Stockton Blvd, Sacramento, CA 95817, USA

Hubertus Haan
HENSOLDT Optronics GmbH, Carl-Zeiss-Str. 22, 73447 Oberkochen, Germany

Cornelis Marinus de Bruijn and Willem Houterman
Wolvega Equine Hospital, Stellingenweg 10, Oldeholtpade 8474 EA, The Netherlands

Margreet Ploeg
Department of Pathobiology, Faculty of Veterinary Medicine, Utrecht University, Utrecht, The Netherlands

Bart Ducro
Department of Animal Breeding and Genetics, Wageningen University, Wageningen, The Netherlands

Berit Boshuizen
Wolvega Equine Hospital, Stellingenweg 10, Oldeholtpade 8474 EA, The Netherlands
Department of Comparative Physiology and Biometrics, Faculty of Veterinary Medicine, Ghent University, Salisburylaan 133, Merelbeke 9820, Belgium

Klaartje Goethals and Catherine Delesalle
Department of Comparative Physiology and Biometrics, Faculty of Veterinary Medicine, Ghent University, Salisburylaan 133, Merelbeke 9820, Belgium

Elisabeth-Lidwien Verdegaal
Equine Health and Performance Centre, School of Animal and Veterinary Sciences, Roseworthy Campus, University of Adelaide, Adelaide, SA 5371, Australia

Anne Mette L. Vinther, Rikke Buhl and Stine M. Andreassen
Department of Large Animal Sciences, Faculty of Health and Medical Sciences, University of Copenhagen, Taastrup, Denmark

Peter M. H. Heegaard and Kerstin Skovgaard
Innate Immunology Group, Section for Immunology and Vaccinology, National Veterinary Institute, Technical University of Denmark, Frederiksberg, Denmark

Pia H. Andersen
Department of Clinical Sciences, Faculty of Veterinary Medicine and Animal Science, Swedish University of Agricultural Sciences, Uppsala, Sweden

E. M. van Grevenhof
Animal Breeding and Genomics Centre, Wageningen UR Livestock Research, Wageningen, The Netherlands

A. R. D. Gezelle Meerburg
Department of Equine Sciences, Utrecht University, Utrecht, The Netherlands

M. C. van Dierendonck
Department of Equine Sciences, Utrecht University, Utrecht, The Netherlands
Behavioral Biology Group, Utrecht University, Utrecht, The Netherlands
Department of Surgery and Anaesthesiology of Domestic Animals, Ghent University, Merelbeke, Belgium

A. J. M. van den Belt
Department of Companion Animal Sciences, Division of Diagnostic Imaging, Utrecht University, Utrecht, The Netherlands

B. van Schaik
GD-Animal Health Service, Deventer, The Netherlands

P. Meeus
Ridderkerk Equine Clinic, Ridderkerk, The Netherlands

W. Back
Department of Equine Sciences, Utrecht University, Utrecht, The Netherlands
Department of Surgery and Anaesthesiology of Domestic Animals, Ghent University, Merelbeke, Belgium

L. Ström and B. Ekesten
Department of Clinical Sciences, Swedish University of Agricultural Sciences, SE-750 07 Uppsala, Sweden

Veronique Saey, Richard Ducatelle and Koen Chiers
Laboratory of Veterinary Pathology, Department of Pathology, Bacteriology and Avian Diseases, Faculty of Veterinary Medicine, Ghent University, Merelbeke, Belgium

Jonathan Tang
Bioanalytical Facility, Norwich Medical School, University of East Anglia, Norwich Research Park, Norwich NR4 7UQ, UK

Siska Croubels and Siegrid De Baere
Department of Pharmacology, Toxicology and Biochemistry, Faculty of Veterinary Medicine, Ghent University, Merelbeke, Belgium

Stijn Schauvliege
Deparment of Surgery and anaesthesiology of domestic animals, Faculty of Veterinary Medicine, Ghent University, Merelbeke, Belgium

Gunther van Loon
Department of Large Animal Internal Medicine, Faculty of Veterinary Medicine, Ghent University, Merelbeke, Belgium

A. Zak, N. Siwinska, M. Slowikowska, H. Borowicz and A. Niedzwiedz
Department of Internal Medicine and Clinic of Diseases of Horses, Dogs and Cats, Faculty of Veterinary Medicine, Wroclaw University of Environmental and Life Sciences, pl. Grunwaldzki 47, 50-366, Wroclaw, Poland

K. Ploneczka - Janeczko and P. Chorbinski
Department of Epizootiology with Clinic for Birds and Exotic Animals, Wroclaw University of Environmental and Life Sciences, pl. Grunwaldzki 45, 50-366, Wroclaw, Poland

P. Szpot and M. Zawadzki
Department of Forensic Medicine, Wroclaw Medical University, ul. Mikulicza-Radeckiego 4, 50-345 Wroclaw, Poland

Nika Brkljača Bottegaro
Clinic for Surgery, Orthopaedics and Ophthalmology, Faculty of Veterinary Medicine, University of Zagreb, Heinzelova 55, 10 000 Zagreb, Croatia

Jelena Gotić
Clinic for Internal Diseases, Faculty of Veterinary Medicine, University of Zagreb, Heinzelova 55, 10 000 Zagreb, Croatia

Jelena Šuran
Department of Pharmacology and Toxicology, Faculty of Veterinary Medicine, University of Zagreb, Heinzelova 55, 10 000 Zagreb, Croatia

Diana Brozić
Department of Animal Nutrition and Dietetics, Faculty of Veterinary Medicine, University of Zagreb, Heinzelova 55, 10 000 Zagreb, Croatia

Karla Klobučar
Faculty of Veterinary Medicine, University of Zagreb, Heinzelova 55, 10 000 Zagreb, Croatia

Krunoslav Bojanić
School of Veterinary Science, Massey University, Palmerston North, New Zealand

Zoran Vrbanac
Department of Radiology Ultrasound Diagnostic and Physical Therapy, Faculty of Veterinary Medicine, University of Zagreb, Heinzelova 55, 10 000 Zagreb, Croatia

Christiane L. Schnabel, Christine L. Wimer, Susanna Babasyan, Heather Freer, Christina Watts, Alicia Rollins and Bettina Wagner
Department of Population Medicine and Diagnostic Sciences, College of Veterinary Medicine, Cornell University, Ithaca, NY 14853, USA

Gillian Perkins
Department of Clinical Sciences, College of Veterinary Medicine, Cornell University, Ithaca, NY 14853, USA

Nikolaus Osterrieder
Institut für Virologie, Freie Universität Berlin, Philippstrasse 13, 10115 Berlin, Germany

Ákos Kenéz
Institute of Animal Science, Faculty of Agricultural Sciences, University of Hohenheim, Fruwirthstraße 35, 70593 Stuttgart, Germany
College of Veterinary Medicine and Life Sciences, City University of Hong Kong, Kowloon Tong, Hong Kong SAR

Tobias Warnken and Karsten Feige
Clinic for Horses, University of Veterinary Medicine Hannover, Foundation, Bünteweg 9, 30559 Hannover, Germany

Korinna Huber
Institute of Animal Science, Faculty of Agricultural Sciences, University of Hohenheim, Fruwirthstraße 35, 70593 Stuttgart, Germany

Index

O

P

R

S

T

U

V

W

CPSIA information can be obtained
at www.ICGtesting.com
Printed in the USA
LVHW062019100122
708212LV00002B/343

9 781647 400767